Applied Concepts in Fractured Reservoirs

Applied Concepts in Fractured Reservoirs

John C. Lorenz
FractureStudies LLC
New Mexico, USA

Scott P. Cooper
FractureStudies LLC
New Mexico, USA

WILEY Blackwell

AAPG
Advancing the World of Petroleum Geosciences™

This edition first published 2020

Registered Office
John Wiley & Sons Ltd, The Atrium, Southern Gate, Chichester, West Sussex, PO19 8SQ, UK

Editorial Office
9600 Garsington Road, Oxford, OX4 2DQ, UK

For details of our global editorial offices, customer services, and more information about Wiley products visit us at www.wiley.com.

Wiley also publishes its books in a variety of electronic formats and by print-on-demand. Some content that appears in standard print versions of this book may not be available in other formats.

Library of Congress Cataloging-in-Publication data has been applied for
ISBN: 9781119055860 [hardback]

Cover Design: Wiley
Cover Image: © John C. Lorenz and Scott P. Cooper

Set in 10/12pt Warnock Pro by SPi Global, Chennai, India
Printed and bound in Singapore by Markono Print Media Pte Ltd

10 9 8 7 6 5 4 3 2 1

For my family and mentors; Mom and Dad, Sean, Ryan, Karen, John, Alvis, and Laurel. It has been an amazing journey, Thanks!

Scott P. Cooper

For Nancy and Ned, Margaret and Norman, who were part of a generation with ideals, principles, and an appreciation for critical thinking; and for Alex who arrived later.

John C. Lorenz

Contents

Foreword

Applied Concepts in Fractured Reservoirs is a much-needed precise and practical treatment of a key topic in the energy industry and beyond. The subject of natural fractures and their impact on hydrocarbon reservoirs has been a mutual interest that I have shared with the senior author of this book, John Lorenz, for decades. In my view this will become a standard reference for geoscientists and engineers working on fractured reservoirs, for example reservoir engineers, geophysicists, geologists, and rock mechanics practitioners. It will take its place among the many other publications by the authors already addressing related issues. The importance of the book lies in the fact that it addresses what is probably the most pervasive feature of rocks: the tendency to break under natural or man-made stresses. The authors put this in an applied context for all involved in exploration and development in the industry and in academia. In that context the book is well organized and clearly illustrated in an easy to grasp collection of applications for fracture studies, for example their impact on reservoir petrophysics, their influence on drilling, and production engineering.

The book is balanced in that it introduces the reader to basic definitions and classifications of fractures and fractured reservoirs at the outset. It then proceeds by outlining a workflow for fractured reservoirs characterization and it goes on to introduce the way fractures impact operational activities. The book allocates a considerable section to discussing the impact of natural fractures on hydraulic fracturing. In my opinion such impact is not fully understood and including it in the book is a timely approach to raise questions, stimulate thoughts, and shed some light on different experimental explanations. The ability to predict the outcome when natural fractures interact with hydraulically stimulated/induced fractures in a reservoir is a challenge not yet fully achieved. Advancement in this area of hydrofracturing is a crucial step in making hydrofracturing more efficient and safer.

John Lorenz and Scott Cooper, who are accomplished researchers and consultants, have produced a valuable resource on the subject of fractured reservoirs, a publication which complements previous texts, and takes the topic to a broader, up-to-date, applied level and scope.

Mohammed S. Ameen (Ph.D., DIC, FGS)
Principal Professional in Geomechanics,
Emerging Unconventional Assets Department,
Saudi Aramco, Dhahran, Saudi Arabia

Preface

This book is a companion to our previous *Atlas of Natural and Induced Fractures in Core*, moving on from the basic recognition of fracture types described in that volume, which must be the foundation of any fracture study, to explanations of how those fractures form, how they are measured, how they can be assessed, and how they affect reservoirs.

This volume is the summary of decades of experience with industry doing applied fracture studies. It brings together numerous fracture-related topics that are not collected elsewhere. We hope that it will be useful to both academia and industry, and that it is not in the vein of the apocryphal third-grader doing a book report on penguins, who concluded that "This book tells me more than I want to know about penguins."

Acknowledgements

Numerous people have contributed to this effort, providing reviews, materials, references, photographs, and insights. We would specifically like to acknowledge and thank Mohammed Ameen, Bruce Hart, Connie Hawkins, Nate Gilbertson, Ron Nelson, Ahmed Ouenes, David Schechter, Joe Simonson, and Norm Warpinski.

Much of the original material used to illustrate this volume has come from detailed, unpublished industry studies, and we thank those companies for allowing us to use the material, sanitized for publication. We would also like to acknowledge all of the astute people who over the years have worked in this amazing field of study and whose published papers were used throughout this text.

We would also like to thank our wives, who, knowing better than we did the size of the undertaking, said "Sure, why not?" when asked if we might carve the needed time from family commitments.

Introduction

One can't begin to write a textbook without self-indulgently explaining how important the subject matter is and how readers therefore absolutely need to know the material. Fortunately, the importance of understanding the origins, characteristics, and effects of natural fractures in hydrocarbon reservoirs is becoming widely accepted, so it is enough for us to note that technology has continued to extend a recognition of the presence and importance of natural fractures in many reservoirs once thought to be un-fractured. Moreover, the increasing use of horizontal and deviated wellbores is providing unique and invaluable information on the close fracture spacings exhibited by many fracture systems, even in relatively undeformed strata.

It was not always so. The default conceptual model of a reservoir before image logs and extensive coring, even in deformed strata, rarely included natural fractures, and there were few data points to indicate otherwise. Cored fractures used to be relatively rare because of the wide fracture spacings typical of the thicker reservoir units most commonly cored, and because vertical core has a low probability of capturing vertical fractures, the most common type in many reservoirs. For example, a vertical, 4-inch (10 cm) diameter core cut vertically through a fractured bed where fracture spacing averages 40 inches (about 1 m) has only a 10% probability of sampling the fracture population (see Lorenz, 1992). The absence of vertical fractures in vertical core used to be accepted as proof that a reservoir was not fractured, but that is like saying that there are no mosquitos about on a summer's evening picnic because you have not been bitten yet.

This absence of good subsurface data, and the slowly maturing study of geomechanics prior to the 1980s, did not support the presence of fractures in the subsurface below about 2000 ft. Statements such as "At depth… most joints are generally closed…" (Heck, 1955) were widely accepted, and even the experts at the respected rock mechanics laboratory of the Exploration and Production Research Division of the Shell Development Company would write that "It is, of course, inconceivable that an open crack could exist at depth…" (Griggs and Handin, 1960). Thus, those fractures that *were* cored were usually considered to be exceptions rather than, as they are, evidence for abundant fracturing.

Substantive, definitive data now document the common presence of natural fractures with significant, permeability-enhancing apertures at the depths of hydrocarbon reservoirs. Moreover, improved understandings of geomechanics and the dynamics of reservoirs (e.g., Ameen, 2018) have allowed for predictions of the behavior of fracture-permeability systems during production. Thus, the effects of natural fractures must be included in assessments of most reservoir permeability systems, especially in unconventional reservoirs.

The default conceptual natural fracture in early reservoir models was a regularly spaced, randomly oriented, open slot of uniform width. The reality is that the fractures of a typical fracture set in a hydrocarbon reservoir are log-normally spaced, systematically oriented, and have irregular apertures. Although a significant variety of fracture types exist in hydrocarbon reservoirs (see Lorenz and Cooper, 2018a), their effects on permeability can be reasonably assessed if the fracture type, degree of occlusion, degree of development, and tectonic setting can be characterized.

For example; shear fractures commonly occur as related, intersecting conjugate pairs whereas extension fractures occur as poorly-connected parallel planes. Shear fractures form vertically and laterally interconnected drainage networks but the individual shear-fracture apertures and therefore permeabilities are irregular. In contrast, sets of extension fractures create highly anisotropic drainage in a reservoir, and are likely to be vertically limited by lithologic discontinuities. Thus, it is important to not only recognize the presence of fractures in a reservoir but also to both correctly identify the fractures by type and to fully characterize them. Fractures in a reservoir must be understood, they cannot be merely counted and oriented.

Correct fracture identifications and characterizations allow the geologist, modeler, and engineer to be accurate in spotting well locations and designing horizontal wellbore azimuths. These data provide anchor points for seismic and petrophysical interpretations as well as basic data for reasonable determinations of fracture-system volumetrics including porosity, permeability, reserves, recovery factors, and production rates. Accurate knowledge of fracture systems allows engineers to design appropriate completion and production strategies, taking advantage of or at least accommodating fracture-controlled drainage anisotropy and stress-sensitive permeabilities. Such in-depth knowledge allows the petrophysicist to more accurately interpret image logs.

This volume is a companion volume to our earlier *Atlas of Natural and Induced Fractures in Core* (Lorenz and Cooper, 2018a). That volume provides a tool for accurately identifying different fracture types whereas this volume discusses how the different types formed, and how they affect reservoir volumetrics. The present volume is divided into three sections. Part 1, *Understanding Natural Fractures: Fracture Types, Dimensions, and Origins*, discusses the origin, the characteristics, the variations among, and distinctions between extension fractures and shear fractures. It also describes microfractures, fractures associated with faults, the effects of the different geomechanical properties of different lithologies on fracture development, fracture domains, and fracture corridors. The important mineralization that can occlude fracture apertures and reduce fracture permeability, as well as the dissolution that can enhance it, are included in the discussions. The characteristics of both individual fractures (length, widths, heights, apertures) and fracture populations (spacings, interconnectivity) of different fracture types are described.

Part 2, *Measuring and Analyzing Fractures in a Reservoir*, deals with techniques for logging and obtaining the maximum amount of data from cores. It includes techniques for distinguishing natural from induced fractures, and discusses the potential uses of different kinds of induced fractures in determining the in situ stress orientations as well as in determining the orientations of fractures relative to the stresses and to each other. We describe the pros and cons of oriented core and techniques for quality-checking a core-orientation survey, as well as techniques for fracture analysis once the fracture data have been collected. Included in this section are the expressions of fractures in image logs and the advantages of correlating image logs to cores, noting the significantly increased value of both the image log once it has been calibrated to core, and of a core that has been oriented using an image log.

Part 2 also offers the observation that there are several valid fold-related fracture models, each appropriate to one of the different mechanisms for folding strata, suggesting that there is no universal model that describes fracture distributions and orientations on anticlines. In addition, we will describe the potential distributions of fractures along wrench faults and within fold and thrust belts.

Part 3 describes *Effects of Natural Fractures on Reservoirs*, offering techniques for estimating volumetrics in fractured reservoirs, and discussing the fracture-related enhancement and/or degradation of reservoir permeability. This section will describe differences in the potentials of different fracture systems to create drainage and permeability anisotropy in a reservoir. We also illustrate the changes in the fracture-related permeability that can be caused when fracture apertures narrow during changes in reservoir fluid pressures and the related effective stresses during production. A series of wells drilled into a fractured reservoir tends to have non-uniform production rates and recoveries, so the effects of fracture corridors and domains in creating reservoir sweet spots are discussed. Finally, we will address the issue of interactions between natural fractures and hydraulic stimulation fractures.

The primary literature on natural fractures is vast and varied, and is replete with inconsistencies and conflicting hypotheses. There are a dozen good, basic structural-geology textbooks (e.g. Fossen, 2010; Pollard and Fletcher, 2005; Mandl, 2005; National Research Council, 1996; Twiss and Moores, 1992) that describe natural fractures and the mechanics of fracturing rock, but even these summaries are not in total agreement. It is interesting to compare the variety of concepts, terminology, and interpretations offered in these texts (see for example Lorenz and Cooper, 2019), and to compare the different experimental, theoretical, and empirical approaches authors have used in describing, assessing, and interpreting natural fractures. Significantly fewer texts (e.g. Nelson, 2001; National Research Council, 1996; Narr et al., 2006) have focused on fractures and their effects on hydrocarbon reservoirs.

In writing a textbook, authors presume that they know something that the readers also really need to know, or that the authors have a different and valuable perspective, or that they can capture the essence of a body of literature and condense it into a useful package. We hope that we have done all three, and that our experience has allowed us to build on the important concepts and information presented in earlier works, to distill them, and to find the useful commonalities.

PART 1

Understanding Natural Fractures: Fracture Types, Dimensions, and Origin

1.1 Introduction

We begin this text with a short discussion of basic fracture nomenclature in order to provide a common understanding and framework for the rest of the volume. Nomenclature would seem to be a rather stodgy lead-off topic, but the whiplash provided by the variety of fracture terminology in the literature (e.g. Pollard and Aydin, 1988; Lorenz and Cooper, 2019) should prevent the discussion from being overly dull and boring. A nomenclature discussion provides a basis for melding data, observations, and concepts from the laboratory, outcrops, and theory, as well as for following the arguments of papers written by different authors.

All fracture types do not have equal effects on a reservoir, so the nomenclature section is followed by descriptions of the range of fracture dimensions and by illustrations of the characteristics that must be accounted for when attempting to model fluid flow through a naturally fractured reservoir. This discussion focuses on shear and extension fractures, which are the most common fracture types in hydrocarbon reservoirs, but it will include some of the other kinds of fractures found in reservoirs. We will also explore the basic mechanics of fracturing rock in shear and extension since that knowledge is useful in extrapolating the limited fracture data obtained from a wellbore into the three-dimensional volume of the reservoir, and can sometimes be used to infer the characteristics or even the presence/absence of fracture systems in a reservoir prior to drilling.

Fracture characteristics depend on the stress conditions at the time of fracturing as well as the mechanical properties of the host rock, which in turn depends on the basic composition, sedimentary heterogeneity, and diagenetic history of the rock. Therefore, we will discuss the variations of fracture characteristics as controlled by lithology, principally the gross-scale differences inherent in fracturing limestone, sandstone, and shale or mudrock. We will also examine what happens when a fracture forms within one stress system and is later reactivated within a reorganized stress system, forming a compound fracture.

Open fracture apertures in a reservoir provide high-permeability pathways that can be enhanced by dissolution and/or restricted by mineralization, so fracture apertures and modifications to fracture apertures are also discussed. Finally, the origins and effects of closely spaced fractures in fracture corridors are examined.

These descriptions and discussions from **Part 1** of this volume and will provide the basic understandings necessary to move on to **Part 2**, *Measuring and Analyzing Fractures in Reservoirs*, and **Part 3**, *The Effects of Natural Fracture on Reservoirs*.

1.2 Nomenclature and Fracture-Classification Systems

1.2.1 Introduction

A fracture is a mechanical discontinuity in a rock, typically planar and commonly associated with a loss of cohesion in the rock across the fracture plane. Many fractures are filled by post-fracturing mineralization, restoring some or all cohesion across the facture plane. Natural fractures are brittle to brittle-ductile strain-accommodation structures that develop when the rock is subjected to a stress anisotropy greater than its strength.

This definition covers an astonishing variety of structures, and numerous authors have divided fractures into various classes with assorted names in efforts to make some sense out of them. Classification, i.e. the identification of populations within which the fractures have distinctive and similar characteristics, is the first step in assessing fracture effects on a reservoir, since all fracture types do not have the same permeability or degree of interconnectedness. Nevertheless, it's useful to remember that most classification schemes, including those for fractures, are artificial constructs, and the boundaries between classes are commonly gradational rather than abrupt.

Applied Concepts in Fractured Reservoirs, First Edition. John C. Lorenz and Scott P. Cooper.

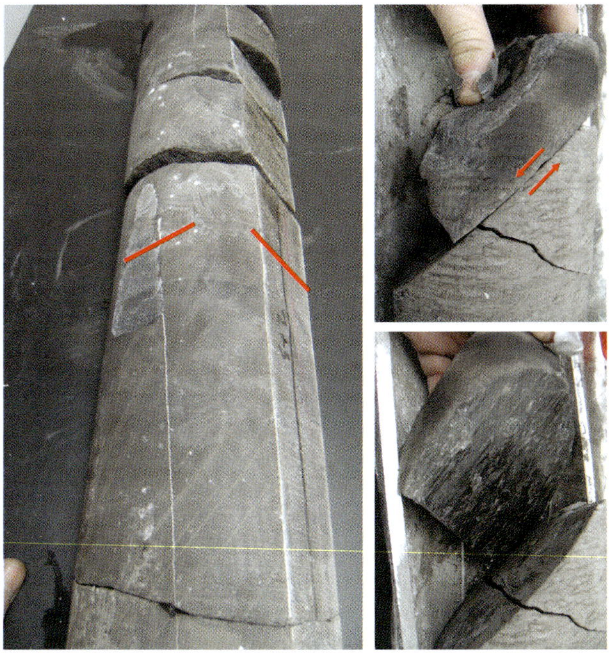

Figure 1.1 *Left*: two calcite-mineralized vertical extension fractures captured by a core cut from a marine sandstone. The fractures do not intersect in the core but have relative strikes (red bars) that will intersect in the reservoir outside the core volume. *Upper right*: a side view of an inclined shear fracture in the same core. *Lower right*: view of the calcite-mineralized and slickenlined surface of this shear fracture. 5.25 inch (13.3 cm) diameter core; uphole is towards the top of all three photos.

The focus of this volume is on *shear fractures* and *extension fractures* (Figure 1.1). Extension fractures form when the opposing fracture walls move apart from each other in the direction normal to the fracture plane, whereas shear fractures form when the opposing walls move in opposite directions but parallel to the fracture plane. Both structures accommodate strain in a brittle fashion under conditions of anisotropic stress, and although there is a gradation between these two basic fracture types, they have fundamentally different characteristics and therefore have significantly different effects on reservoirs. These two fracture types comprise the majority of fractures found in hydrocarbon reservoirs.

We prefer not to not use the terms "joint," an open break in the rock, or "vein," a mineralized joint, since the two terms always seem to generate discussion and require definition during debates on the outcrop. Dictionaries in fact suggest that the term "joint" comes from the verb *to join,* and its original connotation referred to the location where two rock masses come together. Although "joint" may have historical precedence, "fracture" is perhaps more apt since the related verb *to fracture* applies more appropriately to the origin of the structure as a break in the rock mass. "Joint" and "vein" are both too broad (not distinguishing between extension and shear) and too specific (is a mineralized shear fracture a vein?) to be widely useful. Some authors (e.g. Hancock, 1985; Mandl, 2005; Weijermars, 1997) apply "joint" to any fracture discontinuity even if it has shear offset; others restrict it to those fractures that do not have offset parallel to the fracture plane and prefer the term "fault" for any structure with indications of shear. We have used the term "vein" for early diagenetic filled fractures that are insignificant to reservoir permeability (see Lorenz and Cooper, 2018a), and the term also adequately describes the short, wide, mineralized, en echelon structures found in shear zones that form in rock near the brittle-ductile transition. However, "vein" does not serve well as a generic term.

A "fracture"-based lexicon is simple and flexible. It is easily modified with descriptors to indicate any of the numerous and important fracture characteristics such as mode of origin, dip angle, and whether or not the fracture is mineralized (i.e. an "inclined, calcite-mineralized shear fracture"), making it easy to understand. Most fracture types and most other nomenclatures can be fit into a fracture-based nomenclature system (Table 1.1).

Table 1.1 Fracture nomenclature as used in this volume.

Fracture Classifications Used in this Volume	
Mode of Origin	**Modifiers**
Extension	Mineralized, Unmineralized, and Dissolution Dip angle: High, Intermediate, Low Bed-parallel Microfractures
Shear	Mineralized, Unmineralized, and Dissolution Dip angle: High, Intermediate, Low Normal, Reverse, and Strike-slip Bed-parallel Conjugate Microfractures
Special Shear Types	Mixed-mode Deformation Bands Faults En Echelon Tension Gashes
Other Types	
Anticracks	Compaction Bands Stylolites
Fissures	
Veins	
Expulsion	
Ptygmatic	
Beef-filled	

1.2.2 Other Classification Systems

Fracture classifications can also be based on other criteria, and a welter of classification systems that have been used for different purposes and in different contexts exists in the literature. Examination of a half dozen recent structural geology textbooks and the numerous websites and published papers shows some consistency, several conflicts, and a few jarring anomalies. Several of these systems are described briefly here so that a student using the literature can recognize and use the differences and commonalities.

Petrophysicists who characterize fractures from image logs run in a wellbore commonly classify a fracture signature by its electrical characteristics, i.e. whether it is *conductive* or *resistive*. These electrical properties may be used to determine whether a fracture has some open width, which in turn can be used to assess fracture permeability, which is ultimately why we study fractures in hydrocarbon reservoirs. Service companies, however, often leave the permeability interpretations to their client, recognizing that it can be a trap. For example, conductive fracture signatures, which are usually inferred to indicate the invasion of drilling mud into open fracture apertures, can also result from fractures that are completely occluded but where the mineralization includes traces of pyrite. Likewise, resistive signatures can indicate either fractures that are largely occluded by mineralization, deformation-band shear fractures that are not mineralized but that consist of zones of collapsed porosity, or fractures that are filled with a drilling mud that is resistive relative to the formation fluid and lithology.

Nelson (2001) offers several fracture-classification types, including one based loosely on a fracture's morphology and its potential effects on reservoir plumbing, the categories including *open fractures*, *deformed fractures*, *mineralized fractures*, and *vuggy fractures*. Other authors (e.g. Aguilera, 2003) separate fractures into systems based on their relationship to structure, such as *tectonic fractures* (primarily shear fractures related to faults or folds), *regional fractures* (mostly extension fractures that occur over wide areas), and *surface-related fractures* (fractures related to weathering, spalling, stress release, and gravitational forces on valley walls).

One of the common fracture-classification systems is based on Griggs and Handin's (1960) experiments on carefully machined, centimeter-scale laboratory specimens. Their Figure 1, reproduced here as Figure 1.2, shows a series of block diagrams representing a progression from extension fractures to various types of shear fractures as the ductility of the rock increases, all fracturing formed under laboratory conditions of tri-axial compression.

This figure has been reproduced and modified by numerous authors, some adding a block for *tensile fractures*, formed under true tension, on the left side of the figure, obscuring the important distinction between tension and extension. Although most rock is an order of magnitude weaker in tension than in compression, and although tension is easy to create in the laboratory, true tension is rare in the subsurface and is not applicable to the origin of the numerous and widespread extension or shear fractures in hydrocarbon reservoirs.

It is important to remember that Griggs and Handin specified that their figure illustrates a spectrum of fracturing rather than discrete stages. Other authors have added terms for the intermediate stages of fracturing between shear and extension, fractures that show evidence for both extension and shear offset (i.e. *hybrid*, *mixed-mode*, or *oblique extension* fractures; Ramsey and Chester, 2004).

The extension fractures that form parallel to the maximum applied stress and normal to the minimum compressive stress during tests on Griggs and Handin's most brittle rocks have been called *longitudinal-splitting*, *axial splitting*, and *cleavage* fractures, or sometimes *load-parallel extension* fractures. Because they occur under conditions of minimal confining stress these structures are not always considered to be analogous to subsurface fractures where the compressive confining stresses, dictated by the weight of the overburden and any additional tectonic compression, are assumed to be significant. However, advances in geomechanics and an understanding of the significant effects of pore pressure suggest that the axial splitting extension-fracture process can be important in explaining the origin of sets of regional extension fractures.

Fractures are also commonly divided into "Mode" categories based on origin:

Mode I: extension fractures

Mode II: shear fractures where shear was parallel to the fracture face

Mode III: shear fractures distinguished by rotational shear along the fracture face. Mode III, rotational-shear fractures are not common in the rock record and they would be almost impossible to recognize from the limited data available in the subsurface, and we will not discuss them further.

The Mode system is sometimes extended to include a *Mode IV* class of planar, *compaction* or *anti-crack* structures such as *stylolites* and *compaction bands*, where the fracture walls moved towards each other in a direction normal to the fracture plane, requiring volume loss.

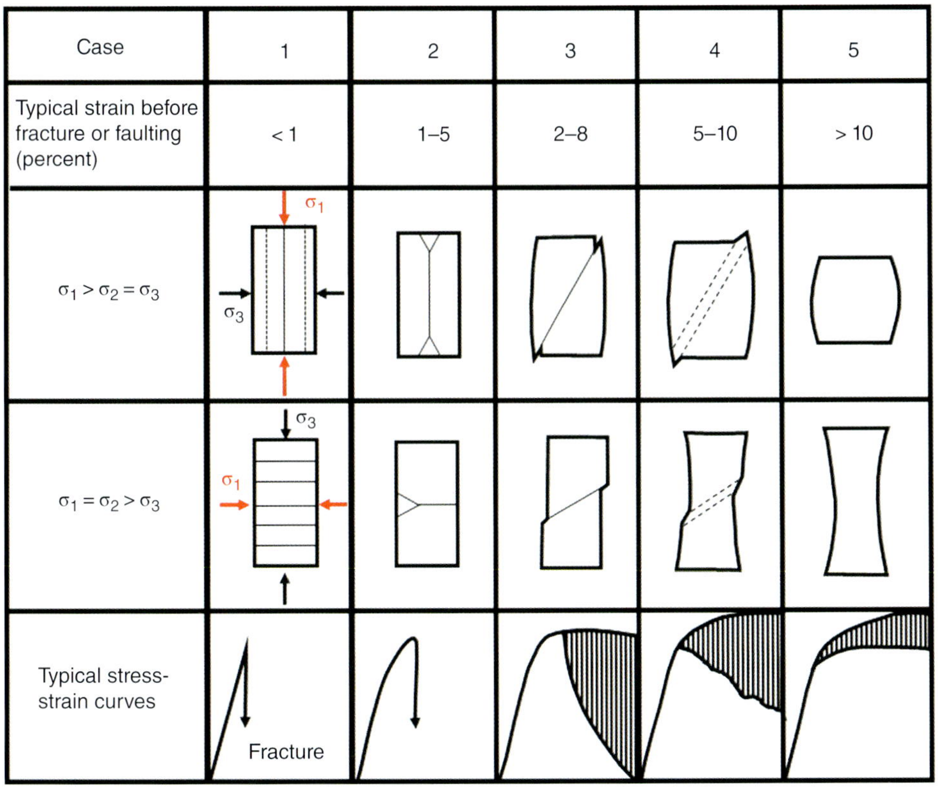

Figure 1.2 Deformation in compression in the laboratory (from Griggs and Handin, 1960, their Figure 1). The original caption reads "Schematic representation of the spectrum from brittle fracture to ductile flow, with typical strains before fracture and stress-strain curves for uniaxial compression and extension. The ruled portions of the stress-strain curves indicate the variation within each case and the overlap between cases 3, 4, and 5." Columns 1 and 2 show extension fracturing, columns 3 and 4 show shear fracturing, and column 5 shows ductile deformation. Griggs and Handin labeled the top row of samples "Compression" and the second row "Extension," suggestive of the type of deformation imposed on the samples. However, that invites confusion with the designation of extension fractures (columns 1 and 2) vs. shear fractures (columns 3 and 4), so we have omitted those labels on the two rows. All test conditions were compressive, as indicated by the formula indicating the stress conditions (σ) for deformation: none of the fractures were formed in tension, by pulling on the ends of the samples. Note that the angular relationships between the three imposed stresses and the resulting fracture planes are identical in the two rows of samples, the only difference being whether the maximum compressive stress (σ_1) was imposed parallel or normal to the long axis of a sample. (The use of σ to designate stress in this figure follows Griggs and Handin's original usage, whereas we have used S for stress in most of this text, reserving σ to denote effective stress.)

1.2.3 Classifications for Fractures in Outcrops and Cores

The nomenclature used for fractures in outcrops and in cores cut from the subsurface is sometimes different from that used in the laboratory. Some authors prefer to use a dual nomenclature in order to distinguish outcrop from laboratory structures. For example, although Griggs and Handin used the term *fault* for laboratory fractures that formed in shear, Jaeger et al. (2007) use *fault* for shear structures in outcrop and *shear fracture* for similar structures formed in the laboratory. The existence of two sets of terms for natural vs. laboratory-created fractures highlights a degree of ambiguity that is common in the application of laboratory results to the formation of natural fractures in the rock record.

Some authors suggest that a fracture plane with any amount of shear offset should be called a *fault*, qualifying the usage by referring to shear structures with small offsets as *microfaults*. Others suggest that the term *fault* should apply only to structures having relatively large offsets and having associated secondary features such as antithetic shear fractures and/or a central core of fault rock consisting of clay smear, breccia, and/or gouge. In general discussions of shear structures, the thresholds for "large" and "small" magnitudes of offset are subjective, but we are happy to use both *fault* and *shear fracture* for structures with large and small offsets respectively, specifying offset magnitudes where possible for discussions of specific fracture sets.

The term *shear fracture*, as for extension fractures, can have appended modifiers for the presence and type of mineralization, dip angle, and for the sense of offset

Figure 1.3 *Left*: Anderson's (1951) three ideal orientations for pairs of conjugate shear fractures and faults. The orientation of the "X" is determined by the orientations of the maximum, intermediate, and minimum compressive stresses (S_1, S_2, and S_3, respectively). Ideally the two sets of the conjugate-fracture pair are equally well developed, but unequal development is common. *Middle and right*: photo and sketch of a fault with antithetic shear fractures illustrating the common occurrence of the unequal development of conjugate shear fracturing.

(normal, reverse, and strike-slip, following Anderson [1951]). Shear fractures commonly occur as conjugate pairs although the two sets of the pair may not be equally well developed in an ideal X pattern in any given outcrop. The maximum compressive stress at the time of shear bisects the acute angle of the X (e.g. Rothery, 1988). One set of the pair may consist of larger, localized shear, as in a fault, with the other set occurring as smaller, antithetic shear fractures intersecting the fault plane (Figure 1.3).

Shear planes where porosity has collapsed and where grains of the host rock have been crushed to create low-permeability bands rather than an open fracture aperture have been referred to as *deformation bands*, *shear bands*, and *gouge-filled fractures*. The first two terms are in common usage in the industry as a shorthand for deformation-band shear fractures.

1.2.4 Expulsion Fractures and Natural Hydraulic Fractures

The tangle of terminology, combined with poorly supported geomechanical models for fracturing in the presence of high formation pore pressures, has created concepts and a lexicon of *expulsion fractures* and *natural hydraulic fractures*. The mechanics of fracturing and pore pressure are discussed later, but the term "natural hydraulic fracture" invites a misleading analogy to the hydraulic stimulation fractures common in the hydrocarbon industry. During stimulations, fluids from an external source (pump trucks) are forced into a formation under pressures well in excess of the combined minimum in situ compressive stress and tensile strength of the subsurface rock.

Expulsion fractures and natural hydraulic fractures do exist, witness igneous dikes and a variety of pre-lithification sedimentary injection structures in the rock record, but the evidence for these mechanisms is weak or nonexistent in most natural fracture systems. Many of the fractures that have been called expulsion fractures and natural hydraulic fractures are more plausibly explained by other mechanisms.

Nevertheless, there are examples of structures with associated clays and remnant hydrocarbons that record the passage of fluids expelled from one high-pressure stratum into or across an adjacent, lower-pressure layer (Figure 1.4). High pore pressure *within* a formation alone, however, does not fracture the rock; rather it makes the rock more susceptible to fracture (see Lorenz et al., 1991; Fall et al., 2015; and later discussions in this book). Like the distinction between tension fractures and extension fractures, there is a subtle but important difference between fracturing rock with pore pressures in excess of the minimum compressive stress (implausible), and making rock susceptible to fracturing by raising the pore pressure such that the rock is effectively unconfined and therefore weak, brittle, and susceptible to fracture.

1.2.5 Other Fracture Terminology

Other terms such as *pinnate fractures, feather fractures,* and *horsetail splays* describe structures that splay off larger shear fractures and faults due to local stresses that

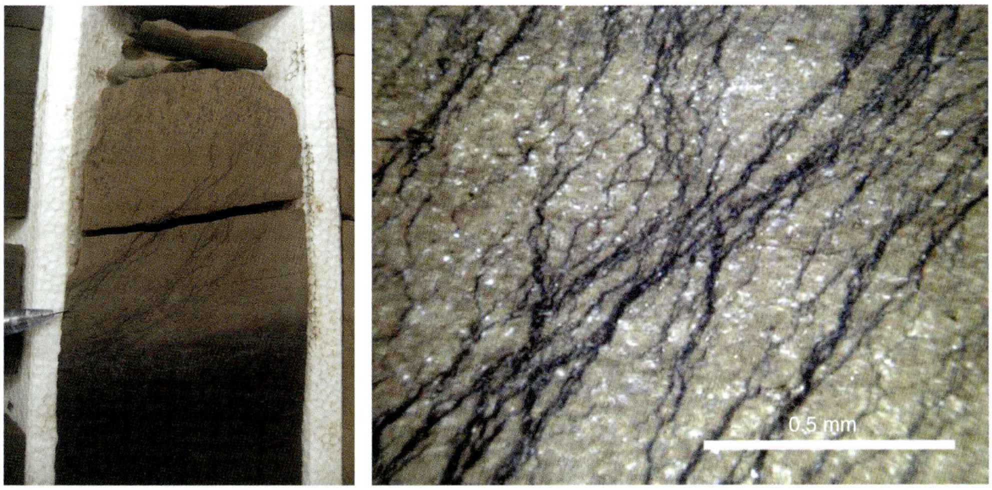

Figure 1.4 Wispy tendrils of clay and hydrocarbon, derived from the underlying, under-compacted organic-rich marine mudrock, mark the episodic passage of over-pressured fluids. *Left*: slabs of 4-inch (10-cm) diameter core, pencil point for scale. *Right*: close-up of the core at the left. The structures are roughly planar in the third dimension (i.e. in and out of the plane of the photograph). They are inclined due to down-slope gravitational creep prior to lithification.

build up along the shear plane during offset, commonly found in strata that have been relatively highly deformed. Occasionally one stubs a toe on undefined terms such as *"hydraulic shear fracture"* and *"traversal fracture."* Terminology must be flexible if a science is to grow, but such terms should be defined when they are introduced, and new terms should not supplant existing terms for the same structures without providing the reason for doing so.

Both shear and extension fractures can occur at a microscopic scale, where they are called *microfractures*. Microfractures may form within grains, between grains, and across several grains, being called *intra-granular*, *inter-granular*, and *trans-granular*, respectively (e.g. Mandl, 2005). Although they can enhance both permeability and porosity in some systems (e.g. Ameen and Hailwood, 2008; Ameen et al., 2012), the role of microfractures in most reservoirs is unclear. Anders et al. (2014) suggest that most unarguably natural microfractures are mineralized or healed, and these should create little effect on a reservoir. Likewise, Loucks and Reed (2016) found that many of the microfractures identified in mudrock cores are induced, often by dehydration. Significant populations of open microfractures in sandstone cores can be created just by the release of the rock from the in situ compressive stresses. Caution should be exercised when extrapolating microfractures observed in a core into a subsurface reservoir.

The term *fissure* generally invokes the image of the wide and open slots across glaciers and lava flows, typically open to the surface of the earth and tapering downward. Warpinski (1991) used the term to distinguish between natural fractures (his fissures) and hydraulic stimulation fractures in a paper assessing the interactions between the two. We have used fissure to refer to fractures, both extension and shear, that have wide, irregular, dissolution-enhanced apertures and that are filled with allochthonous materials derived from an associated subaerial exposure surface (Lorenz and Cooper, 2018a). Fossen (2010) suggests that fissures are extension fractures that are wider than usual but without quantifying the threshold for "wider." Fissures are not easy to recognize in the small samples of a reservoir offered by image logs and cores although they might be expected to be present in karsted strata. If they are filled with relatively permeable materials, they can provide good, albeit localized, permeability in a reservoir.

The term *vein* is used commonly in several geologic contexts including, notably, to describe tabular ore-bearing zones in hard-rock mining, and by some authors to describe any mineralized fracture. In the sedimentary strata that form most hydrocarbon reservoirs, the term may best apply to short but relatively wide fractures that have elliptical, low-aspect-ratio apertures, and that are typically occluded with the same basic mineral that comprises the host strata, most commonly in carbonates. Veins can occur as early diagenetic structures, but they are probably best known as systems of en echelon structures, sometimes referred to as *tension gashes* or *gash fractures*, that form within shear zones in strata that were relatively ductile at the time of shear (Figure 1.5). Veins typically have little effect on a reservoir because they are short, poorly interconnected, and occluded.

Compaction bands form as planar *anticracks* or *closing fractures* (as opposed to most fractures that open to form

Figure 1.5 Quartz-filled en echelon veins in silicified sandstone, southwestern Algeria. In contrast to the discrete shear planes, en echelon veins form within wider, tabular zones of shear in rock that is near the brittle-ductile transition, thus they are more common in relatively ductile lithologies such as limestone than in siliceous sandstone. In this example, the rock is presently very brittle, but at the time of fracturing the rock was deeply buried within an aulacogen and rendered relatively ductile by the high temperatures and confining pressures at depth. The trains or arrays of en echelon veins in this formation form thrust-oriented, reverse-dip-slip conjugate pairs (i.e. in the third orientation shown in Figure 1.3). As with conjugate shear-fracture Xs, the bisector of the acute conjugate angle of the conjugate en echelon trains records the maximum compressive stress.

an aperture between the fracture walls) due to localized, planar compaction in a rock (e.g. Pollard and Fletcher, 2005; Holcomb et al., 2007). The bands can be difficult to distinguish from deformation-band shear fractures, especially in the small samples provide by core, and they are not common in most outcrops. Like shear bands, they form permeability baffles and barriers in a reservoir.

Stylolites also form along planar zones and under compression, but they are related to localized pressure dissolution and form by chemical removal of material along the anticrack rather than by compaction. Stylolites are common in carbonate reservoirs, and short, related extension fractures may extend into the host strata on either side of the structure, forming localized high-permeability streaks in a reservoir (Nelson, 1981).

1.2.6 Sets, Systems, Domains, and Systematic Fractures

A group of fractures with consistent characteristics forms a *set* of *systematic fractures*, and the lateral and/or vertical extent of that zone of similar fractures defines a fracture *domain*. A fracture *system* consists of one or more fracture sets. Fracturing may also occur as groups of irregular, *non-systematic* fractures. Assessing the effects of fractures on a reservoir requires an understanding of not only the characteristics of individual fractures (orientation, height, length, aperture or lack of it) but also the characteristics of the system (interconnectivity, spacing, distribution with respect to lithology, etc.).

Fractures that are locally well developed in linear domains have been called *fracture corridors* or *fracture swarms* that, unless plugged with mineralization, can form high-permeability zones in a reservoir. Fracture corridors are often genetically related to the zones of high stress around local structures (fold hinges, incipient faults, fault tips, e.g. Souque et al., 2019).

Fracture orientations and other characteristics are dictated by the in situ stresses at the time of fracturing, thus all fractures that formed within a given lithology and within the same stress field should be similar, forming a systematic fracture set and a relatively simple fracture system. The members of an extension-fracture set will have similar strikes, dips, surface characteristics (*fractography*), mineralization, distributions relative to lithology, and terminations.

Some sets of similar fractures can be classified by their geometric and therefore possibly their genetic relationship to a structure, i.e. systematic fracture sets may be classified as *cross fractures* or *strike fractures* if they have consistent strikes relative to a fold axis. Stearns and Friedman (1972) described six specific fracture sets with consistent orientations relative to a fold axis, grouped into two systems called "Pattern 1" and "Pattern 2," each consisting of three systematic fracture sets.

Similarly, Price (1966) defined several patterns for fractures related to folding, labeling fractures relative to the directions "a" (normal to the fold axis), "b" (parallel to the fold axis), and "c" (vertical). Fractures striking normal and parallel to the fold were inferred to have originated in tension and were called *cross joints* or *a-c joints*, and *b-c* or *longitudinal joints*, respectively. Additional *oblique joints* were inferred to have originated in shear.

Multiple fracture sets may be distinguished and classified by origin and/or timing. For example, Engelder (1985) suggested a local four-fold fracture classification based on tectonics and mechanics, where *tectonic*, *hydraulic*, *unloading*, and *release* fracture sets are members of a spectrum related to changes in stress and pore pressure.

Since extension fractures form as sets of unidirectional, parallel fractures, the presence of two sets of extension fractures in a reservoir records two fracture events. (Some authors have suggested that the formation of one extension-fracture set can alter the local stress system sufficiently such that a second, orthogonal fracture set forms within one regional stress event, but the mechanics of this theory are not compelling and there is little evidence for it in outcrops.) The component sets within superimposed fracture systems typically have their own unique characteristics and they may not contribute equally to reservoir permeability. Other fracture systems may consist of a restricted domain of fractures related to a local structure, such as a fold, superimposed onto a set of more widespread regional fractures (e.g. Cooper et al., 2006; Ameen et al., 2012).

In contrast to extension fractures, a set of shear fractures commonly consists of two intersecting, related, systematic, *conjugate fracture subsets* that form penecontemporaneously. Except for orientation, members of the two subsets may have similar characteristics and contribute equally to reservoir plumbing.

The stress domains that control fracturing, and thus the resulting fracture domains, may be as small as the rock immediately adjacent to a fault or along the hinge of an anticline, or they may extend laterally for many kilometers within uniformly stressed strata. The transitions between domains of different fracture sets may be gradual, recording regional changes, or they may be abrupt where the domain limits are defined by structural or lithologic discontinuities.

Irregular, *non-systematic* fractures can form where the stresses are low and/or nearly isotropic. Fractures that propagate in such systems are easily diverted by sedimentary or other mechanical heterogeneities in the rock. Complex structure can also result in non-systematic fracture systems, with variable strikes and irregular or curved individual fracture planes. Non-systematic fracture systems related to weathering and/or stress release during uplift can be superimposed onto systematic fracture sets related to regional tectonics, forming combined fracture systems that must be filtered carefully to determine which aspects of the system might be analogous to a related subsurface reservoir.

1.3 Fracture Characteristics and Dimensions

1.3.1 Introduction

In the absence of data, the default conceptual model for fractures in a reservoir typically consists of several sets of intersecting generic vertical, open fractures of uniform lengths, apertures, and heights, each with the same impact on permeability. Multiple fracture sets are modeled to be oriented at 90° to each other, and three mutually orthogonal fracture sets form the classic "sugar cube" model of natural fractures in a reservoir. These convenient albeit necessary simplifications can be misleading since the published data consistently show that fracture systems can consist of a single fracture set although multiple sets are common, and that sets of shear and sets of extension fractures have significantly different characteristics and therefore each affects a reservoir in a different manner. Moreover, fractures within any given set are rarely if ever uniform in their dimensions and effects on permeability.

In this section, we will describe differences and similarities between shear and extension fractures as well as their common distributions and dimensions. Where possible, the illustrations and data in this volume are from subsurface examples in order to emphasize the characteristics of natural fractures in situ.

Although fractures within a population are not all uniform in size, many populations have systematic size distributions. Subsurface data usually cannot be used as direct model inputs, but even the limited and truncated fracture datasets that can be obtained from the subsurface provide a foundation for fracture modeling.

The different parameters that can be measured for fractures must be used as interrelated data. For example, if only the strikes of two fracture sets are considered, other significant differences between them may not be recognized. The differences are significant to reservoir production and can include apertures, the expected degree of fracture interconnectivity, and the orientation of fracture planes to the in situ stresses. Shear and extension fractures also have different potentials for vertical extent within a reservoir, impacting the intra-reservoir connectivity across sedimentary heterogeneities. Fracture strike and dip must be used in conjunction with observations of fractography, measures of aperture, and strikes, among others, when building a conceptual model of fracturing and its implications.

1.3.2 Fracture Distribution Patterns

Most fracture populations have numerous small fractures and fewer large ones, more narrow fractures than wide ones, and more close fracture spacings than large spacings. Such distributions are important considerations when assigning fracture attributes to models, since, for example, an assumed average fracture aperture may capture neither the disproportionate effects of the few widest and longest fractures in enhancing reservoir permeability, nor the drainage into those fractures provided by the more numerous small fractures. A detailed understanding and realistic concept of a fracture population is important in correctly populating fracture models, in extrapolating from outcrop data into the subsurface, and in projecting data obtained from a wellbore out into the adjacent formation.

Most histograms of the distributions of fracture heights, widths, lengths, and spacings show patterns that are inferred to reflect either power-law or log-normal distributions. Whereas the shapes of both types of curves tail off to the right where there are increasingly fewer

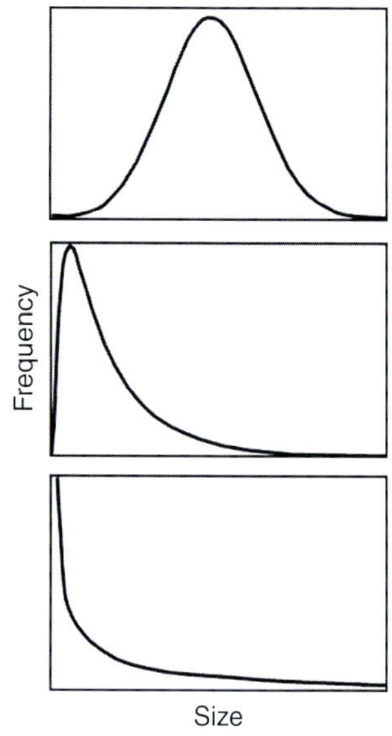

Figure 1.6 The difference between a normal population distribution (*top*), a log-normal distribution (*middle*), and a power-law distribution (*bottom*). Most of the sizes in a normal population would fall into an intermediate range, with relatively few small fractures and relatively few large fractures. However, power-law and log-normal population distributions are more typical of natural fracture populations, with numerous small fractures and increasingly fewer fractures with each larger size category. In a log-normal population, the trend of increasing frequency as size decreases reaches a peak, and then reverses for the smallest fractures (*middle*), whereas frequency continues to increase for ever-smaller fractures in a power-law distribution (*bottom*).

fractures with larger dimensions, the curves are distinct at the left side: if a population fits a power-law distribution, the number of fractures continually increases as the dimension gets ever smaller, whereas if a log-normal distribution applies, the number of fractures increases as the dimension decreases but only up to a point, after which the frequency decreases as size decreases (Figure 1.6).

Statistical analysis is a topic too broad to cover here, and we refer the reader to texts with in-depth coverage such as Davis (2002), and Jensen et al. (2007). In brief, a power-law distribution has one quantity changing proportionally to another, independent of the size. For example, the area of a square grows proportionally to an increase in the length of the sides (Length x Width = Area). Power-law relationships will be linear on a log-log plot and may be the best way to model a distribution if the data indicate a regular change in values regardless of scale. Power-law distributions are common in the natural sciences including geology (Gale, 2014; Gale et al., 2014; Zeeb et al., 2013).

A log-normal distribution is similar to a power-law distribution in that there are numerically more smaller values and a long tail of fewer larger values, and log-normal data will also plot as a line in a log-log plot. However, the frequency trend reverses and decreases at the lower end, where the relationship between size and frequency does not continue as it would in a power-law distribution. Log-normal distributions are also common in natural systems (Limpert et al., 2001).

The true distribution of a fracture population is not always apparent, particularly where the size of a sampled fracture population is limited as from core or an image log. Even where a dataset is relatively robust, the distinction between a power-law and a log-normal distribution can be masked by the recording and plotting techniques, and/or by the scale of measurement (see Ortega et al., 2006).

For example:

1. Failure to log, or the inability to see, the smallest fractures in a population may create an artificial drop-off at the left of a distribution curve, making a power-law distribution appear to be log-normal.
2. In contrast, a log-normal size distribution can be masked by the bin size chosen for the histogram used to display the data, if a limited number of small fractures is grouped with more numerous fractures of a larger size, obscuring the leftward drop-off in the curve that defines a log-normal distribution and making the data appear to fit a power-law distribution.

Even if a fracture population has a power-law distribution where the smallest fractures are most numerous, that part of the population may not contribute significantly to reservoir flow. Anders et al. (2014) and Laubach et al. (2016) suggest that the smallest fractures in a fracture system, i.e. grain-size microfractures, are typically mineralized. Other authors, however, including Ameen et al. (2012), Zeng and Li (2009), and Zeng (2010) report open, grain-scale microfractures that influence reservoir production, contributing up to 1% of the total porosity and 25% of the permeability in reservoirs.

Our empirical experience has been that a log-normal distribution from meter-scale to centimeter-scale fracture dimensions is common, but that there is a jump between centimeter-scale fractures and sub-millimeter, grain-scale microfractures. Although microfracturing commonly precedes macrofracturing as a fracture system develops, the grain-scale microfractures coalesce into macrofractures that are centimeters to meters in scale, with little gradation between them.

Gale et al. (2014) suggest that fracture-aperture distribution patterns can be different in various lithologies, describing log-normal distributions in shales that are different from the power-law distributions common in sandstones and carbonates. Even populations of microfractures can be variably distributed: Hooker et al. (2009; 2014) describe two populations of microfractures

Figure 1.7 *Left*: diagram of the ideal fractography of an extension fracture (adapted from Kulander and Dean, 1985). From the point of origin (1), the fracture propagated both to the left and the right along the plume axis (3), leaving rays or plume branches (2) that diverged from the main fracture plane near the distal edge of the fracture to form twist hackle (4). Arrest lines or ribs (5) suggest incremental fracture propagation. *Right*: a subtle plume structure ornaments the face of an extension fracture (parallel to the plane of the photograph) in a siltstone. The plume records a fracture that originated at an inhomogeneity near the upper edge of the photo and grew both downward and laterally. (A set of younger, narrow, closely spaced, calcite-filled extension fractures, striking normal to the plane of the photograph and marked by faint vertical lines, intersects the plumed fracture face. The absence of interaction between the rays of the plume and the narrow orthogonal fractures suggests that the plumed fracture formed first.)

in the same core, one with a power-law size distribution and one with a log-normal distribution.

1.3.3 Fractography

Assessments of the effects of a natural-fracture system on reservoir permeability begin with identification of fracture type. This in turn starts with recognition of the distinctive fracture-face markings, the *fractography* that is present on many fractures. The markings on shear fractures are entirely different from those found on extension fractures, and fractography can be used to differentiate these two primary fracture types (see Bahat, 1979; Hancock, 1985; Petit, 1987; Kulander et al., 1990; Ameen, 1995; Doblas, 1998). Fractography forms as a fracture initiates and propagates, recording some of the conditions of fracturing and the direction of fracture propagation. The inferences derived from fractography were developed in large part by analogy to similar features found on the surfaces of fractures in materials such as glass and ceramics in the laboratory (e.g. Kulander et al., 1979)

Although not always present, fractography can be used to help assess fracture-controlled reservoir plumbing since an extension-fracture set typically consists of poorly interconnected parallel fractures whereas shear fractures commonly form as networks of conjugate pairs. Moreover, extension fractures commonly form parallel and normal to the in situ stresses whereas shear fractures are more commonly oblique to those stresses, impacting interpretations of how the fractures will behave during changes in stress caused by reservoir production, as well as the potential interactions between natural fractures and hydraulic stimulations (see discussions and examples in Part 3). Where fracture strike data are absent, fractography may provide the only indicator for assessing fracture type and orientation relative to the in situ stresses.

Plume structure (Figure 1.7) is diagnostic of extension fractures, and it is best developed in fine-grained and well-cemented rock. Some of the early authors suggested that plumes form on shear fractures, but most recent publications, most laboratory experiments, and most field evidence suggest that plumes form on extension-fracture faces. Weinberger and Bahat (2008) suggest that different plume patterns can form at different rates of fracture propagation, with well-developed plumes forming during rapid propagation. Slower propagation rates, along with coarse-grained and/or poorly cemented lithologies, may preclude the formation of plumes on extension fractures. Plumes can also be obscured by mineralization or removed by dissolution after fracturing, so the absence of a plume does not exclude an origin in extension.

Plumes typically consist of rays forming systematic arcs that diverge from a central axis on a fracture plane. The central plume axis is often parallel to bedding and located near the middle of a bed, and the divergent rays of the plume record the direction of fracture propagation. Plume axes may also be located nearer the upper or lower bedding surfaces of a layered rock rather than the middle of a bed, or they may follow bedding planes. Some plumes have less definitive and less linear axes, wandering irregularly across a fracture face and suggesting that the fracture grew under conditions of relatively low stress anisotropy.

The parabolic ribs or arrest lines that commonly also occur on plumed fracture faces are typically interpreted as the record of pauses in fracture propagation, although more closely spaced Wallner lines with similar shapes may also form due to other processes involving propagation rates and associated sonic waves (see Frechette, 1972; Kulander et al., 1979). Both ribs and Wallner lines are concave towards the point of origin of the fracture, which is often located at an obvious flaw in the rock such as a fossil, clast, or even an older, intersecting fracture.

Arrest lines indicate the position of a fracture front when the driving stress anisotropy was temporarily reduced below that needed for fracture propagation. They can also result from the extra volume created within the new fracture width which reduced the local pore pressure, making the rock temporarily less susceptible to fracture propagation until the fluid pressures within the aperture was re-established and the rock was again prone to fracturing. Arrest lines do not mark cyclic injections of high-pressure fluids into the fracture.

Twist hackle may form at the distal edges of an extension fracture, most commonly being found where the fracture has propagated to and terminated against a bed consisting of a different lithology. The common interpretation of twist hackle is that the fracture plane changed orientation by a few degrees in a zone of slightly altered stresses at the edge of propagation.

Shear fractures have a wider variety of fractographic markings. Doblas (1998) listed 61 kinematic indicators for shear fractures that can be combined into 11 groups, but many of these indicators occur in rock that is more structurally deformed than that typically of interest to the hydrocarbon industry. The more limited list of fractographic markings that are common on shear fractures in reservoirs includes slickensides, slickenlines, slickencrysts, and steps, while gouge and breccia can develop along larger shear fractures and faults.

Slickenlines form when the two fracture faces slide past each other and their irregularities score linear patterns on the opposing faces. Multiple shear events in different directions may be recorded by superimposed oblique slickenlines, the youngest obscuring or even destroying older lineations. A single set of slickenlines indicates two possible directions for the shear offset, and the 180° ambiguity usually cannot be resolved without correlation points across the shear plane.

Slickensides (Figure 1.8) are created where larger magnitudes of shear offset and/or higher magnitudes of compressive stress normal to the fracture plane during shear create a layer of comminuted rock on the fracture face. The layer of comminuted rock and the fracture face may even have been metamorphosed by the high pressures and local high temperatures created during shear, reducing permeability across the fracture plane. Slickensides may also be marked by cuspate chatter-marks or other of the shear markings illustrated by Doblas

Figure 1.8 Accretionary, congruent steps formed of comminuted host rock and lighter-colored calcite mineralization ornament the face of this shear fracture, indicating that the pictured block moved to the viewer's right and the missing block moved to the left on a strike-slip shear fracture. The magnitude of offset cannot be determined. Handle of a pocketknife, left-center, for scale. (For another example of slickensides see Figure 3.4.)

(1998), and the layer of comminuted rock may form accretionary steps on the fracture surface. The short, abrupt risers of the steps typically face the direction of offset ("congruent steps"; Hancock, 1986; Petit, 1987) and the risers strike normal to the slickenlines found on the longer treads of the steps, although Gay (1970) did experimental work that suggested accretionary steps on slickensided fracture faces can be asymmetric in either direction.

Shear, especially in carbonates and especially where there was a significant compressive stress normal to the fracture plane during shear, can cause minor dissolution of the host rock adjacent to the shear plane. Dissolution is evident from the thin films of insoluble residue that line the faces of these fractures (see Figure 2.27). The process is similar to that which produces stylolites but with typically smaller amounts of dissolution, and shear prevents the formation of stylolite teeth. As with the residues that accumulate along stylolites, insoluble residues on shear planes can be barriers to flow across the fracture faces.

Shear-fracture surfaces in mudstones can be shiny, seemingly polished, and slickenlined, since shear motion aligns the clay platelets in the rock parallel to the fracture face. Such surfaces should also inhibit permeability normal to the fracture plane although the effect may be less important since the matrix permeability of mudrocks is low to begin with.

Shear fractures may be mineralized after shear has taken place, so undeformed crystalline mineralization with no evidence for shear does not preclude an interpretation of a fracture set as originating in shear. On the other hand, the mineralization itself may show evidence for shear in several forms. Mineralization may preserve an impression of the slickensided or slickenlined fracture face on which it was precipitated, or the mineralization itself may have been sheared by offset while or after it was being precipitated in the fracture aperture. In the case of multiple, concurrent events of shear and mineralization, multiple layers of mineralization are commonly truncated and lens-shaped in cross sections cut normal to the fracture plane. Slickencrysts (asymmetric, crystalline deposits), may form with an asymmetric fish-scale pattern (see Lorenz and Cooper, 2018a) when small-magnitudes of shear offset are concurrent with mineral precipitation (Figure 1.9).

Evidence for small-scale shear offset is important, especially where the sampling of a fracture population is small or truncated as from a core or an image log, since shear fractures with even millimeter-scale offsets can form as networks of intersecting conjugate pairs, creating permeability systems that are distinctly different from those formed by fracture sets consisting of parallel extension fractures. Small-scale shear is commonly

Figure 1.9 Stepped shear-fracture faces. *Top left*: asymmetric calcite slickencrysts marked by slickenlines were drawn out by strike-slip shear between the faces of this shear fracture. The pictured block moved to the viewer's left and the missing block moved to the right. *Top right*: non-congruent shear steps ornament the face of this normal dip-slip shear fracture, part of a pair of dip-slip conjugate shear fractures. A similar non-congruent stepped fractography was produced experimentally by Paterson (1958) on shear fractures in the laboratory. The steps are composed of unaltered host rock and indicate that the pictured block moved downward while the missing block moved upward, against the apparent step asymmetry. Bedding offsets show that the magnitude of shear offset was on the order of a millimeter. The short risers on these steps formed as connectors between the fracture segments that form the step treads when the rock broke along the shear plane during road construction. *Bottom*: a vertical, strike-slip fracture ornamented with small non-congruent steps similar to those shown in the upper right photo, but with an orientation that indicates minimal left-lateral offset. The high points of the steps are slickenlined, having been modified by a small amount of additional shear. Continued left-lateral offset would have destroyed these steps and created new ones with a reversed asymmetry out of comminuted rock.

recorded by asymmetric steps, superficially but misleadingly resembling those created out of comminuted rock on slickensided fracture faces. These steps consist of unaltered rock, and the long treads of the steps are small en echelon fractures that formed along and within a narrow zone of shear. The abrupt risers that connect these treads, non-intuitively facing *away* from the direction of offset ("non-congruent" steps), are created when the rock breaks open along the shear zone, connecting the en echelon segments. Small, sometimes lunate steps may be scattered across such fracture faces. The high points of the steps may be secondarily slickenlined if shear continues, but shear of any significant magnitude converts the steps to comminuted rock, forming accretionary steps with the opposite, "congruent" sense of asymmetry relative to the direction of shear.

In contrast to these indicators of small shear offset, gouge and breccia indicate shear fractures with large offsets. Antithetic shear fractures oriented oblique to the shear plane may even form as a shear fracture grows to become a fault.

As described by Anderson as early as 1905, shear-fracture fractography will record normal dip-slip, reverse dip-slip, or strike-slip offset (depending on which of the three axes were the maximum, minimum, and intermediate stresses) in the ideal condition where the three compressive stresses are vertical and horizontal. However, many shear fractures show raking slickenlines that record oblique slip, having formed in more complex stress systems or where the stress system changed after fracturing and the shear plane was reactivated within the reoriented stress field. Reoriented stress fields can develop as a fold tightens or as a fold migrates through a formation, and raking shears may be expected within such structurally complex settings. Ideal shear fractures can also be tilted with bedding during post-fracture deformation so that the fractography records raking offset relative to the present orientation of the fracture.

1.3.4 Fracture Dip Angles

Fracture dip angle is a simple, easily understood characteristic in non-complex structural settings. Most extension fractures are vertical or nearly so, since they form in the plane defined by the maximum and intermediate compressive stresses and since the vertically acting weight of the overburden provides the maximum compressive stress in such settings. Most vertical extension fractures are normal to bedding, not because bedding controls dip but because bedding is commonly horizontal.

Topographic relief and structural complications can create stress systems that are inclined relative to vertical which can result in inclined extension fractures with intermediate or even low dip angles. Inclined extension fractures may also have originated as vertical, bed-normal planes that later became tilted along with bedding during folding, and bed-normal but non-vertical extension fractures may form during folding due to extension on the outside of curvature (Figure 1.10). Occasionally fractures that are inclined relative both to vertical and to bedding form during folding of a layered formation where bed-parallel shear (flexural-slip) at the sedimentary contacts above and below a fractured bed sets up a local bedding-oblique extensional stress system.

Most extension fractures are planar, but the planes can be curved if the rock was being twisted as it fractured. Extension fractures with plume markings on curved faces occur within strike-slip structural settings where impingements at asperities and variations in throw along irregular faults create complex stress conditions that change during faulting.

The ideal dip angles in simple structural settings, described by Anderson (1951), are vertical for strike-slip shears, 60° for normal dip-slip shears, and 30° for reverse dip-slip shears (Figure 1.3). Intermediate-angle, hybrid extension-shear fractures can form as conjugate pairs in any of the three configurations, but the intersection angles are less than the 60° ideal angle (Hancock, 1985). In the laboratory, the acute conjugate intersection angle for conjugate shear-fracture pairs decreases as the confining pressure on the rock decreases (e.g. Paterson, 1958). The conjugate pairs of normal and reverse dip-slip shears have parallel strikes and opposing dip directions, a useful relationship to remember when reconstructing fracture systems from the one-dimensional samples afforded by cores and image logs.

As with extension fractures, shear planes with non-ideal dip angles can also form where the stress system is inclined, as in complex structural systems, and can be found where the strata were folded or tilted after fracturing. The dip angles of shear fractures that cut across multiple lithologic units can change due to variations in the mechanical properties by layer (Figure 1.11). Finally, certain kinds of extension and shear fractures can have horizontal dip angles, commonly parallel to bedding, as described in Section 1.5, "Other Fracture Types."

1.3.5 Fracture Distributions

Fractures in heterogeneous formations are distributed neither uniformly nor randomly, so stochastic models predicated on random or probability-based fracture distributions make less than full use of geologic data and theory (e.g. Loosveld and Franssen, 1992). Fracture systems occur in domains where the fractures have similar characteristics, and domains can vary laterally as well

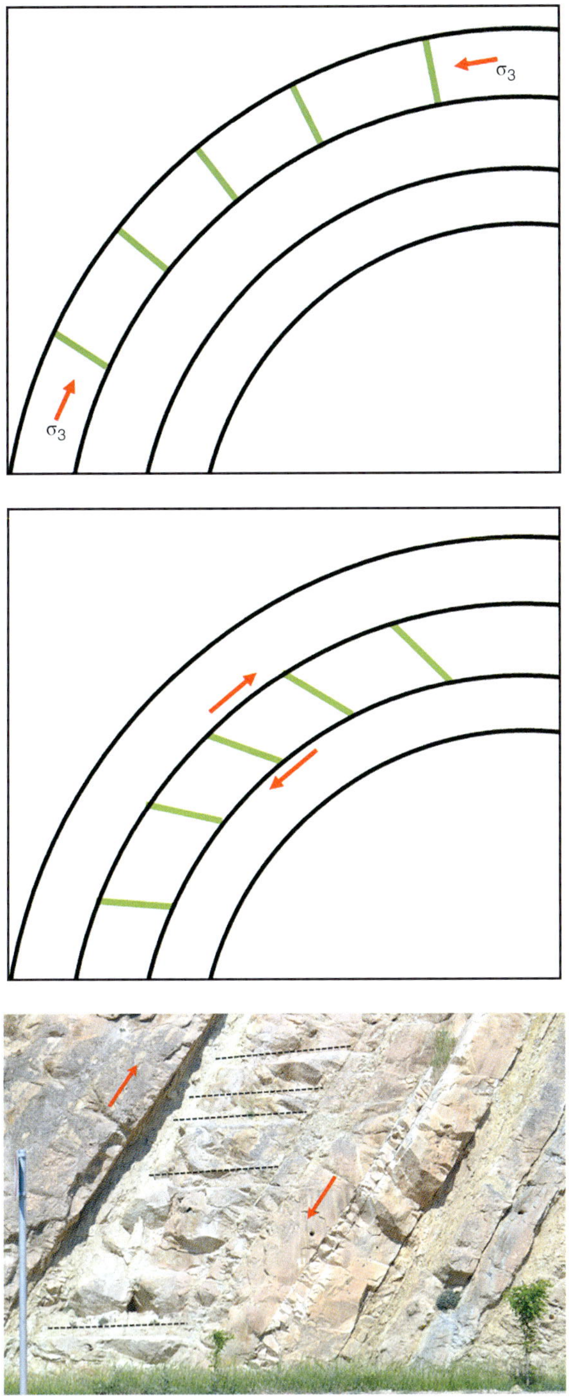

Figure 1.10 *Top*: bed-normal, inclined extension fractures can form due to extension of the outer layers of strata in a fold (the layer is not in tension, but rather the minimum compressive stress σ_3 is reduced to form a stress anisotropy large enough to break the rock), and they can form as vertical fractures prior to folding, becoming tilted as the strata fold. *Middle*: more rarely, extension fractures that are inclined relative both to vertical and to bedding can form on a fold due to oblique extension within a bed between two bed-parallel shear zones. *Bottom*: an outcrop example of this type of extension fracturing (parallel to the black lines) in tilted limestones of the Eocene Pila Spi Formation in northern Iraq.

as vertically. Strain can be accommodated by different fracture types in the different lithologies of a heterogeneous formation, called "strain partitioning." Fractures are typically more common in, and can even be limited to, the more brittle beds of a formation (Figure 1.12); if all units are fractured, extension fractures may be more widely spaced in the more ductile beds than in the brittle beds, or the ductile units may contain shear fractures while extension fractures occur in the relatively brittle lithologies (e.g. Lorenz et al., 2002; Lorenz and Cooper, 2018b, 2018c).

The rule of thumb that fracture intensity is a function of ductility/brittleness can be useful, but beds that are brittle today would not necessarily have been brittle during strain events that occurred in the geologic past. The fracture susceptibility of rock can change over time due to diagenesis, cementation, burial depth, temperature, compaction, pore pressure, confining stress, and strain rate. An extensive but proprietary fracture database compiled from core cut from marine strata shows the presence of three extension-fracture sets with different strikes in a reservoir: one fracture set is restricted to the dominant mudstones, one set is restricted to the interbedded limestones, and one set cuts indiscriminately across both lithologies. This system of three fracture sets records an evolution in the mechanical properties and related fracture susceptibility of the three lithologies during the course of three separate fracture events.

In another example, a coarse-grained arkose of the Abo Formation in New Mexico is cut by an early set of normal dip-slip conjugate shear fractures, and a younger set of extension fractures that strike nearly normal to the shear fractures. These two fracture sets record a change in rock properties and/or the stress conditions between the two fracture events (Lorenz and Cooper, 2018b).

Since mechanical properties can vary by lithology, the mechanical stratigraphy that controls fracturing in many sedimentary formations correlates to conventional stratigraphy. In more heterogeneous formations an irregular mechanical stratigraphy may control a similarly irregular fracture distribution. Where the mechanical contrasts between units of a heterogeneous formation are low, for example where all units are heavily cemented regardless of grain size or composition, they may form an overarching mechanical unit composed of multiple sedimentary units, and fractures may extend across the otherwise heterogeneous package (Figure 1.13).

Fracture distribution and intensity depend not only on lithology but also on structural setting. For example, extension fractures may be more closely spaced along the more abruptly folded hinge of an anticline, where percent strain was higher than on the flanks of the fold. Similarly, shear-fracture intensity commonly increases

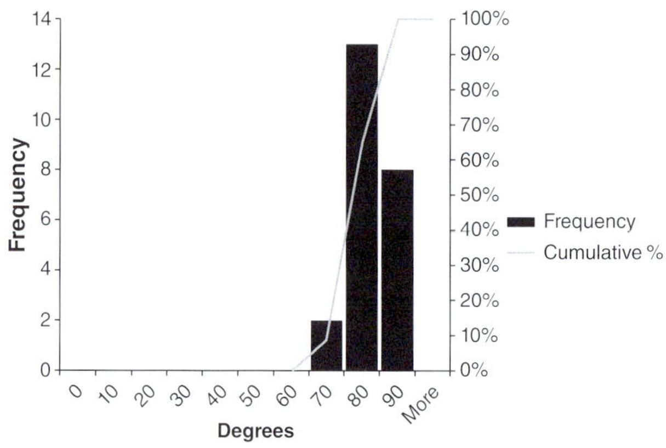

Figure 1.11 Fracture dip angles: *Left*: histogram of the non-ideal dip angles of 23 strike-slip shear fractures sampled by a horizontal core (unpublished data). *Right*: the effect of lithology on dip angle: an inclined dip-slip shear fracture in core becomes steeper where it crosses a layer of stiffer rock in a 4-inch (10-cm) diameter core. Fractures can transition from extension fractures to shear fractures and back again to extension fractures as they extend from one bed to another in a heterogeneous lithology.

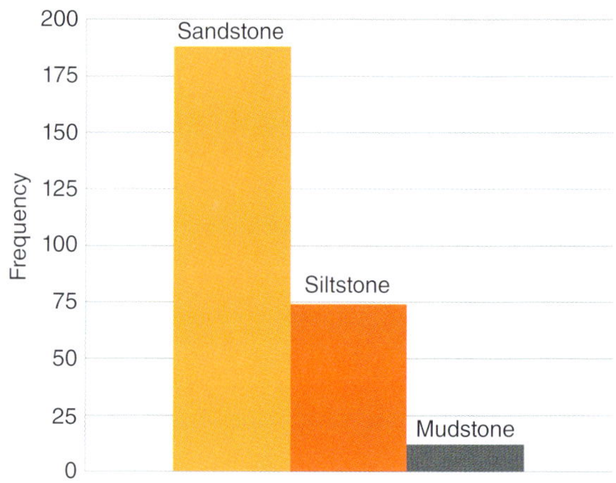

Figure 1.12 *Left*: the frequency of extension fractures by lithology from 4,200 ft (1280 m) of vertical core from the Cretaceous Mesaverde Formation, Colorado. Sixty-eight percent of the 274 cored fractures occur in sandstones, which comprise only about 30% of the cored section, suggesting that the sandstones are intensely fractured compared to the mudstones, which contain only 4% of the cored fractures but comprise some 50% of the section (from Finley and Lorenz, 1988). *Right*: short, strata-bound fractures that are restricted to the limestone layers in a thin-bedded limestone-marl sequence (Cretaceous Twin Bridges Formation, shallow marine, New Mexico).

near faults, particularly in the hanging walls of normal dip-slip faults (Figure 1.14) (e.g. Nelson, 2001; Withjack et al., 1990). Fracture intensity increases both with the brittleness of the strata and with percent strain, so it can be thought of as fitting into a diagram where brittleness increases on one axis and with degree of deformation on the other axis: fracturing should be best developed in brittle rocks near a fault and least well developed in ductile strata distant from a fault.

Gross and Eyal (2007) described a field example from a folded, heterogeneous, layered carbonate formation that illustrates the combined effects of lithology and strain. The more ductile layers on the broad fold are cut by strata-bound shear fractures whereas strata-bound extension fractures formed in the more brittle layers. With increasing strain, the fractures coalesce into fracture swarms and faults that cut across all lithologies (Figure 1.15).

Turcott (1986) was among the early authors to describe natural-fracture distributions as fractal, i.e. that the distributions of fracture dimensions are similar at numerous scales of observation, and fractal fracture distributions have been reported from outcrops (e.g. Li et al., 2018). However, Loosveld and Franssen (1992) suggested there may be distinctions between the distributions of extension and shear fractures, citing published data for deformation-band shear fractures that are fractally distributed and documenting extension fractures in two

Figure 1.13 *Left*: parallel-striking extension fractures with irregular heights and spacings have an irregular distribution in a heterogeneous fluvial sandstone. Some fractures are vertically through-going, others are limited by some of the internal bedding planes (Cretaceous Mesaverde Formation, Colorado). *Right*: tall extension fractures cut vertically through some 120 ft (40 m) of homogeneous eolian quartz sandstone (Permian Coconino Sandstone, Arizona).

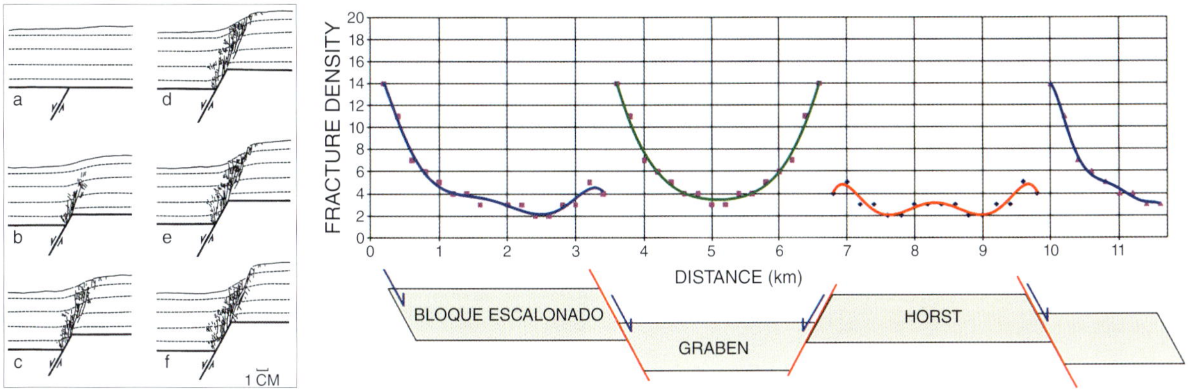

Figure 1.14 *Left:* Withjack et al. (1990) documented the enhanced fracturing that forms as part of a fault-related process zone in their experimental models. Most of the fractures form in the hanging wall. (Reprinted from Withjack et al., 1990, with permission from AAPG, whose permission is required for further use). *Right*: a subsurface fracture model based on outcrop studies across a 7.5 mile (12 km) transect in a normally faulted terrane in the Talara Basin, Peru, highlights the expected increases in fracture density and concentration in the hanging wall near normal faults in an extensional setting (adapted from Roldan et al., 2013).

well-exposed outcrop examples that are not. Odling et al. (1999) also reported that the sizes of shear fractures, but not strata-bound extension fractures, are "often" fractally distributed.

To be truly fractal, all fracture dimensions should follow the fractal pattern, so this concept may not apply to fractures where various dimensions were controlled by different processes. For example, extension fractures heights and spacings are controlled in part by bedding thickness whereas width and length are controlled largely by percent strain. Moreover, whereas a fractal fracture distribution might develop in a homogeneous system, the presence of structural or lithologic heterogeneities such as a pre-existing fracture set or a strong crossbedding fabric introduces local controls on fracture distributions, so the concept must be used carefully.

1.3.6 Fracture Heights and Terminations

Fracture-height data provide insights into the vertical continuity and interconnectivity of a fracture system within and between reservoirs. Unfortunately, many subsurface fracture-height datasets are severely truncated since "vertical" fracture planes are not always oriented parallel to the axis of a vertical wellbore or core, thus the fractures may exit a core before terminating (Figure 1.16). Horizontal wellbores and cores capture an even smaller percentage of a fracture set's full vertical height data since most fracture heights exceed the diameter of the wellbore or core. Where complete fracture heights are not captured, the partial/minimum fracture heights that can be measured offer insights into the population of heights in a fracture system, especially when

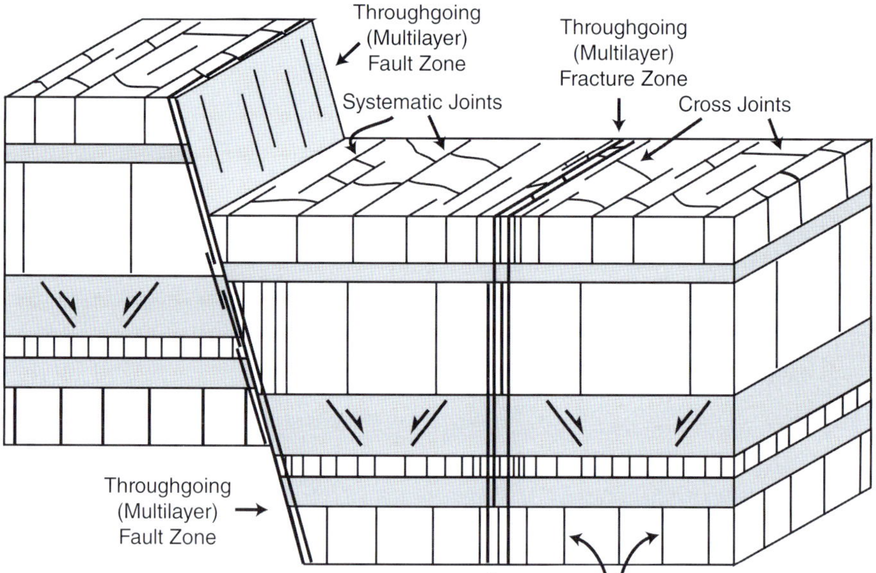

Figure 1.15 Schematic illustration of the distribution of shear and extension fractures in a layered carbonate formation, and the coalescence of some of those fractures to form throughgoing, interconnected fracture systems (from Gross and Eyal, 2007).

Figure 1.16 A calcite-mineralized, inclined extension fracture in a Paleozoic marine shale, Algeria. The measurable height of this fracture is less, by an unknown amount, than its total height since it exits the core both upward and downward before terminating. Nevertheless, the truncated height is still useful as a minimum-height data point.

combined with the related data on vertical termination locations and types that *are* frequently captured in a core.

The heights of fractures in a developing extension-fracture system are initially controlled by the amount of strain imposed on the formation, and circular fractures commonly grow both vertically and laterally in proportion to that strain. Heights are initially likely to be log-normally distributed, with a few taller fractures growing and accommodating strain at the expense of the numerous short fractures. However, the stress differentials and related strains that are required to propagate extension fractures are relatively small, so height growth is commonly arrested by bedding planes and the lithologic contrasts which create mechanical barriers to fracture growth in sedimentary rock. Vertical growth stops once the fractures propagate to one of these boundaries, but continued strain causes the smaller fractures to grow. Well-developed extension-fracture sets commonly become "strata-bound," where fracture heights become consistent, correlating nearly 1:1 to bed thickness (Figure 1.17).

If full fracture heights have not been captured by a core or image log, the observation that fractures consistently terminate at lithologic boundaries can still suggest that fracture heights are related to bed thicknesses, and moreover that these fractures are likely to be extension fractures, providing a constraint on the conceptual fracture-permeability model. In contrast, fractures that consistently cut across minor lithologic boundaries are likely to be shear fractures and taller than bed thickness (e.g. Odling et al., 1999). The difference between the two fracture types impacts inferences on the vertical, fracture-controlled permeability in a reservoir. In the absence of fractography (i.e. where the data are from image logs), the vertical relationship to bedding is one of the criteria that can be used to differentiate extension from shear fractures.

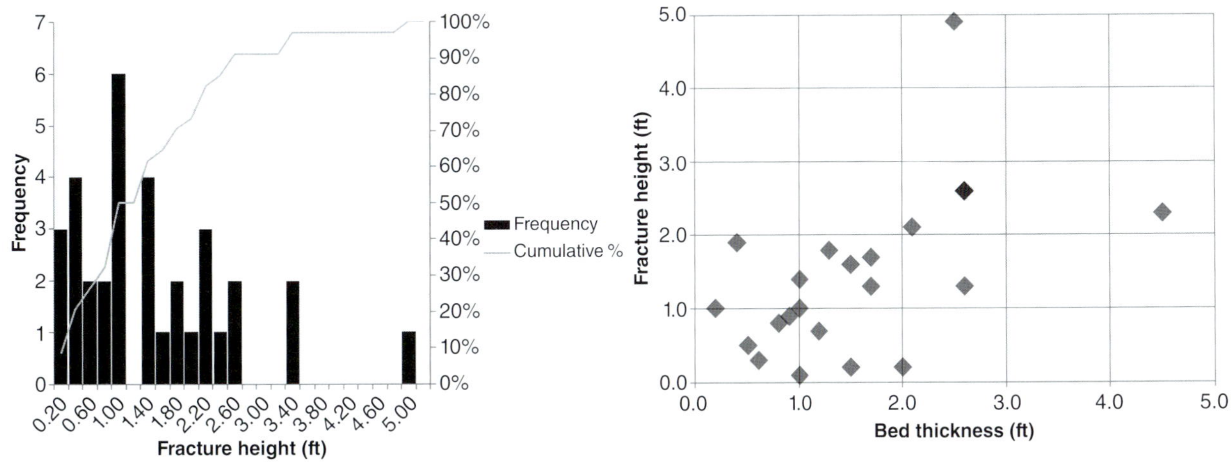

Figure 1.17 *Left*: histogram showing the height distribution of bed-normal extension fractures in vertical core cut from a formation consisting of interbedded marine sandstones and calcareous mudrocks. The fractures occur preferentially in the mudrock layers, terminating at the contacts with sandstones. The height population is truncated since the full heights of about 16% of the fractures were not captured by the core (n = 34; min 0.10 ft, max 4.9 ft, ave 1.41 ft). The fracture-height histogram appears to have a log-normal distribution, but it is controlled by bedding thickness. *Right*: a cross plot shows a loose correlation between fracture heights and the thicknesses of their host layer for this dataset. Only a few fracture heights exceed the thickness of their host bed (n = 22). Darker points indicate overlapping data. The cross plot dataset is smaller than the fracture-height dataset since host bed thickness could not always be determined (unpublished data).

1.3.7 Fracture Lengths

Many fluid-flow models require information on fracture length, and lengths cannot be obtained from subsurface data. Outcrop pavements provide the only direct measurements of fracture lengths, provided that the dimensions of the pavements exceed most of the fracture lengths. Even if the length data are truncated by the limits of an outcrop, they still offer useful minima in the same way that truncated fracture height data are useful.

Determining fracture lengths on pavements can be tricky since the geometry of fractures in the third dimension is usually unknown and it is not always obvious whether closely spaced, overlapping fracture tips (Figure 1.18) represent the blind terminations of two offset fractures or whether the overlaps indicate fracture segments that join into a single plane in the third dimension (see Vermilye and Scholz, 1995, for a summary of cautions when measuring fracture lengths in outcrop; see Ortega and Marrett, 2000, for a list of criteria useful in determining the probable lengths of interconnected fractures on pavements displaying fracture segments).

Maximum reported extension-fracture lengths logged with a tape measure on an outcrop are on the order of hundreds of meters; longer fractures are measurable on larger pavements with remote imagery, but the shorter fractures may not be measurable using this technology. Published outcrop-length histograms (Figure 1.19) suggest that, as with widths, spacings, and sometimes heights, and regardless of whether the longest fractures of a set are meters or hundreds of meters long, distributions of lengths within sets of extension-fracture populations show either power-law or log-normal distributions (e.g. Segall and Pollard, 1983; Laubach, 1992; Lorenz and Laubach, 1994; Wennberg et al., 2007).

In terms of reservoir engineering and fluid flow, the "lengths" of the fractures in a well-developed set of parallel extension fractures are probably effectively infinite within a given fracture domain due to closely spaced and locally hooking overlaps of the fracture planes, and to interconnectivity in the third dimension. A set of short, poorly developed fractures, however, is less likely to be interconnected along strike. Well tests can often provide assessments of effective fracture lengths, but effective lengths can be qualitatively estimated from measured outcrop fracture lengths in combination with related fracture parameters such as fracture spacing, aperture, and lateral termination types.

Where two sets of extension fractures are present in an outcrop, their lengths may have been controlled by different factors. The lengths of the older set were probably controlled by stress and strain magnitudes, i.e. fractures lengthened and new fractures initiated, as long as the rock was subject to stresses capable of fracturing the rock. In contrast, the lengths of the younger fracture set may have been controlled in part by the imposed stress and in part by the spacing of the older set, especially if the older set was open or poorly mineralized at the time the younger set formed (e.g. Laubach 1992). However, if the fractures of the older set are narrow and completely healed by mineralization, they may not have created significant mechanical weakness planes in the

Figure 1.18 *Left*: two extension fractures exposed on a shale bedding surface hook towards each other where the tips overlap, suggesting that they propagated towards each other. Hooks commonly connect such fractures, extending their effective length. Discoloration along the fracture planes suggests fluids flowed along the fractures. *Right*: a bedding-surface pavement of the Frontier Sandstone in Wyoming cut by numerous bed-normal extension fractures. The fractures are uniformly distributed in the 10 ft (3 m) thick shallow-marine sandstone, and the fracture population consists of fractures of varying lengths and spacings. The inset depicts the scan line (red dashed line) that was laid out across this outcrop, and the positions and lengths of the fractures intersected by the line. The fractures shown with arrow tips indicate truncated measurable lengths where the fractures extended under cover, or extended to and presumably beyond the edge of the outcrop. Only the fractures that crossed the scan line are portrayed. Minimum, average, and maximum lengths of the fractures in this outcrop are 6 m, 32 m, and 135 m (20 ft, 104 ft, and 443 ft), respectively, and the length population has a log-normal distribution (see Lorenz and Laubach, 1994).

rock, and the younger fractures may cut indiscriminately across them.

Most of the compiled data on shear-fracture lengths are measurements of faults, and present cross plots between fault length and fault offset rather than histograms of fault lengths (e.g. Marrett and Allmendinger, 1990), so the distribution of shear-fracture lengths is not apparent. Moreover, the "length" of a shear fracture depends in part on whether it formed in normal dip-slip, reverse dip-slip, or strike-slip. Shear fracturing typically involves displacement and volume constrictions at the end of a fracture, where additional strain-accommodation structures including splay faults, anticlines, and fracture corridors are common, adding to effective fracture lengths and increasing the probability of fracture intersection.

Few datasets report and compare the lengths of shear fractures and extension fractures. Vermilye and Scholz (1995) provided information on mineralized extension fractures up to 80 ft (25 m) long and exposed on several large pavements. They also found relationships between width and length for extension fractures, varying from outcrop to outcrop but consistent within a given fracture domain, the ratio of width to length varying between 1:1,000 and 1:8,000. The authors compared these ratios to the consistently-lower reported ratios between the lengths of faults and their offset magnitudes, which are on the order of 1:30 to 1:250. Vermilye and Scholz implied that the fault ratios can be extrapolated down-scale to shear fractures of the same general size as their extension fractures.

1.3.8 Fracture Widths, Apertures, and Mineralization

Fracture width and the remnant unmineralized aperture within that width control the permeability of individual fractures. With a few caveats, direct width data are relatively easy to obtain from cores, although measurable widths are rarely equivalent to effective widths. The capacity for fluid flow along an ideal fracture that consists of a slot between smooth, parallel walls is neatly proportional to the cube of the open fracture width (e.g. Warren and Root, 1963; Reiss, 1980), and individual *un*mineralized extension fractures have geometries that are perhaps close to that ideal, wedging out only near the fracture tips. Most *un*mineralized extension fractures also have relatively smooth, planar walls, even where ornamented with plumes, and the width of an unmineralized extension fracture can often be reasonably captured with a single measurement.

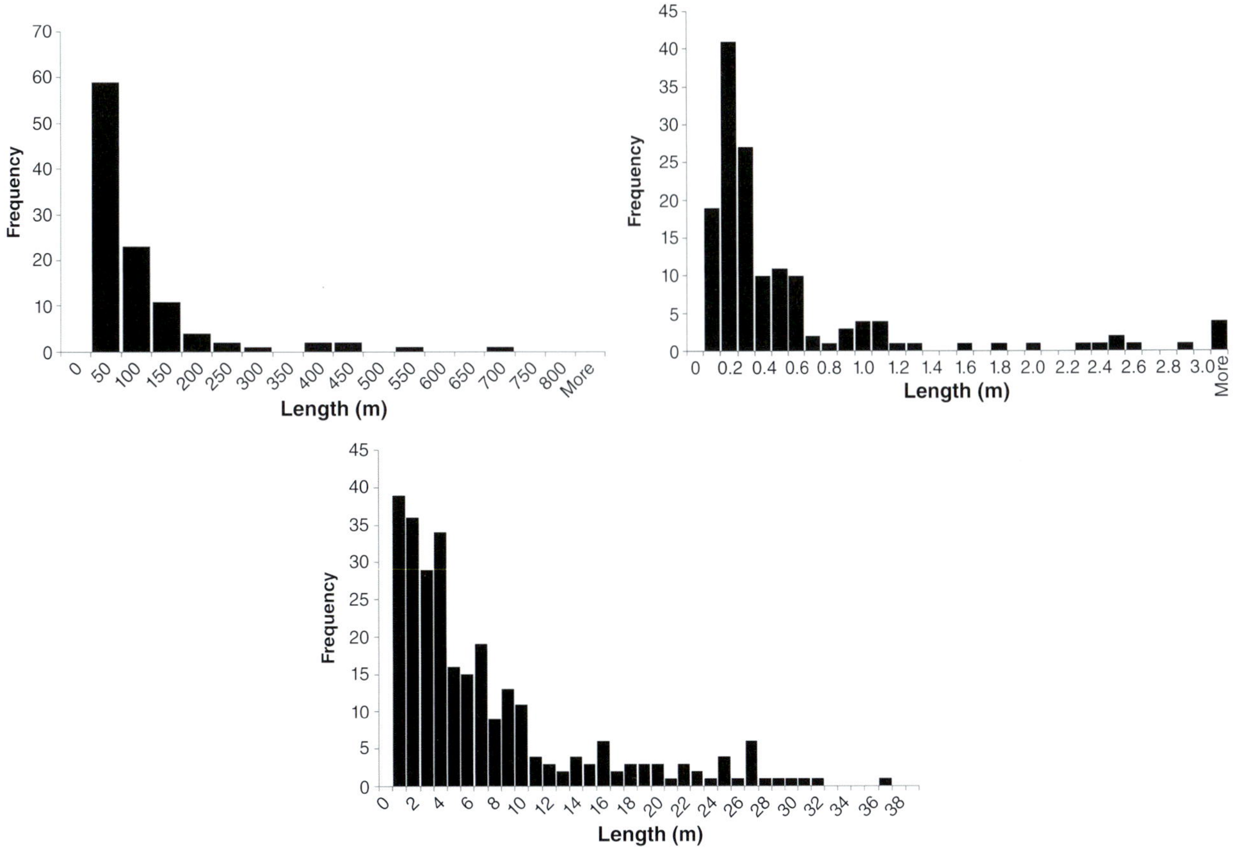

Figure 1.19 Distributions of fracture length populations follow log-normal or power-law patterns at scales ranging from meters to hundreds of meters. *Top left*: Cretaceous sandstones of the Frontier Formation in Wyoming; 107 lengths measured from Google Earth (adapted from Laubach et al. 2016). Minimum, average, and maximum lengths are 6.2 m, 90 m, and 600 m (20 ft, 300 ft, and 2,000 ft), respectively. *Top right*: 142 lengths of "fracture segments" measured in Triassic rock (from Vermilye and Scholz, 1995). The segments are commonly linked end-on to form longer composite fractures over 12 m (40 ft) long. *Bottom*: unpublished length data from Cretaceous sandstones of the Mesaverde Formation in Colorado, n = 279, maximum length 37 m (120 ft). The length distributions for extension fractures in granites can also follow similar patterns (Segall and Pollard, 1983).

However, even if they started as parallel-plate slots, natural-fracture widths have usually been reduced by mineralization and/or enhanced by dissolution one or more times over the course of geologic time, and present aperture geometries rarely resemble slots of uniform width. Extension fractures that are mineralized or that have been subjected to dissolution have irregular widths (Figure 1.20), and shear-fracture widths are inherently irregular whether they have been mineralized or not.

The necessity of modifying the width parameter for use in assessing fluid flow along irregular natural fracture planes, to account for roughness and tortuosity, is well recognized, and techniques have been proposed to address it (e.g. Pryak-Nolte et al., 1980; Zimmerman and Bodvarsson, 1996; Barton, 2007; Hooker et al. 2014). Additional difficulties arise from the fact that populations of widths within a fracture set are not uniform, commonly having, once again, log-normal distributions, with many narrow fractures and increasingly fewer wide apertures (e.g. Gale, 2004). The width distributions are systematic and can be mathematically characterized within each fracture set, but no formula yet describes a more universal frequency-width distribution for extension-fracture populations.

Fracture widths are useful in calculating percent strain, but remnant open aperture is the important parameter when assessing the contribution of fractures to reservoir porosity and permeability. The relationship between fracture width and open aperture, or the more easily estimated percent remnant porosity within the mineralized fracture width, as described in Part 2, can be a useful measure in determining which of the fractures in a set contribute most effectively to reservoir system permeability. Some datasets show that the widest fractures in a set also have higher percentages of remnant aperture, suggesting that they are the most important contributors to reservoir permeability. More commonly, however, cross plots between width and remnant porosity show a

Figure 1.20 Irregular open fracture apertures. *Left*: an incompletely mineralized inclined fracture in a Cretaceous chalk from Tunisia, pencil point for scale. Calcite bridges the fracture width but there are irregular, millimeter-scale, connected, remnant voids that will accommodate significant fluid flow. (The small right-stepping offsets and the dip angle suggest these are segments of a dip-slip, en echelon, shear fracture system formed with a narrow shear zone rather than along a single shear plane, a variation on the en echelon pattern shown in Figure 1.5). *Middle*: the face of a vertical extension fracture in a Cretaceous sandstone from Colorado, captured by an inclined core, is covered with isolated but closely spaced calcite crystals. The fracture has relatively planar host-rock walls so width is easily measured, but the effective aperture and estimates of the fluid flow capacity through that aperture are not as easily quantified. Flow through the fracture will be channeled and turbulent. *Right*: a shear fracture in an eolian sandstone (Pennsylvanian, Wyoming) has an irregular, discontinuous, partially mineralized width. The millimeter-scale shear offset is indicated by the displaced oil-stained layers. The fibrous tan material within the aperture at the top of the photo is Lost Circulation Material (commonly known as "LCM" and in this case consisting of shredded cedar bark), added to the drilling mud to minimize the mud loss into the fractured reservoir while cutting the core. All three cores are four inches (10 cm) in diameter.

roughly inverse correlation within a wide scatter of data (Figure 1.21.1), suggesting that although the wider fractures were the most effective conduits for fluid flow early in their history, they were more quickly and more completely plugged by mineralization. Some cross plots show more irregular data distributions, without a trend, typically because there has been significant dissolution along the fracture planes (Figure 1.21.2) or because several fracture populations with different characteristics have been combined in the plots. Still other fracture populations have a relatively narrow distribution of widths but significant variability in the degree of mineralization within those widths (Figure 1.21.3).

The three fracture populations provided to illustrate common width and remnant-fracture-aperture distributions for fracture sets (Figures 1.21.1, 1.21.2, and 1.21.3) were measured in different cores, cut from different lithologies in three different tectonically inactive, mid-basin settings. The techniques used to acquire the data are described in Part 2. The first and third cores are horizontal, the second is vertical. Two of the cores are longer than those of most core programs and therefore the datasets are more comprehensive, offering relatively complete characterizations of the actual fracture properties in the subsurface.

The first example (Figure 1.21.1) shows measurements characterizing a regular set of parallel, partially mineralized, vertical extension fractures in a calcareous marine mudrock. The more irregular widths and apertures of the second example (Figure 1.21.2) come from vertical-extension fractures in a dolomite formation where significant dissolution along the fracture planes left highly irregular fracture widths and remnant apertures that were further altered by incomplete remineralization. The third example (Figure 1.21.3) documents widths and apertures in a set of parallel, narrow, incompletely mineralized right-lateral strike-slip shear fractures, with stepped surfaces and sub-millimeter offsets, in a marine sandstone. An opposing subset of what could be a conjugate shear pair is not present in this core since the strain in that direction was accommodated by a nearby larger, more localized, left-lateral strike-slip fault with a strike close to the ideal of 60° clockwise from the strikes of the right-lateral shear fractures. The potential for fracture widening as a core is retrieved from the subsurface is discussed in Part 3, but it is not considered to be an issue in these three cores where the fracture widths are bridged or partially bridged by mineralization and therefore fixed.

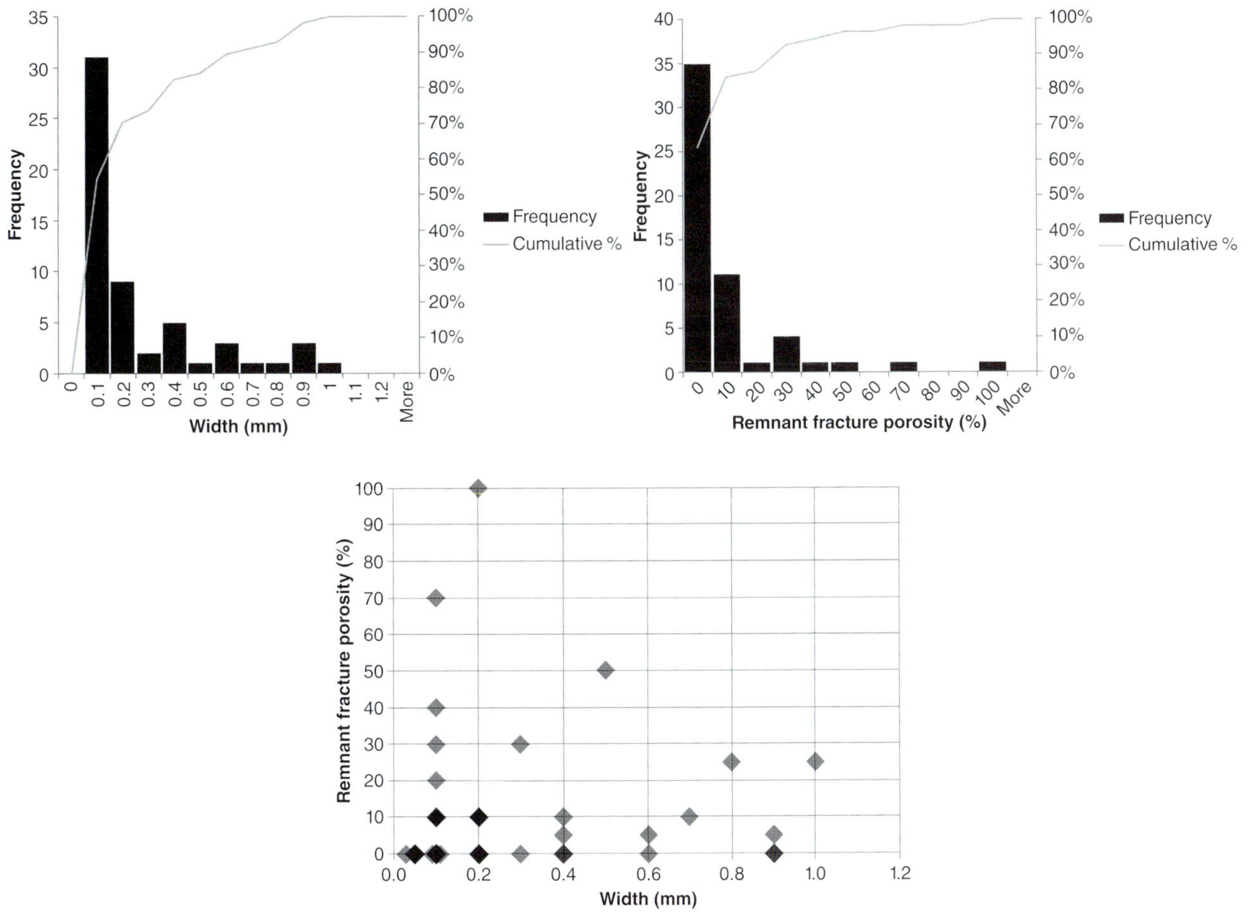

Figure 1.21.1 Histograms and a cross plot showing the distributions of fracture widths, and the remnant fracture porosities within those widths, for a set of vertical extension fractures captured in 470 ft (143 m) of horizontal, 4-inch (10 cm) diameter core cut from a calcareous marine mudrock. *Upper left*: fracture widths (min 0.03, max 1.00, ave 0.25 mm). *Upper right*: remnant fracture porosities (min 0, max 100, ave 8.8%). *Bottom*: fracture width to remnant fracture porosity cross plot (n = 57). Darker points highlight overlapping data (unpublished data).

Characterization of the irregular widths and apertures of shear fractures is problematic. The relatively uniform distribution of shear-fracture widths plotted in Figure 1.21.3 merely reflects the necessity of assigning one number to the width of each fracture. Although the plot indicates a generally restricted range of widths within the fracture population, it does not adequately characterize the degree of variability of widths along the individual fractures, since the points of contact between the irregular opposing fracture walls alternate with open void spaces twice the size of the assigned width. (If "zero width" is defined as intact rock, then any break in a rock must have some width. Functionally, most unmineralized cracks in rock no matter how narrow provide more permeability than intact matrix rock.)

The difference between shear and extension fracture widths results from the dissimilar mechanisms by which width forms in the two types of fractures. Width in an extension fracture, where fractures open normal to the fracture walls, is a direct record of the amount of strain the rock has accommodated, and opening extension fractures increases the volume of the rock.

In contrast, strain is accommodated in a shear fracture system by fracture-parallel offset along a plane which does not by itself increase the volume of the rock because it does not produce any width: no width results from shear along a perfectly planar and smooth fracture regardless of the magnitude of offset. Width in a shear fracture is created by the offset of asperities along the fracture plane. Regardless, the widths of both fracture types are commonly altered by dissolution and mineralization.

1.3.9 Fracture Spacing

More papers have probably been published on fracture spacing than on any other fracture dimension, in part because spacing is important, impacting connectivity and system permeability, in part because spacing is a useful measure of strain and the degree of fracture

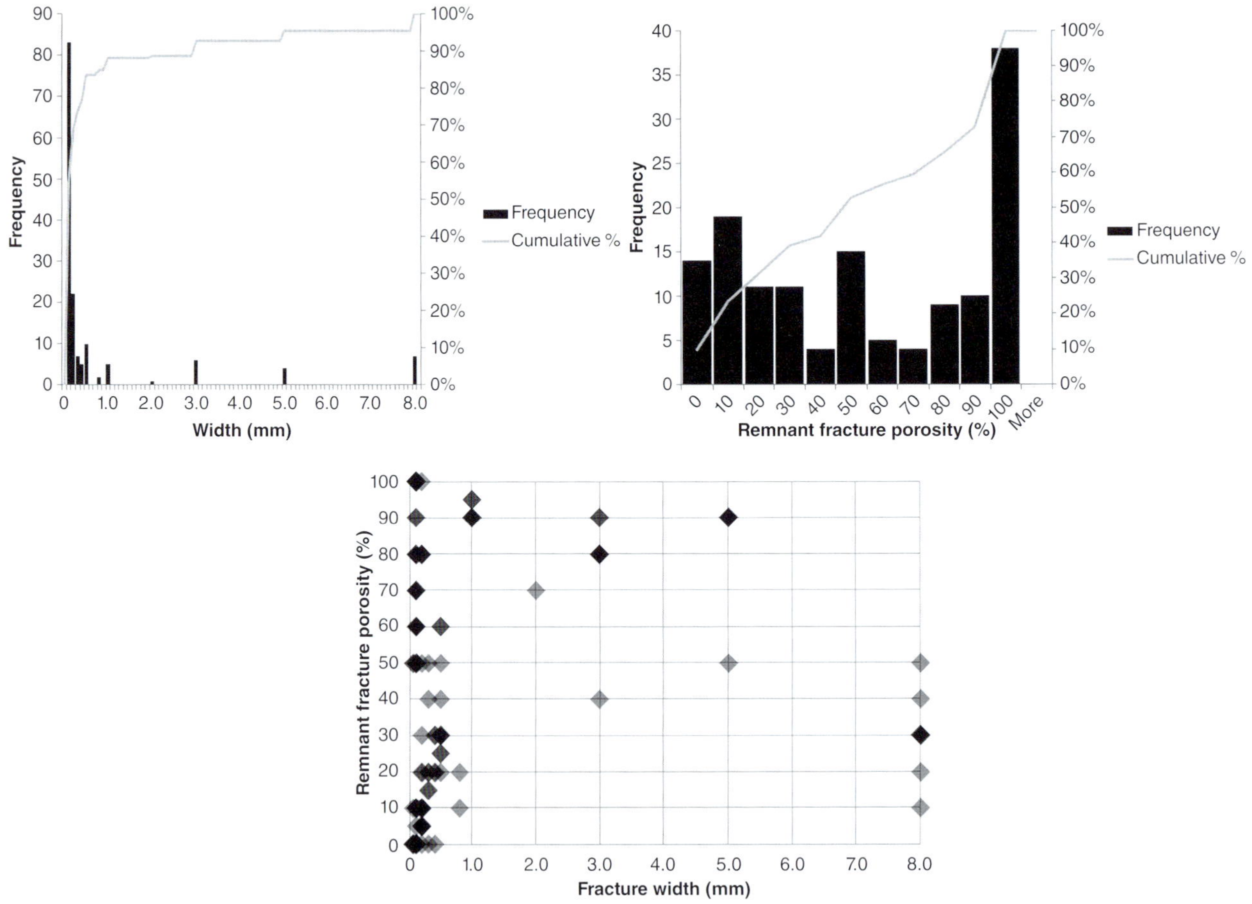

Figure 1.21.2 Histograms and a cross plot showing the distributions of fracture widths, and the remnant fracture porosities within those widths, for a set of vertical extension fractures captured by 960 ft (290 m) of vertical, 4-inch (10 cm) diameter core cut from the Ordovician Arbuckle dolomite in Kansas. *Upper left*: fracture widths (min 0.05 mm, max 8.00 mm, ave 0.82 mm). *Upper right*: remnant fracture porosities (min 0%, max 100%, ave 54%). *Bottom*; the cross plot of fracture widths and remnant fracture porosities (n = 152). The non-systematic patterns in these figures reflects the chemical history of the fractures which includes dolomitization of the host rock, fracturing, fracture mineralization, dissolution along the fractures, and remineralization of the dissolution-enhanced fractures. The core is not oriented so fractures could not be segregated by strike, but the rough groupings shown in the remnant porosity histogram and the cross plot may indicate the presence of two fracture sets (unpublished data).

development, and in part because spacing is easy to measure. Fracture spacing can be obtained from outcrops (from remote imagery and from scan lines measured directly on the outcrop), and it can be measured in the subsurface from horizontal cores and image logs.

Fracture measurements from horizontal cores suggest that like many outcrop fracture spacings, both extension and shear fractures in the subsurface commonly have log-normal spacing distributions (Figure 1.22). In contrast, three of the four references cited by Loosveld and Franssen (1992) and using data from faults, conclude that spacings are fractal and can be extrapolated down to the scale of shear fractures, suggesting power-law distributions.

Another published shear-fracture spacing dataset (Lorenz et al., 2002) was acquired from horizontal core taken from the Spraberry Formation, a Permian deep-marine sandstone in Texas. In this fracture population, the angle between the conjugate shear-fracture pair is on the order of 45°, somewhat less than the ideal 60° conjugate angle, and the shear offsets were minimal. Both aspects suggest conjugate hybrid shears, i.e. shear fractures displaying the characteristics of both extension and shear. The bisector of the conjugate angle for this pair is parallel to the strike of associated extension fractures, supporting that interpretation. The measured spacings for 21 pairs of NNE-SSW striking shear fractures, and for 27 pairs of ENE-WSW-striking shear fractures in the same intervals of the same core are both essentially log-normally distributed.

As with other fracture parameters, however, spacing is not as simple as it might seem. The concept of spacing is intuitive and comfortable when fractures are regularly distributed and extend top-to-bottom within

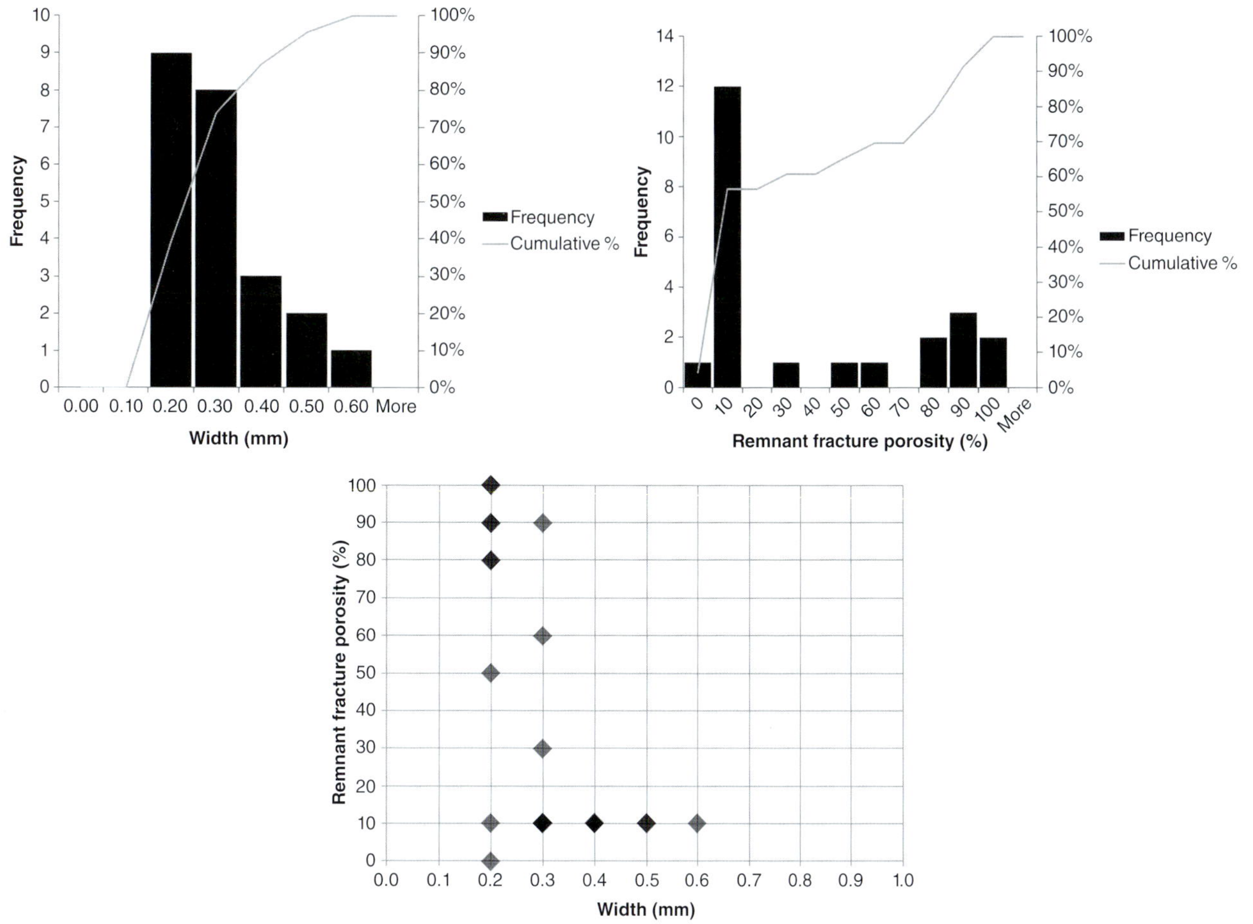

Figure 1.21.3 Histograms and a cross plot showing the distributions of fracture widths, and the remnant fracture porosities within those widths, for a set of narrow, high-angle shear fractures captured by 110 ft (33 m) of oriented, horizontal, 4-inch (10 cm) diameter core cut from a deep marine sandstone. *Upper left*: fracture widths. *Upper right*: remnant fracture porosities. *Bottom*: the cross plot of fracture widths to fracture porosities for these fractures (n = 23). Darker points highlight overlapping data (unpublished data).

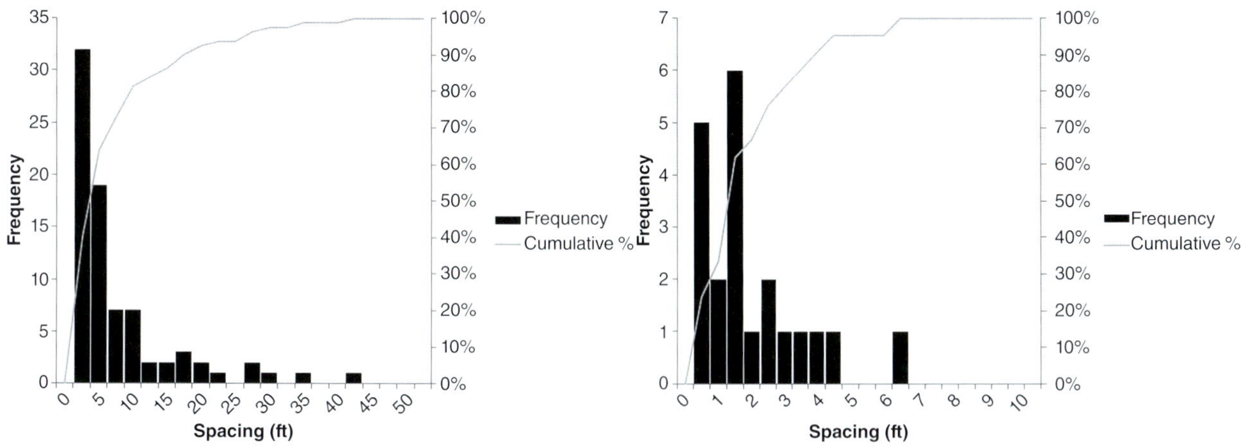

Figure 1.22 Histograms of fracture-normal fracture spacings measured in two horizontal cores. *Left*: vertical extension fractures in a marine mudrock (n = 80; min 0.01, max 40.3, ave 6.4 ft). *Right*: parallel strike-slip shear fractures in a marine sandstone (n = 21) (unpublished data).

Figure 1.23 *Top and lower left*: outcrops showing the relatively regular spacings of two sets of bed-normal extension fractures in the Permian, eolian, Cedar Mesa sandstone, Utah. The two fracture sets are both marked by plumes on the fracture faces and cut top-to-bottom in the homogeneous, 30 ft (10 m) thick unit. As shown by the histograms in Figure 1.24, one set is more regularly spaced than the other, and neither is as uniform as the photos seem to suggest. *Lower right*: the spacing histograms are derived from remote imagery and do not have the resolution to capture the most closely spaced fractures such as this one from the same outcrop, shown in plan view looking down on the bedding surface.

well-defined layers (Figure 1.23), but it becomes more nebulous when applied to fractures that have a range of heights and that are irregularly distributed in a heterogeneous reservoir (see Figure 1.13). Moreover, like length, spacing can change depending on the scale at which it is measured, i.e. small but potentially important fractures may not show on remote imagery, whereas measurements made from an outcrop at a scale of a few tens of meters may not extend far enough to capture a representative sample of larger, more widely spaced fractures.

Additionally, the effective fracture spacings that control fluid flow often do not equate to the precisely measured spacings from an outcrop scan line or captured by a horizontal core. Those exact, quantitative spacing measurements do not incorporate the range of heights, lengths, and apertures of the sampled fractures, which dictate that some fractures contribute more to fluid flow than others. The log-normal distributions of many fracture-spacing populations also mean that a mathematical average does not capture 1) the pervasive but small-scale contributions of the numerous small fractures; 2) the more dramatic but localized contribution of the relatively few large fractures; or 3) the clusters of closely-spaced fractures that are effectively single conduits at the scale of a reservoir.

An example of measured spacing vs. effective spacing comes from low-permeability Cretaceous sandstones in Colorado, where the spacings of WNW-ESE striking extension fractures in both outcrop and horizontal cores range from several inches (a few cm) up to 17 ft (5.2 m), averaging about 3 ft (1 m) (Lorenz and Finley, 1991; Lorenz and Hill, 1994). Although carefully measured and precise, the measured average spacing was useless to the reservoir engineer whose models required an effective fracture spacing on the order of 10 ft (3 m) to match the observed drainage anisotropy and the carefully tested production rates from wells drilled into these reservoirs (P. Branagan, personal communication, 1988). The measured fractures in this example are log-normally spaced, and the clusters of closely spaced fractures have spacings on the order of 8 ft (2.4 m), closer to the engineering-determined 10 ft (3 m) effective fracture spacing. In addition, some of the measured fractures are occluded by mineralization and therefore less effective as permeability conduits. Taking clustering and individual fracture effectiveness into account, the geologically estimated effective fracture-spacing estimate for the reservoir is compatible with the engineering model. Thus, exact spacing measurements are important guidelines for modeling, but only rarely can or should they be put directly into a model. Actual and average fracture spacings are still important measures of strain when they can be combined with width, and spacing provides a key constraint in calculating reservoir volumetrics, as discussed in Part 3.

The oil-field rule of thumb is that fracture spacing is equivalent to bed thickness, and in the absence of data this rule offers a starting point for estimating fracture spacing, but it should not be used if actual data are available. The rule of thumb is based in part on the bent-beam model for fracturing (e.g. Price, 1966), in which tension created by stretching the outer arc of a beam during flexure is relieved at regular intervals by fractures having spacings that are dictated by both the thickness of the beam and the degree of curvature. Folding rock creates local strains, but it takes less stress to fracture a rock than to fold it and therefore rock commonly fractures before it folds: many fracture sets on anticlines are inherited, formed prior to folding (e.g. McQuillan, 1973; Bergbauer, 2007). Nevertheless, fold-related strains do commonly result in fracturing, and fold-related fractures may form parallel and normal to the axes of folding, often being superimposed onto pre-fold fractures (e.g. Cooper et al., 2006). Fractures that formed prior to folding can also be reactivated, opened, and extended during folding.

Bed thickness is important, but it is only one of the several controls on fracture spacing and it is not usually the primary control. Published cross plots between fracture spacing and bed thickness, including early studies such as that by Ladeira and Price (1981), show significant variations in the ratio of fracture spacing to bed thickness, often as a function of lithology. Ladeira and Price showed that there can be different trends for different lithologies, i.e. sandstone and limestone, and different trends even for different formations with similar lithologies. Moreover, the Ladeira and Price thickness-spacing relationships are not linear; fracture spacing increases with bed thickness but only up to a given thickness, above which fracture spacing in the outcrops does not change significantly.

There are also innumerable examples where a given bed contains two sets of extension fractures of different ages (e.g. Hanks et al., 1997) and where the two sets have different spacing characteristics. Since two distinct spacing populations occur in the same bed (Figure 1.24), bed thickness may influence but it cannot be the only, or even the primary, control on spacing.

Numerous spacing datasets have been published. Among them, Bai and Pollard's (2000a) data suggest that the spacing-to-thickness ratios can vary over two orders of magnitude for different formations, while McQuillan (1973), Narr and Suppe (1991), and Huang and Angelier (1989) suggested that the ratio between fracture spacing

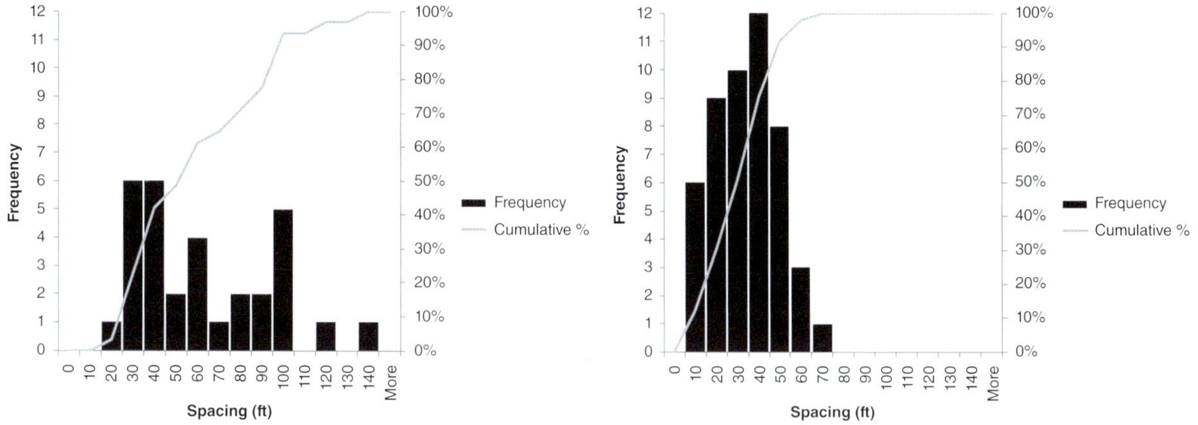

Figure 1.24 Spacing distributions measured from remote imagery for the two mutually crosscutting extension-fracture sets in the Cedar Mesa Sandstone shown in Figure 1.23. *Left*: the N-S striking fractures (n = 31) have a debatably log-normal spacing distribution. *Right*: the distribution of spacings in the nearly orthogonal E-W striking fractures (n = 49) has affinities to both log-normal and normal distributions. The remote imagery could not capture the closer spacings illustrated in Figure 1.23. The difference in the spacings of two fracture sets in one bed highlights the fact that bed thickness is only one of the factors controlling extension-fracture spacing. Although there are few abutting relationships between the two fracture sets, the more regular spacings of the E-W set (right) may indicate that it propagated in a homogeneous medium, creating heterogeneities that affected the spacing distribution of the N-S set. (Unpublished data.)

and bed thickness in sedimentary strata is quantifiable and reasonably consistent as long as measurements are from the same lithology in the same formation. Wennberg et al. (2007) presented data showing that bed thickness is a poor control on spacing where the strata have been subjected to multiple strain events. The lesson is that the spacing-to-thickness ratios can be quantified at specific locations and for specific formations, but that the ratios change with lithology and with structural setting.

There is point at which fractures, or at least extension fractures, stop developing in a rock despite the continued imposition of stress, identified as the point of fracture saturation (e.g. Bai and Pollard, 2000a, 2000b; Wu and Pollard, 1995). Bai and Pollard suggest that for the formations they studied this threshold lies at a ratio of spacing-to-bed thickness between 0.8 and 1.2. There are, however, many examples where fracture spacing is much less than bed thickness, so saturation is not universal, or at least not at these ratios. Saturation also does not occur in formations where the fractures become healed by mineralization soon after they form and where the host rock and fracture filling have similar mechanical properties, such as systems of closely spaced, calcite-mineralized, hairline fractures in chalks. When healed, the fractures in such settings are no longer weakness planes in the rock, so new fractures continue to form as long as the rock is being strained.

1.3.10 Fracture Strike

Strike is perhaps the least ambiguous of the numerous fracture dimensions. Most fractures stay in approximately the same plane over their height and length, so a small sample of a fracture from core is usually a reliable representation of the strike of the entire fracture plane. The same small sample of one fracture can even provide a reasonable characterization of the orientations of other fractures of the same set.

Strikes tend to be more regular and more tightly constrained in rock that fractured under conditions of high stress anisotropy, and in fine-grained, well-cemented, and/or thin-bedded strata (Figure 1.25). In contrast, strike variability among the members of a set commonly increases, and the planarity of individual fractures decreases, where fracturing occurred under conditions of low stress anisotropy or in strata that are thicker, more heterogeneous, less well-cemented, and/or coarser-grained.

The two sub-sets of a set of conjugate shear fractures also typically have systematic strikes. Shears form conjugate pairs where the strata are allowed to extend in only one direction normal to the axis of compression, and fractures of a pair will have parallel strikes, but opposing dips if they are normal or reverse dip-slip offset. If they are strike-slip pairs, ideally, they should be vertical and have strikes that intersect at an angle of about 60° (Figure 1.26). However, if the strata are allowed to extend in two directions while being compressed in the third, a more complicated system consisting of two intersecting "orthorhombic" conjugates (Reches, 1983; Reches and Dietrich, 1983), or even more than two "polymodal" shear pairs may form (Healy et al., 2015).

Reservoirs containing only one set of extension fractures have elliptical drainage patterns and anisotropic horizontal permeability ratios, but secondary and even tertiary sets of extension fracture can be superimposed

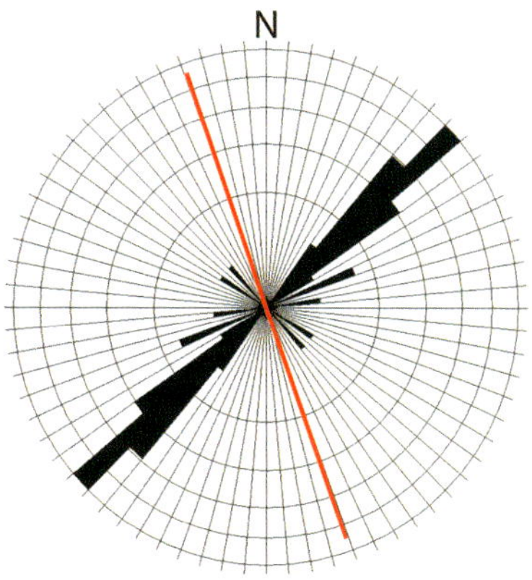

Figure 1.25 *Left*: parallel, bed-normal, vertical extension fractures in thin bedded Cretaceous chalks and marls of the Niobrara Formation, Colorado, create a pervasive fracture fabric in the strata. A two-track dirt road across the middle of the photo gives the scale. Cross fractures are present but rare, and most of them are surficial features that are not present in the equivalent subsurface strata. *Right*: the parallel strikes of a set of vertical extension fractures as measured in a horizontal core. The red line shows the wellbore azimuth through the cored interval. Fractures striking nearly parallel to the wellbore have a low chance of being cored and would be under-represented by this core if they are present (n = 53, unpublished data).

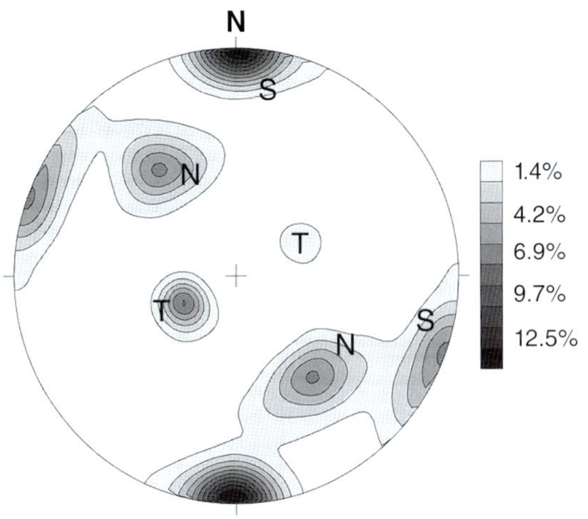

Figure 1.26 A stereoplot of 360 poles to planes measured for deformation-band shear fractures in sandstones of the fluvial Jurassic Morrison Formation of New Mexico (from Olsson et al., 2004). The clusters of poles define three conjugate pairs, with normal dip-slip (N), reverse dip-slip (T for 'thrust'), and strike-slip (S) senses of offset. The orientations of the superimposed pairs were controlled by stress axes that were oriented NE-SW, NW-SE, and vertical. Sequential development of the three conjugate pairs records changing stress magnitudes along these three axes, but not a rotation of the principle axes. The scale bar gives the percentages of points per unit area.

onto the first fracture set if a stress system becomes reoriented by changes in the tectonic or local structural systems. Superimposed fracture sets reduce anisotropy and enhance permeability above matrix values in the direction normal to the first set of fracture planes. If both younger and older fracture sets are equally well developed and minimally mineralized, the resulting horizontal drainage can be almost radial.

Younger cross-fracture sets can also form when only the stress magnitudes, and not orientations, change: for example, an existing minimum in situ compressive stress can ramp up in magnitude to the point where it exceeds the intermediate or even the maximum compressive stress during the development of a thrust system (e.g. Olsson et al., 2004). If the older fractures are tightly cemented and do not form significant mechanical discontinuities in the rock, or if they are held together by a high-magnitude stress normal to the fracture planes, then the two fracture sets may be mutually cross-cutting and it will be difficult to tell their relative ages. Older fractures can also be reactivated during the formation of a younger set, resulting in ambiguous or even contradictory cross-cutting and abutting relationships. An older fracture set is not always or even usually the best-developed set of fractures in the rock.

In contrast to extension fractures, shear fractures commonly form an inherently interconnected system if they form as a conjugate pair, but the intersection angle is not always the 60° mechanical ideal. Hancock and Bevan (1987) reported that the intersection angle between the strikes of a pair of conjugate strike-slip shear fractures decreased with distance from the Zagros thrust front, eventually morphing into extension fractures striking normal to the thrust front and parallel to the bisector of the conjugate angle of the shear pairs.

1.3.10.1 Fracture Orientations Relative to the In Situ Stresses

Fracture strike relative to in situ stress system helps determine fracture permeability. Fracture orientations record the in situ stress orientations at the time of fracturing, but fracture planes may be tilted or rotated if the host strata are deformed after fracturing, so the relative orientations between fracture planes and the stress system may change over time. The relative orientations of fractures should be noted in a reservoir (see Section 2.4.8 in Part 2) since the behavior of fractures and fracture-related permeability is related not only to the inherent fracture characteristics but also to fracture orientations within the stress field (see Section 3.5 in Part 3).

We will not discuss stress characterization in detail since this volume is devoted to fractures, but a range of techniques exists for measuring and computing the present-day in situ stress orientations and magnitudes (see for example Hill et al., 1994; White et al., 2002; Fairhurst, 2003; Barree et al., 2009; Amadei and Stephansson, 2012). A fracture-characterization study that is designed to assess the effects of fractures on reservoir permeability must include measurements of fracture orientations relative to the in situ stresses, and should include assessments of the compressive stress magnitudes and the reservoir pore pressure in order to evaluate the potential behavior of the fractures within the stress system as it changes during production.

1.3.11 Discussion

From these descriptions of fracture dimensions, it should be apparent that although individual fractures are discrete structures with definitive dimensions, those dimensions cannot always be exactly measured, and even where they can be, they are not always directly applicable to a reservoir model. There is enough variability between the numerous fractures of a set that, except for strike, any given fracture is probably not fully representative of the system. Fractures and their effects on a reservoir cannot be assessed in isolation but must be considered within the stress and lithologic systems in which they occur. A fracture is not just a geometry. The most intensely developed fracture set, with long, tall, wide, and closely spaced members, may in fact be fully mineralized and may therefore have less effect on permeability than a superimposed system comprised of fewer, smaller, but more open fractures.

Models are limited by the constraints of mathematics; most cannot handle the full range of data provided by a fully quantified set of fracture dimensions, so simplification and upscaling are required. The geologist, engineer, seismologist, and petrophysicist must work together to use the available fracture measurements, data, and inferences from all four disciplines to make best estimates of the effective fracture dimensions for lengths, widths, heights, apertures, and spacings, as described by Nelson (2020).

1.4 The Mechanics of Fracturing Rock in Extension and Shear

1.4.1 Introduction

Many of the early outcrop studies were conducted in structurally complex settings (the Alps, the Appalachians), where fractures are a manifestation of obvious rock deformation and where shear is common. Because of this, authors investigating less-deformed strata noted with puzzlement that fractures can be well developed in flat-lying, undeformed strata, where they show no evidence for shear (e.g. Hodgson, 1961). However, laboratory tests and theoretical work were also beginning to suggest that rock does not need to be folded or faulted in order to fracture (e.g. Paterson, 1958), and that stresses in the subsurface are not isotropic (Hubbert and Willis, 1957). It became apparent that stress differentials capable of fracturing rock may be common, even in relatively undeformed terrains. Laboratory experiments also showed that shear and extension fractures are stages along a spectrum of brittle deformation (Griggs and Handin, 1960), and that the pressure of fluids in the pores of a formation ("pore pressure") plays an important role in the mechanics of fracturing (Secor, 1965).

Nevertheless, there were still were questions about issues such as the exact role played by pore pressure, and the source of stress anisotropy in the subsurface. The planar breaks of fractures in rock have deceptively simple geometries, suggesting simple mechanics, but there are many ways to break rock and there are still ongoing discussions regarding some of the basic aspects of fracturing such as the difference between tension and extension in breaking rock in the subsurface, and the mechanics of breaking and dilating rock within compressive stress systems.

Ideas for the origins of the stresses under which rock fractures have been limited only by imagination. Hodgson (1961) listed nine possible sources of stress that might account for pervasive fracturing in flat-lying strata of the Colorado Plateau. Some of these are not currently considered to be viable, and some of the present theories are not included in Hodgson's list. The theories have included stresses created by changing rock temperatures during burial and uplift, by cyclic glacial loading at the earth's surface above fractured strata, and by lateral extension and compression during uplift and basin subsidence. All of the mechanisms create stresses, but some of them do not provide the stress anisotropy needed to fracture rock, and the anisotropies provided by others are incompatible with the fracture geometries in question.

Turning from the conditions of fracturing to the susceptibility of rock to fracture, the mechanics of fracturing rock would seem to be simple: rock is weak and can easily be broken in tension by pulling on it or bending it. Rock can also be broken in extension or shear by compressing it, and in torsion by twisting it. These mechanisms are well documented in the laboratory, but it is not always apparent which laboratory results apply to the genesis of fracturing in the subsurface. The susceptibility of rock to fracture changes significantly with variations in the fundamental, intrinsic properties of the rock, which are related to its original composition as modified by compaction and diagenesis. To complicate matters, fracture susceptibility also changes with variations in external conditions such as temperature, stress magnitudes and strain rates. Rock is stronger by an order of magnitude in compression than it is in tension, therefore it is sometimes assumed that rock containing extension fractures must have failed in tension, but that is rarely the case since tension is rare in the subsurface except at the scale of the grains that make up rock. Tension, a true pull on the rock, is not the same as extension, where rock extends and dilates in the direction of the least compressive stress.

A note on terminology: most authors place shear and extension fracturing into a general category of "brittle" failure (e.g. Paterson, 1978; Paterson and Wong, 2005), and some do not even feel it necessary to specify whether they are describing shear or extension fractures in their papers. However, the two basic fracture types have significantly different effects on a reservoir and need to be distinguished, even though Hancock (1986) and Hancock and Bevan (1987) showed that extension fractures and conjugate shear pairs form the end members of a spectrum of brittle fracture geometries. They documented in the outcrop what Griggs and Handin (1960), among others, demonstrated in the laboratory. The failure spectrum can be extended past shear fracturing and into the realm of ductile failure, where the rock deforms by pervasive deformation without discrete failure planes, but ductile deformation does not create discrete permeability pathways in a reservoir, and it is not covered here.

The pressure of fluids filling the pore spaces within a formation is a supremely important component in the mechanics of breaking rock in compression. Although a high internal pore pressure increases the susceptibility of a rock to fracturing, under geologic conditions excessive pore pressure within a confined rock does not break it by driving it apart in tensile failure. The hydrocarbon industry routinely fractures rock by hydraulically injecting fluids into a formation under high pressure, but this process is entirely different from the effects of pervasive high fluid pressures within a formation, and it is not an analog to natural extension fracturing: debates of "natural hydraulic fracturing" must be approached carefully.

The following discussions are built on the empirical characteristics of fractures in core and outcrop, supported and constrained by experimental laboratory results and by theoretical considerations. Conclusions based on these three views of fracturing overlap, but they are not always mutually supporting. For example, the theoretical considerations derived from Mohr diagrams can be limiting: whereas these diagrams represent shear failure well, if extension factures are represented at all they are often considered to be anomalies, forming only in the special circumstances represented by the end cap of the failure envelope on the negative, tensile side of the vertical axis. In contrast, extension fractures are common in rock. As Jaeger et al. (2007, page 2) put it: "Joints [extension fractures] are by far the most common type of geological structure," more common than would be suggested by the special, tensile, and nebulous end-cap conditions required for extension fracturing as depicted on most Mohr diagrams.

Laboratory observations show that rock breaks easily in tension, created by pulling on "fixed grips" attached at the ends of a specimen. Competing observations of extension fractures that form when a sample is compressed in one direction while confined in the other two axes in the laboratory were largely discounted as being unrelated to fractures that formed in deeply buried rock. However, except for special, local conditions such as the tips of propagating extension fractures and igneous injections, and probably within the microcracks that form in individual sand grains as the precursors to macrofractures (Paterson, 1978; Paterson and Wong, 2005), true tension is difficult to generate in confined rock at depth, and in fact only compressive conditions have been measured in situ (e.g. Brown and Hoek, 1978; Amadei and Stephansson, 2012). Therefore, we start these discussions with the premise that almost all subsurface fractures, both extension and shear, must have formed when all three stress axes were compressive.

Because rock is stronger in compression than in tension, it requires significantly more applied force to fracture laboratory samples when they are being pushed than when they are being pulled. Moreover, tests show that rock becomes stronger when it is subjected to compression in all three axes, as it is the subsurface, and that its strength increases with growing confining stresses. In fact, theory in the mid-20th century suggested that because of the large confining stresses at depth, open fractures could not exist in the subsurface, and one could read lines such as:

"At depth, however, most joints are generally closed…" (Heck, 1955)

"It is, of course, inconceivable that an open crack could exist at depth…" (Griggs and Handin, 1960).

Thus, the cores that recovered rock with open natural fractures from strata in the deep subsurface (Figure 1.27) created a conundrum; rock should not be fractured at depth. Strata at depth are under significant confining compressive stress due to the weight of the overburden and should theoretically be strong and not susceptible to fracturing. Moreover, if a fracture did form at depth,

Figure 1.27 A vertical, incompletely mineralized extension fracture in sandstone, recovered from a depth of 20,481 ft (6,243 m) in the Cretaceous Frontier Formation, Wyoming (courtesy Connie Hawkins). Since deeply buried strata are under significant compressive stress in all three directions due to the weight of the overburden, and in the absence of an understanding of the effects of pore pressure, it was not obvious how fractures could form, open, and stay open long enough to become mineralized at such depths.

or if one formed in shallow strata and was subsequently deeply buried, it should be held shut by the large confining stresses. The answer to the puzzle lay in laboratory tests that measured the importance of pore pressure, which acts against confining stresses at depth, allowing fractures to remain open and reversing the strength increase caused by high confining stresses. A more recent quote recognizes the presence and significance of pore pressures at depth: "We assert that almost no macro-scale brittle deformation in the upper crust takes place in the absence of elevated pore pressures…" (Turner et al., 2017).

1.4.2 Origins of Geologic Stress Systems

1.4.2.1 Stresses in a Tectonically Quiescent Basin

Simplistically, a stress is a force applied to a rock in a certain direction and with a certain strength or magnitude. In the absence of tectonically imposed stress asymmetry, the maximum compressive stress imposed on a subsurface reservoir is typically the weight of the overlying rock, and this stress is vertical due to the downward pull of gravity. The magnitude of a vertical stress is calculated, not measured, by using the densities of the different layers of rock, their thicknesses, and the pull of gravity (see Appendix 1.A). As a rule of thumb, it increases by 1 psi per foot of depth, or 22.6 MPa/km.

If reservoirs were composed of strong, perfectly rigid strata, the weight of the overburden would be entirely supported by the strength of the reservoir rock in the same way the bottom brick in a stack of ten square bricks supports the weight of the overlying bricks without deforming, and the weight of the overburden would not create additional stress in the rock. The lateral stresses imposed on such a reservoir, just as on the sides of the brick at the bottom of the stack even if it is one of many side-by-side brick stacks, would be zero.

If on the other hand the bottom brick of the stack is replaced by a plastic bag full of water, then the weight of the bricks overlying the plastic, water-filled bag must be supported entirely by the incompressibility of the water and an increase in fluid pressure in the bag equal to the weight of the bricks. The pressurized fluid, which has no inherent strength, tries to escape laterally, creating a fluid pressure inside the plastic bag equal to the weight of the overlying bricks, but the bag is constrained by adjacent bricks, imposing a lateral pressure equal to the weight of the overlying bricks onto the sides of the bricks at the bottom of the adjacent stacks.

Rock at depth is neither perfectly rigid nor entirely fluid. When rock is under compression in one direction, vertically from the weight of a column of overlying strata, the strength of the rock supports part of the load, but since it is not perfectly rigid it would expand laterally to some degree if it were not laterally constrained by the surrounding rock. The amount of expansion depends on lithology. Some of the overburden weight imposed on the buried rock is supported by the strength of the semi-rigid rock framework, and some of the weight is supported by an increase in the horizontal stresses created by the attempted lateral expansion of the rock. Replace the bag of water with a rubber block at the bottom of the stack of bricks and part of the weight of the stack will be supported by the strength of the rubber, but the rubber would also expand laterally if it weren't prevented from doing so by the presence of adjacent, similar brick piles with basal rubber blocks. Thus, the vertical weight of the stack of bricks creates horizontal, lateral stresses where the rubber blocks are prevented from expanding sideways by the presence of adjacent rubber blocks.

The horizontal compressive stress in a formation is usually considered as two stresses oriented at right angles to each other in the horizontal plane, and in this hypothetical case they are equal to each other since the rock at the base of the stratigraphic column, equivalent to the rubber blocks, would expand equally in all directions. The horizontal stresses are passive; they are the result of the inactive weight of overlying strata rather than the product of tectonic activity. Like the rubber block, and unlike the bag of water, the rock has some strength which carries part of the overburden so that the horizontal stresses are less than the weight of the overburden. The amount of potential lateral expansion of rock, and the lateral stress magnitudes generated by preventing it, are governed by Poisson's ratio, a basic property of the rock that relates the amount of vertical shortening to the amount of lateral expansion that the rock would undergo in the direction normal to an imposed compressive stress. Poisson's ratio varies by lithology but for different sedimentary rock it can have values ranging from 0.1 for stiff rock to 0.5 for more ductile rock.

Layers of very ductile rock with even higher Poisson's ratios such as halite or coal have little or no strength and therefore have no inherent structure that would resist lateral expansion and carry part of the vertical stress; they behave nearly as a fluid over geologic time scales. In layers composed of these lithologies, the entire weight of the overburden is supported by the lateral constraint provided by the surrounding rock. Remove those constraints and these rocks flow, albeit slowly. The walls, roofs, and floors of tunnels in salt mines creep slowly into the caverns and are a constant problem. This property is being used to permanently bury nuclear waste at the Waste Isolation Pilot Plant near Carlsbad, New Mexico. Layers of such ductile strata are highly stressed since the horizontal stresses are nearly equal to the maximum, overburden stress, and the *differential* between the maximum and minimum stresses is low.

In contrast, stiffer, stronger rocks have a low Poisson's ratio and their rigid frameworks support some of the overburden weight. Their potential for lateral expansion under the weight of the overburden is less, creating lower horizontal stress magnitudes within the formation at depth. The lower horizontal stress magnitudes result in a greater differential between the vertical and horizontal stresses, and rock compressed in this way can fracture if the stress differential exceeds the compressive strength of the rock.

Extension fractures strike normal to the minimum compressive stress since it takes less energy to overcome the minimum stress when opening a fracture in that direction than it would to counteract the two larger stresses. Extension fractures formed under these stresses are oriented in the vertical plane since the minimum stress is one of the two horizontal stresses. Dip-slip shear fractures also accommodate vertical shortening and lateral expansion of the rock in the direction of the minimum compressive stress, and ideal, normal-dip-slip shear fractures have 60° dip angles, 30° oblique to the vertical compressive stress.

1.4.2.2 Other Potential Sources of Horizontally Isotropic Stress

There are a number of mechanisms that can cause changes in the volume of rock and that are therefore theoretically capable of creating significant stresses, although relating them definitively to specific fracture sets is not easy since only a few of them increase the stress differential that is needed to fracture rock. The volume change of a rock caused by changes in temperature (the coefficient of thermal expansion) during burial and uplift, as well as the significant difference between the thermal coefficient of rock and that of the fluids filling the pores, or even differences in the thermal expansion of the different mineral components of a formation, are sources of stress that may be capable of fracturing rock (e.g. Robertson, 1988; Billingsley, 2005; English, 2012). English (2012) suggests that the magnitude of the contraction of strata that should be created by cooling during uplift and exhumation is sufficient to create tensile stresses in the strata, the effect depending on the amount of cooling and being more pronounced in stiffer, well-cemented sandstones and carbonates. The theoretical strain magnitude caused by cooling should be sufficient to break the rock, but it is not anisotropic. This mechanism probably does not account by itself for the common sets of parallel extension fractures, but it may play an important role in changing the magnitudes of anisotropic stresses, enhancing the stress differential needed to break rock.

Heating a quartz grain can change its volume by nearly 1%, although it takes hundreds of degrees of change to do so and the amount of linear expansion varies along the different crystallographic axes. A quick search of the internet for household plumbing tolerances related to domestic water heaters suggests that water expands by 1.25% when temperature is raised from 90° to 140°F (from 32° to 60°C). Measurements presented by Russell and Hoskins (1973) show that test samples of granite change their volume by nearly .01% over the course of the small temperature change from 0°F to "room temperature," and Haxby and Turcotte (1986) note that the volume change in sandstone is about 20% larger than that for limestone, setting up the potential for bedding-plane slip during thermal changes. Warpinski (1989) suggested that the stresses generated by temperature changes can be similar in magnitude to tectonic stresses.

Thermal expansions and contractions have magnitudes that are consistent with fracturing rock, but as yet they include no obvious mechanism for producing the stress anisotropy necessary to fracture rock systematically, creating fractures with preferred strikes and dips. Nevertheless, the mechanism may contribute to fracturing within an anisotropic stress field created by other processes.

The repeated subsidence and uplift of the earth's crust caused by cyclic glacial loading was also calculated to be a theoretically plausible source of stress magnitudes capable of fracturing Paleozoic shales in eastern North America (Clark, 1982). This mechanism has not found a widespread following since most of the fractures in the affected strata appear to predate glacial conditions (e.g. Engelder and Geiser, 1980), and since more conventional mechanisms for stressing and fracturing the rock have generally been found to be more plausible.

Similarly, the daily vertical displacement of the earth's surface created by the orbit of the moon, i.e. earth tides, probably creates enough stress to break rock (see Hodgson, 1961), and in fact samples in the laboratory become significantly weaker and more susceptible to fracturing when they have been cyclically subjected to stresses well below their failure strength (e.g. Haimson, 1978). The magnitudes and orientations of the stresses generated by this mechanism are not well constrained, and like the others listed, it has not been widely invoked as an explanation for pervasive extension or shear fracturing in reservoir strata.

1.4.2.3 Stresses in a Tectonically Active Basin

Under the stress conditions described so far, there is no reason for either extension or shear fractures to have preferred strikes if the two horizontal stresses are equal. Creating a horizontal stress anisotropy capable of producing preferentially oriented fractures requires the influence of external, usually tectonic stresses, and in turn, the typical occurrence of systematic fracture sets with uniform strikes shows that horizontal stress anisotropies are and have been common and widespread in the subsurface. Specific fracture sets in a basin can often be related to specific sources of stress anisotropy, i.e. to specific tectonic events and structures.

Just about any tectonic process that moves rock produces a stress anisotropy. Tectonic processes can asymmetrically enhance compression in one horizontal direction, and/or they can reduce it preferentially in one direction. These active stress systems are superimposed onto the passive isotropic horizontal compressive stresses created by Poisson's ratio and the overburden weight. Sources of anisotropic tectonic compressive stress can include the following, and multiple sources can be superimposed to form complex stress systems (e.g. Tavener et al., 2017).

1. Basin-margin thrusting and the bulldozer-type indentation of a thrust belt into the strata filling the adjacent basin increase the horizontal compressive stress in the basin in the direction parallel to the direction of thrust movement. The stress anisotropy in the undeformed strata immediately in front of the thrust front is significant since the horizontal compressive stress is sufficient to lift strata along inclined thrust planes in these strata as it forms successive thrust sheets. The stress anisotropy diminishes with distance from the thrust front, but it is still measurable in strata hundreds of kilometers distant from the front (Figure 1.28) (Craddock and van der Pluijm, 1989; van der Pluijm et al., 1997). Extension fractures that formed in thrust-related anisotropic stress systems around the world strike normal to the thrust fronts, and associated shear fractures have dynamically compatible orientations (e.g. Parker, 1942; Hancock and Bevan, 1987; Cooper, 1992).

   The fractured foreland strata can subsequently become incorporated into the thrust sheets as the system develops, and although many fracture sets found in thrust sheets are directly related to thrusting and thrust-related folding, others are inherited, having formed in the strata prior to their inclusion into the thrust belt (e.g. Winslow, 1983). In addition, thrust sheets commonly become extended laterally as they propagate towards the basin along arcuate fronts, reducing the horizontal compressive stress parallel to the thrust front, enhancing the stress anisotropy, and locally creating additional thrust-normal extension fractures (e.g. Crosby, 1969; Cooper, 1992).

   Compressive plate-margin interactions can produce stress anisotropies in a fashion similar to that of thrust belts, which can affect the stress magnitudes and trajectories within strata for a significant distance into the adjacent plates (e.g. Tavener et al., 2017).

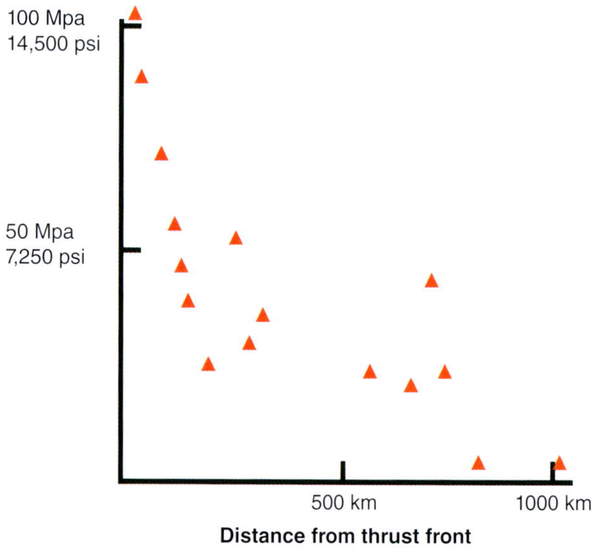

Figure 1.28 The record of thrust-related differential stresses, measured from calcite twin lamellae in strata in front of the Appalachian, Ouachita, and Sevier thrust systems, show significant stress anisotropies adjacent to the thrust front, diminishing, but still measurable for hundreds of kilometers out from the thrusts (adapted from Craddock and Van der Pluijm, 1989).

2. In contrast to stress anisotropies created by horizontal compression, lateral extension can create a stress anisotropy due to a *reduction* of one of the horizontal compressive stresses. The horizontal compressive stress in one direction is reduced (but it does not become tensile), resulting in a stress anisotropy that can create extension fractures and dip-slip conjugate shear fractures striking normal to the minimum compressive stress. The weight of the overburden and the maximum horizontal stress may be minimally affected, but the strata are extended parallel to the minimum compressive stress under the influence of the weight of the overburden. Structural settings where the minimum compressive stress is diminishing include zones of continental-scale rifting, down-slope extension of strata on gentle depositional slopes such as the Gulf of Mexico, and uplifted basins where the compressive stresses are reduced unequally due to the removal of lateral constraints and/or erosion of the overburden. Strata above a rising salt diapir or uplifted basement block can also be extended laterally at a smaller scale. Teufel et al. (1993) describe radial stresses and fractures created by doming of chalk layers, the dome probably being the result of an underlying salt diapir in the North Sea.
3. Local stress anisotropies are created when strata are folded, and those stresses can be superimposed onto regional stresses. Theoretically, the horizontal compressive stress oriented normal to a fold axis is reduced on the outer arc of the fold due to extension and increased on the inner arc of the fold due

to volume constraints (e.g. Bengtson, 1981). This mechanism accounts for fracture sets and some of the fracture corridors striking parallel to and localized along fold hinges (e.g. Stephenson et al., 2007), and the associated stress changes can often be calculated given ideal folds and knowledge of essential pieces of data such as the thickness of the folded strata, the type of fold, and the position of the neutral line separating extension from compression within the folded strata.

However, many geologic folds have shapes that are less than ideal, and parameters such as the location of a neutral plane in the fold are usually poorly constrained. Moreover, fracture predictions based on such calculations assume the unlikely conditions of a homogeneous, unfractured, and unconfined layer prior to folding. Finally, there are multiple ways of folding strata including layer-parallel compression and the draping of strata over a rising thrust front or basement block; the fold-related stresses and the fracture-fold relationships are different for each mechanism.

Folds can also migrate through a formation, producing multiple fracture sets (e.g. Rawnsley et al., 2007), and some folds have localized, fractured hinges with flat, unfractured dip-panel flanks whereas others consist of strata with continuous curvature across the fold and evenly distributed fractures, complicating fold-related fracture models.

Simple fracture-fold mechanisms are undoubtedly valid locally, but not all documented fracture patterns in outcrop conform to theoretical fracture strikes and the ideal distributions described by Sterns and Friedman (1972). Studies including that of Cooper et al. (2006) and Bergbauer (2007) suggest that many of the fracture sets in folded strata are inherited, while at the other end of the spectrum, Jamison (2016) documents the sequential development of multiple fracture sets related to evolving stress conditions as folding matures.

4. Residual, locked-in stresses, formed when a rock at depth becomes cemented under anisotropic compressive stresses, can break the rock when the external compressive stress is reduced or relieved during uplift and erosion. The elastic rebound of the individual component grains against each other as the external compressive stress is removed breaks the cement holding the grains together, creating fractures. The resulting fractures will have preferred strikes since initial deformation, and therefore the elastic rebound, are greater in the direction of the previous maximum compressive stress. Fractures created this way are commonly shorter and more irregular than the older fractures formed at depth (e.g. Wise 1964; Friedman, 1972; Gallagher et al., 1974; Lajtai and Allison, 1979; Engelder, 1993; Lorenz, 2006). The elastic strain energy stored in each sand grain is relatively small, and the mechanism is more applicable to rock composed of elastic materials such as quartz sand than to more ductile rock such as limestone. Moreover, the amount of elastic energy that can be stored is limited by the pressure threshold where elastic deformation changes to grain cracking or pressure dissolution.
5. Anisotropic horizontal extension and compression can theoretically result from the uplift and/or subsidence of strata since uplift should extend the strata horizontally to accommodate a somewhat longer arc at the earth's surface as the radius from the earth's center increases. Likewise, subsidence should compress the strata within a shortening arc. A layer in a subsiding basin can even theoretically go from compression, as the concave-downward arc of strata near the curved earth's surface shortens during subsidence, into extension as the layer passes through and beyond the chord of the earth's circumference defined by the basin margins (Price, 1974). The horizontal stresses in an oblong subsiding basin would be anisotropic since the short dimension of the basin reaches the chord of the earth and goes from compression to extension before the long axis of the basin. Although easy to calculate and theoretically plausible, this mechanism is difficult to apply to basins with complex subsidence histories and to those with unpinned or irregular margins. Some aspects of the mechanism have found local application, but it remains largely theoretical.
6. A related mechanism was proposed by Gross and Engelder (1995), who showed that stresses capable of fracturing rock could be set up in a formation consisting of alternating layers of rock with high and low Poisson's ratios as the overburden stress is reduced by erosion during uplift. Removing the overburden reduces the vertical compressive stress which in turn reduces the lateral compressive stresses created by the Poisson's-ratio effect. Stresses and tractions would be created at the interfaces between the weak, high-Poisson's-ratio rock and the stiffer, low-ratio layers, producing tension and fracturing. S. Brown (personal communication, 2003) ran laboratory tests of interlayered rubber sheets and brittle rock slabs and produced fractures in the brittle rock layers in this fashion. If such a mechanism does not generate a horizontal stress anisotropy capable of preferentially orienting fractures by itself, it could at least amplify the stresses in a system where an anisotropy already exists. However, the common geometry of extension fractures that pinch out rather than widen at bed boundaries suggest that bed boundaries were not the point of origin for fractures in many basins.

7. There are special cases for surface conditions where true tension produces fractures in rock that is shrinking in volume, such as cooling lava flows (e.g. DeGraff and Aydin, 1987). If horizontal contraction is isotropic, the fractures form as hexagonal pillars and columnar jointing, but if there are additional influences, for example if the lava flow cools while lying on a slight incline so that the down-slope pull of gravity imparts an asymmetry to the horizontal stresses, the resulting cooling fractures will have preferred orientations parallel to the strike of the slope. Surface cliff exposures are also prone to gravity-assisted tensional spalling, but such fractures cannot be extrapolated into the subsurface.

With so many potential sources of stress and stress anisotropy it is no surprise that most strata are cut by at least one fracture set even in flat-lying, unfolded, unfaulted strata.

1.4.3 Rock Susceptibility to Fracture: Basic Concepts

1.4.3.1 Introduction

When anisotropic external stresses are applied to brittle and brittle-ductile rock, the initial macroscopic deformation response is typically elastic strain, defined as deformation where the rock remains capable of springing back to its original, pre-deformation shape when the stresses are released. Rock under stress begins to "yield" when it deforms beyond its elastic limit, where release of the confining stress at any point after this limit does not bring the rock back to its original shape, since deformation has started to damage the rock structure and some or all of the developing strain has become permanent. Yield also marks the point in the stress-strain curve where rock starts to lose its strength, i.e. where it weakens and starts to lose its ability to support the applied stress (Figure 1.29).

Continuing along the stress-strain curve, laboratory samples are considered to "fail" when they lose all ability to support an applied stress anisotropy: failure occurs somewhere beyond the point of initial yield, when the applied stress reaches the compressive strength of the rock, and, if failure occurs as a fracture, where the specimen loses cohesion across the new fracture plane. The length of the interval along the stress-strain curve between the onset of yield and final failure varies with ductility among other things; it is extended in more ductile rock, which may never fracture if the rock is fully ductile, whereas yield and failure occur at nearly the same point for more brittle rock. Paterson and Wong (2005) define a "brittle fracture" as one which is "not preceded by any appreciable amount [less than a few percent] of permanent deformation...," noting that microscopic

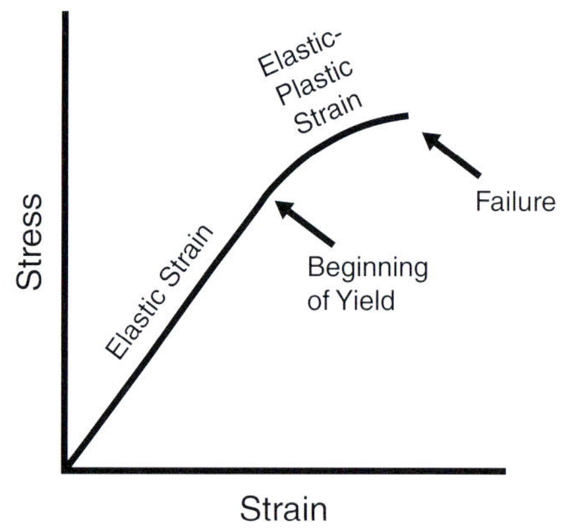

Figure 1.29 A simple, ideal stress-strain curve, showing initial elastic deformation (the straight line), the yield point, the interval of elastic-plastic strain, and failure when the ultimate compressive strength of the rock is reached.

cracks and other inelastic deformation mechanisms may, however, begin to form prior to macroscopic failure.

The present-day stress magnitudes and orientations in a formation can be measured or calculated (see Fairhurst, 2003; Dixit et al., 2017), but reconstructing either the mechanical properties of the rock, or the stresses, at some time in the distant past when a specific set of fractures formed is problematic since the stresses and the mechanical properties both change over time.

Rock fails in extension when it is relatively brittle, it fails in shear when it is somewhat less brittle, and it fails plastically when it is ductile (Figure 1.30). In addition to the mode of failure, ductility also controls the amount of strain prior to failure, but strength controls

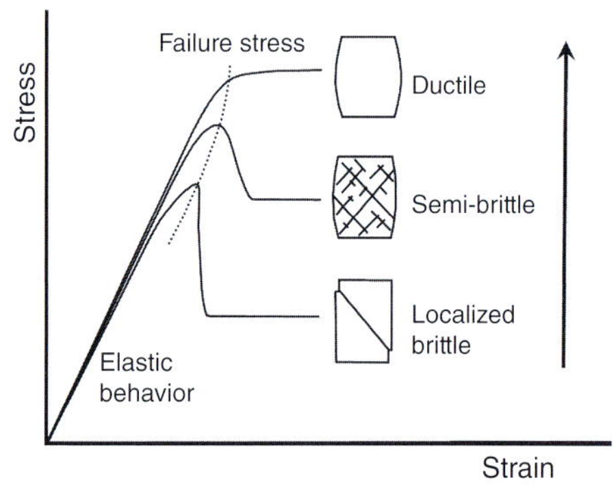

Figure 1.30 This generalized laboratory stress-strain chart shows the effect of ductility, controlled in large part by lithology, on failure. Most rock initially deforms elastically when stressed, and brittle rock fails (either in shear as shown, or in extension), at relatively low levels of strain. Semi-brittle rock commonly undergoes more strain before failure, and ductile rock may fail at an even higher percentage of strain or even without final failure and fracture. The samples begin to fail at a "yield" point where the straight, elastic part of the curves begin to roll over prior to ultimate failure (adapted from Faulkner et al., 2008).

the magnitude of the stress anisotropy that is required to reach failure. In order to fracture, rock must be weak relative to the imposed stress anisotropy. That is a common condition since geologic stress systems are capable of moving continents and raising mountains whereas unconfined rocks can be broken with a hammer. There are numerous combinations of the stress magnitudes, rock ductility, formation temperatures, strain rates, etc., that can produce fractures in a formation.

Still, the default answer to the question "Is deeply buried rock fractured?" used to be "No," since 1) laboratory results showed that rock becomes stronger and more resistant to fracturing as the compressive stresses confining a specimen increase; 2) the compressive stresses on rock increase rapidly with burial depth (the vertical stress increases at about 1 psi/ft, or 22.6 MPa/km); and 3) prior to the invention of image logs and the collection of large fracture datasets from core, little information was available to document the fracture systems that are common, even pervasive, at the depths of hydrocarbon reservoirs.

With the acquisition of more and better subsurface data, and with the knowledge of pore pressure effects in counteracting the in situ confining stresses, it has become apparent that rock can fracture, and that the fractures can be open at considerable depths, if the pore pressure in the fractured formation is high. Elevated pore pressure can turn otherwise ductile lithologies into brittle ones that are susceptible to fracturing, and even inherently ductile rock such as shale and anhydrite can fracture in the subsurface under the right conditions.

Before getting into the important discussion of pore pressure however, we will present and discuss some of the basic laboratory data pertaining to rock fracture. These data illustrate how the intrinsic geomechanical properties of rock such as ductility, strength, and elasticity (which are in turn governed by a rock's mineralogical composition and sedimentary fabric) affect the susceptibility of rock to fracture. Laboratory data also illustrate how basic factors in a geologic system extrinsic to the rock itself, such as temperature, strain rate, and confining stresses, influence a rock's susceptibility to fracture.

Complex subsurface geologic systems can only be approximated in the laboratory, yet the results of laboratory experiments are essential to understanding the mechanics of fracturing in the subsurface. Correlations between the intrinsic mechanical properties of rock, the extrinsic conditions of fracturing, and fracture susceptibility can be readily quantified in laboratory settings, but these are not perfectly analogous to geologic fractures where the multiple, poorly constrained, and interdependent factors that control fracturing have changed simultaneously at estimated rates and over approximately known times during the course of millions of years, where strain rates are significantly slower, and where three unequal stresses are the common condition (as opposed to most "triaxial" laboratory tests where the minimum and intermediate compressive stresses are equal). Efforts to quantitatively reconstruct geologic systems are hampered by the wide range of possible values for the mechanical properties of a given rock during compaction, cementation, and diagenesis as well as by the uncertainties in other parameters including timing, burial depths, formation temperatures, and pore pressures. Nevertheless, such reconstructions are invaluable in constraining the problem.

For example, Figure 1.31 shows a set of extension fractures and an intersecting set of dip-slip conjugate shear fractures that formed in the same rock unit but at different times. Abutting and intersecting relationships show that the shear fractures pre-date the extension fractures. We can say that the rock properties and/or the stress conditions must have been different during

Figure 1.31 Two views of a coarse-grained, arkosic fluvial sandstone (the Permian Abo Formation in New Mexico) containing two fracture sets. Each view is parallel to the strike of one of the fracture sets, which have nearly orthogonal strikes. The red arrows show correlation features (a tuft of grass and a talus boulder) for comparing the two photos. *Left*: looking towards the northwest, parallel to the strike of a well-developed set of closely spaced vertical extension fractures. *Right*: looking towards the northeast, parallel to an equally well-developed set of dip-slip conjugate shear fractures with faces marked by non-congruent shear steps. Stress conditions and/or rock properties were patently different during the two fracturing events, but it would be difficult to quantify the parameters at the time of either fracture event (see Lorenz and Cooper, 2018b).

the two fracture events since the rock fractured in two dissimilar modes, and based on laboratory analogies we can plausibly infer that either the minimum effective confining stress was lower (due to shallower burial? due to higher pore pressure?), or that the rock was more brittle (due to a greater degree of cementation? due to lower porosity?) at the time of extension fracturing than it was during shear fracturing. However, we cannot assign definitive numbers to the confining stresses or to the mechanical properties of the rock at either stage of fracturing. We have the fracture geometries as well as the present-day rock properties at surface conditions as end-point anchors for reconstructing a fracture-genesis model, but the quantitative values for conditions and mechanical properties during either fracture event are not well constrained.

One important component of the fracture-susceptibility question is the strength of rock, i.e. how much applied stress is required to break it. Counterintuitively, rock strength alone does not control its susceptibility to fracture: fracture susceptibility, and type, are also controlled by ductility. Rock that is both brittle and weak is prone to fracture, initially deforming elastically when strained but breaking/fracturing before it undergoes significant deformation, and breaking under a low stress differential. Brittle rock can also be strong, however, breaking at the same low level of deformation but requiring a larger stress differential to achieve the same small amount of deformation (Figure 1.32). As stresses develop in a heterogeneous formation, fracturing should begin earlier in the weak/brittle strata than in interbedded strong/brittle rock, and fractures should form last in the strong, ductile strata which can be strained significantly before breaking, and where a large stress differential is required to achieve that large amount of strain.

Prior to fracture, a brittle rock undergoes primarily elastic deformation and can recover its original shape and volume if the deforming stress is released before the rock fractures. In contrast, when a predominantly plastic or ductile rock deforms it may undergo a small amount of preliminary elastic deformation, but it quickly reaches a yield point, beyond which the deformation is plastic and permanent.

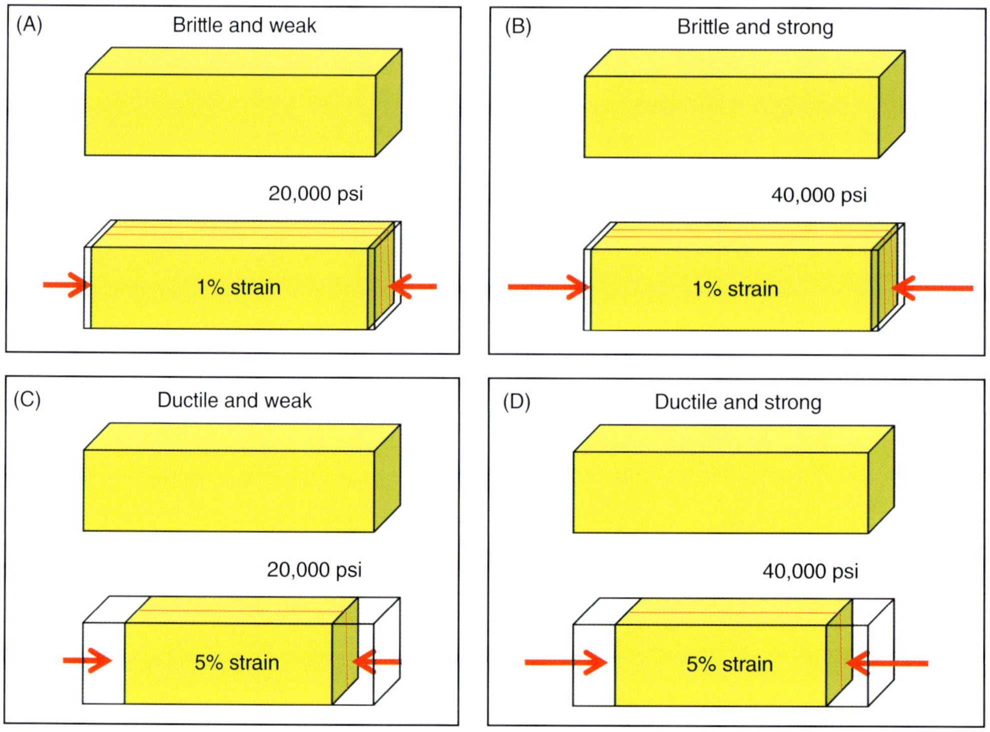

Figure 1.32 The difference between ductility and strength, illustrated by four end-member combinations. (A) Rock that is both brittle and weak fractures before it has been significantly deformed (commonly less than 1% strain) and under a relatively low compressive-stress differential. (B) Similar but stronger brittle rock fractures at the same minimal level of deformation, but it does not fracture until the imposed stress has increased to a higher magnitude. (C) Weak, ductile rock fractures when the compressive-stress differential is relatively low, similar to that which fractures weak brittle rock, but this rock deforms by perhaps 5% before the resistance to deformation and the imposed stress levels increase to the relatively low magnitude where fracturing occurs. (D) Strong but ductile rock undergoes the same large amount of strain before fracturing, but it takes a higher stress anisotropy to reach that amount of deformation. The red dashed lines indicate the formation of load-parallel extension fractures, but as a rock becomes more ductile the mode of fracturing commonly changes from extension to shear.

1.4.3.2 Intrinsic Controls on Fracture Susceptibility

Laboratory tests have shown that the susceptibility of a rock to fracture changes with variations in its intrinsic mechanical properties. These properties are typically tested one at a time in order to constrain and easily interpret the results. For example, when testing the effects of porosity, all samples typically come from the same formation in the same quarry, and they are deformed under the same conditions of confining stress, strain rate, and temperature so that ideally, percent porosity is the only variable between successive tests.

The compositions of common sedimentary rock types overlap, but published measurements of rock properties suggest that rocks within each general type have similar mechanical properties and that the types can be roughly grouped by those properties. For example, in general, ductility decreases (brittleness increases) from halite to gypsum to shale to limestone to sandstone to dolomite (Figure 1.33). In contrast, the compressive strengths of limestone and sandstone are similar to each other, but stronger than halite and weaker than dolomite (see the tables of measured properties for different rock types in Lama and Vutukuri, 1978; Johnson and DeGraff, 1988) (Table 1.2). Limestones tend to be more ductile than sandstones, but a poorly cemented sandstone can nevertheless be ductile compared to a relatively brittle dolomitic limestone.

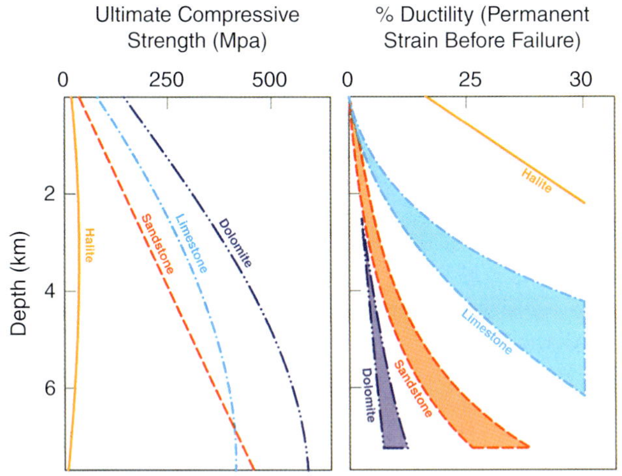

Figure 1.33 The theoretical compressive strengths (*left*) and ductilities (*right*) of different lithologies increase with depth due to increasing temperature and confining pressure, but the mechanical properties of various lithologies change at different rates. The effects of pore pressure are not included. Using standard gradients, the in situ temperature at a depth of 8 km would be on the order of 250°C (480°F) and the in situ compressive stress would be in excess of 180 MPa (26 000 psi). (Reprinted from Handin et al., 1963, with permission from AAPG, whose permission is required for further use.)

Table 1.2 An example of the mean compressive strengths for selected sedimentary rock types (from Johnson and Degraff, 1988).

Lithology	Compressive strength (Mpa)
Limestone	41.5
Limestone	142.5
Limestone	143
Sandstone	77.6
Sandstone	80.8
Sandstone	90.5
Sandstone	167.7
Mudstone	50.1

The fundamental chart presented in Griggs and Handin (1960) (reproduced in Figure 1.2) shows how fracture susceptibility, measured by the percent strain sustained by a rock prior to fracturing, and fracture type (shear vs. extension), both depend on the ductility of a sample. Laboratory data show general trends where rocks become stronger as well as more brittle, and more susceptible to fracturing, when, for example, they are finer grained, when they are better cemented, when they are composed of a more brittle mineralogy, and when they are denser and less porous. The rock-mechanics literature contains numerous stress-strain-yield/failure curves for different types of rock under different conditions (Figure 1.34). Nelson (1985a, 1985b, 2001) provides curves comparing factors such as the strengths of sandstones with dissimilar compositions and the strengths of samples from the same formation but with different porosities. Only a few such curves are reproduced here, enough to illustrate the effects of some of the intrinsic rock properties on fracturing when other factors are kept constant.

Outcrops of heterogeneous formations, where the interbedded layers have different fracture susceptibilities, but which have been subjected to the same stress conditions, can have interlayered fracture systems with different characteristics (Figure 1.35). Such systems have been called "dynamically compatible" systems by Hancock and Bevan (1987).

The size of the rock mass being deformed also affects its properties, primarily its mechanical strength. Most samples tested in the laboratory are carefully selected to be free of heterogeneities and flaws, and they are relatively small, sizes being measured in inches and centimeters and limited by the dimensions of the testing apparatus. However, flaws are inherent in natural rock, and more and larger flaws are unavoidably included as sample size increases. Flaws range from sedimentary discontinuities to fractures to faults, and are numerous in rock masses

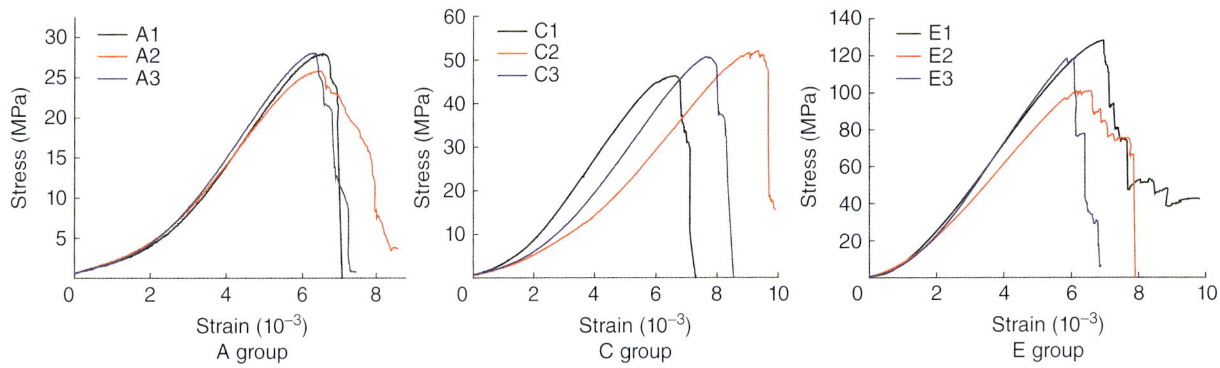

Figure 1.34 Stress-strain curves plotting the strength of sandstones show similar strengths for samples from the same beds (the three curves within each graph), and that most sandstones behave similarly (the shapes of all nine curves are generally congruent), but that sandstones from different beds and different formations (Groups A, C, and E) can have significantly different strengths: the horizontal scales of the three graphs all show failure at about 0.6–1.0% strain, but the *vertical* scales show that strengths of the E-group samples are on the order of four times stronger than the A-group samples (from Li et al., 2018).

Figure 1.35 Interbedded heavily fractured dolomite (the gray upper layers) and poorly fractured sandstone (the brown lower layers) show that the brittle dolomite was more susceptible to fracturing than the relatively ductile sandstone under the same conditions of stress and strain. Shattuck Member of the Permian Queen Formation, New Mexico.

the size of reservoirs. Flaws concentrate stresses and provide weakness planes that can be exploited during strain, weakening a large rock mass and lowering its compressive strength relative to the measured strengths of unflawed samples (e.g. Goodman, 1989; Cundall et al., 2008). Many fractures originate at flaws since it takes less stress anisotropy to initiate a fracture at a flaw than in unflawed rock, and fracture susceptibility increases and strength decreases with the size of a rock mass. However, Paterson and Wong (2005) suggest that weakening of a rock mass as its size increases is minimal above the meter scale, and that increasing confining stresses that close flaws in the rock can negate much of the size effect.

1.4.3.3 Extrinsic Controls on Fracture Susceptibility

The Griggs and Handin chart (Figure 1.2) shows that fracture mode changes from extension to shear as a function of changes in ductility for samples of different lithologies, but the same chart can be used to represent fracture responses for samples of unchanged lithology where increasing ductility is caused by increases in the confining stresses, increases in the temperature, or decreases in strain rate. Almost any inherently brittle lithology can be turned into a ductile rock, and any ductile rock can be made brittle, by changing the conditions within which it is deformed.

The figures below illustrate the effects of some of these external changes; for example, the strength of rock increases significantly as confining stress increases (Figure 1.36), and many authors have noted that the failure mode of laboratory samples changes from extension to shear when they are tested under increasing magnitudes of confining stress (e.g. Janach, 1977). Increasing the temperature at which a sample is deformed also decreases its strength, but the effect is typically minimal until temperatures reach those in excess of common reservoir temperatures.

Fracture susceptibility is also affected by the rate of strain, i.e. how rapidly a rock is deformed (e.g. Sano et al., 1981; Costin, 1987). Rapidly imposed strains can

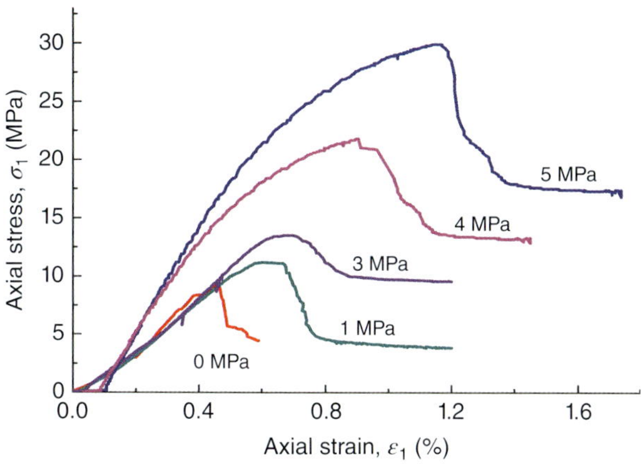

Figure 1.36 Even inherently weak mudrock becomes stronger when subjected to increasing confining stresses. Here, samples of mudrock become stronger by a factor of three or more as confining stress increases from 0 to 5 MPa (0 to 725 psi) (from Zhao et al., 2014).

make rock both stronger and more brittle (Figure 1.37), and the two factors have counteracting effects: rapid deformation increases the strength of rock making it less susceptible to fracture while at the same time decreasing the ductility which makes it brittle and more susceptible to fracture. The change in fracture susceptibility depends on which of the two factors is more important in a given formation, but if changes in brittleness outweigh changes in strength, then rapid deformation results in a higher probability of fracturing.

Most studies addressing this are either theoretical or have focused on laboratory tests of materials other than rock such as cement, metal, and ceramics, and in fact Paterson and Wong (2005) suggest that rate has a negligible effect until entering the realm of rapid, "shock" strain rates. Most geologic processes are relatively slow so strain rate is not commonly a factor, but rupture during strike-slip or normal faulting can impose high strain rates on rock in localized geologic settings, and the episodic, rapid offsets along faults can impart abrupt strains on the adjacent strata.

Some laboratory experiments examine changes in the mechanical properties of a rock as several parameters are changed together over the course of the tests, but the experiments are still tightly controlled compared to geologic systems. In the example shown in Figure 1.38, when porosity and confining stress are varied at the same time, the rock becomes stronger both as porosity decreases and as confining stress increases. The figure also shows that changes in porosity have greater effect on the rock strength at high confining stresses than they do at low stresses.

The presence of water in rock samples can also affect strength and ductility, the magnitude of the effect depending on lithology. For example, laboratory tests show 5–17% reductions in the strength of various sandstones in the presence of water (Baud et al., 2000). The tests performed by Wong et al. (2015) showed that reductions in the uniaxial compressive strength and stiffness of quartzite samples in the presence of water are negligible, but that shale samples can be weakened by up to 90% under the same conditions.

All these extrinsic factors affect the strength and ductility of rock and therefore its susceptibility to natural fracturing. The effects have been quantified in the laboratory where they can be isolated and tested under controlled conditions. Applying these precise laboratory results to the poorly constrained conditions of the subsurface over

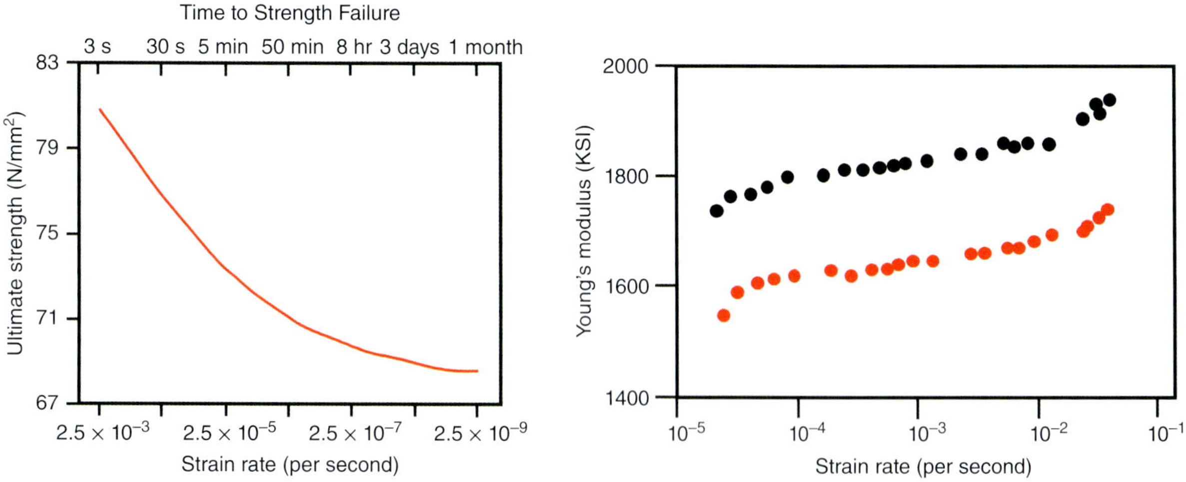

Figure 1.37 *Left*: laboratory tests show that the strength of samples from the Laurencekirk sandstone in the UK increases as strain rate increases (adapted from Sangha and Dhir, 1972). *Right*: the stiffness (brittleness) of samples of New Albany Shale increases as strain rate increases (from Chong and Boresi, 1990). (Strain rate increases in the opposite direction in the two charts.)

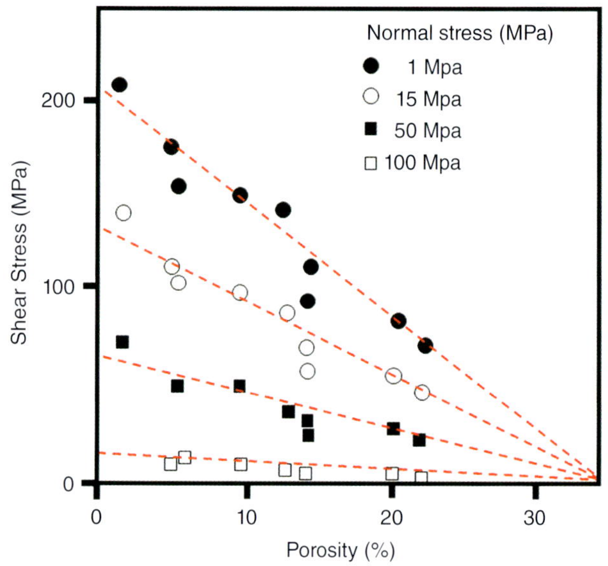

Figure 1.38 Two variables were changed during these tests, one intrinsic to the rock (porosity), and one extrinsic (confining stress). The data show that failure/fracture occurs at higher stress differentials both as the porosity decreases and as the confining stresses increase, and that there are different failure-porosity curves for different confining stresses (adapted from Jizba, 1991a, 1991b).

the course of geologic time is less definitive. Nevertheless, the tests provide important insights into the behavior of rock under different conditions.

1.4.3.4 How Rock Breaks: Grain-Scale Cracking, Yield, and Failure

When a brittle rock is subjected to a stress anisotropy, the initial response is elastic deformation accompanied by the formation of a few scattered microfractures. Reports vary for different lithologies, but most agree that most of the microfractures are oriented within 30° of parallel to the imposed maximum compressive stress. The mechanics of microfracturing are obscure, but Paterson and Wong (2005) suggest that it is "most likely that the fractures originate in response to local tensile stresses around flaws or cracks on a microscopic scale…," while shear is likely along microfractures that are more inclined relative to the maximum stress. Desayes et al. (2000) suggested that the processes controlling microfracturing and macrofracturing are different, but that they are integrally related.

In low-porosity "compact" rock such as marble and granite, these microfractures commonly originate along the edges of the tightly fitting component grains. In contrast, initial microfractures are not as pervasive in porous rock such as sandstone where some of the initial strain can be accommodated by porosity reduction and, in weakly cemented rock, by grain slip and rotation. However, there are more potential sites for microfracture origin in porous rock, including the cement between grains and the stress risers created by pores (e.g. Dunn et al., 1973; Wei and Anand, 2008). Microfractures also commonly originate at the point loads created by grain-to-grain contacts ("Hertzian" fractures, first described by Hertz, 1881).

Microfractures in the rock become more numerous as the applied stress increases (Figure 1.39). Yield in brittle and brittle-ductile rock occurs when enough microfractures develop to noticeably affect the mechanical properties of the rock (e.g. Kranz, 1983; Costin, 1987). As the rock approaches failure, the microfractures "localize," beginning to form preferentially along certain planes in the rock, and failure occurs when the microfractures coalesce along those planes to form one or more macrofractures.

Dunn et al.'s (1973) thin sections of strained laboratory samples show microfractures that are oriented parallel to the direction of the applied maximum stress. As stress increases, the microfractures form preferentially along planes that are oblique to that stress, ultimately coalescing to become shear fractures (Figure 1.40) along those planes. The formation and location of microfractures has also been documented in the acoustic emissions produced by developing microcracks (e.g. Ingraham et al., 2013). Microcracks coalescing into fractures are best documented for shear, but similar microcrack zones also form in front of developing extension fractures in the laboratory (Janach, 1977; Delaney et al., 1986; M. Ingraham, personal communication, October 2018).

Some theoretical fracture models postulate that randomly oriented microfractures ("Griffith cracks") are inherent in any unstressed rock mass, and that those cracks which are favorably oriented parallel to the in situ compressive stresses should become preferentially extended into macroscopic fractures as the rock is deformed. Other models suggest that extension fractures originate at sheared, pre-existing cracks that are inclined to the stress axes, creating tension at the crack tips and opening oblique "wing" cracks that then propagate in extension in the direction of the maximum compressive stress. Such precursor flaws may be inherent in some rocks, in others rocks the microfractures that form when the rock is initially stressed can serve as Griffith cracks. Larger stress risers such as fossils or clasts can also act as the initiation points for macrofractures.

Paterson (1978) and Paterson and Wong (2005) summarize several papers that suggest that, as a rock is strained, it undergoes four stages of behavior on the way to developing microfractures and, ultimately,

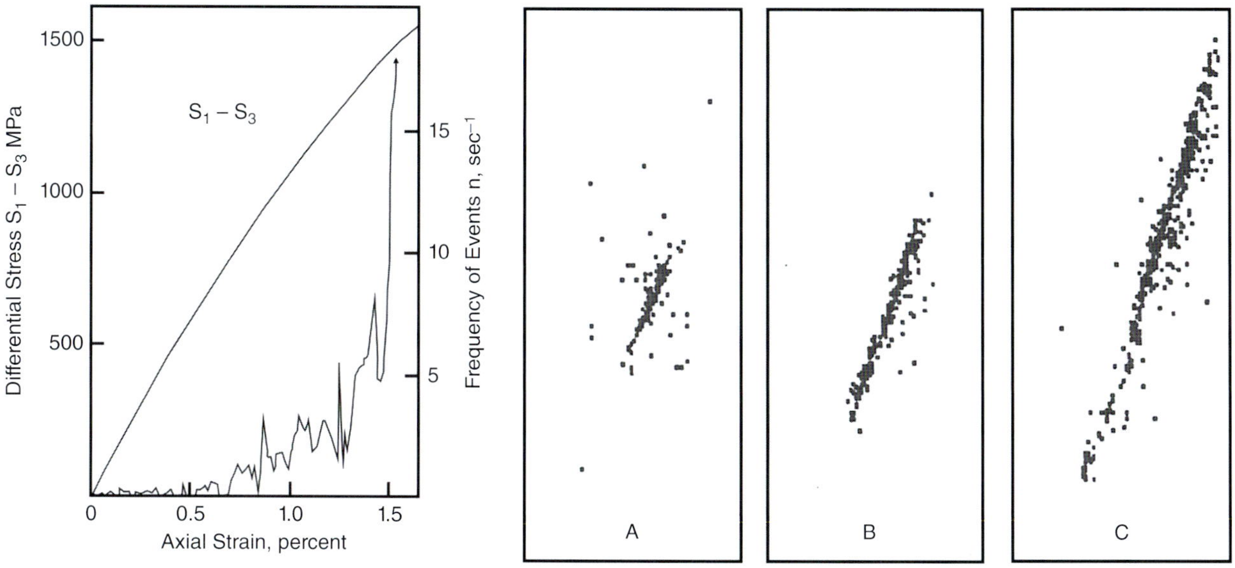

Figure 1.39 *Left*: graph showing the increase in frequency of acoustic emissions created by the formation of microfractures in a sample (vertical scale, right side) as the differential stress (S_1–S_3) imposed on the sample increases (vertical scale, left side). The horizontal axis of the figure shows the axial strain (deformation) in the sample as differential stress increases (adapted from Scholz 1968). Acoustic emissions indicating the formation of microfractures start almost as soon as stress is imposed on the sample, but occur in significant numbers only as the differential stress level approaches the yield and failure points. *Right*: plots of the locations of acoustic emissions within laboratory samples as they are strained. The sequential stages (A, B, C) show that as a sample is strained, the acoustic emissions recording the formation of microfractures are initially scattered within the sample (A) but that with increasing strain they begin to localize, forming more exclusively along an inclined, planar process zone (B), and that they eventually coalesce to form a shear fracture at the point of failure (C) (Westerly Granite, adapted from Reches and Lockner, 1994).

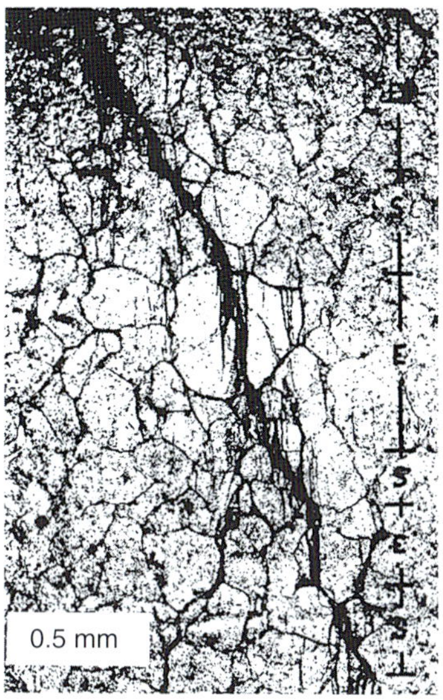

Figure 1.40 Thin sections cut from rock deformed in the laboratory have shown the concentration of extension microfractures, oriented parallel to the maximum applied compressive stress but forming a process zone of en echelon cracks that is inclined relative to the stress axes. The zone eventually becomes a through-going fracture as the compressive strength of the rock is exceeded (from Dunn et al., 1973).

macrofractures. Different lithologies behave somewhat differently, but the four general stages are:

1. As stress is first imposed on the rock, the initial response is the closure of any existing microfractures.
2. With these cracks closed, the rock and its component grains behave elastically up to an elastic limit, although some of the existing microfractures may shear and a few new microfractures may form during elastic deformation.
3. Once the elastic limit of the rock is reached, large numbers of microfractures begin to form. They are uniformly distributed, they are predominantly oriented parallel to the applied stress, and they are relatively stable so they do not link together.
4. In the final stage, the growing number of microfractures "localize," forming preferentially along certain planes and becoming unstable, linking together into macrofractures as the rock mass loses strength. The microfractures may be axial or inclined, linking to form either axial-splitting or shear macrofractures, respectively. Higher confining pressures lead to more pervasive microfracturing as well as microfractures that are more likely to be inclined relative to the maximum stress.

1.4.3.5 Extrapolation to the Subsurface

The application of laboratory tests to the fracturing of larger rock masses at depth is not always clear cut. For example, Paterson and Wong (2005) suggest that shear is the dominant mode of fracturing in compressive tests on the laboratory, which runs counter to the observation that extension fractures dominate most fracture systems at depth. The apparent discrepancy is probably due to the fact that conditions promoting extension are more common in the subsurface than those typically designed for in the laboratory, rather than to any greater susceptibility of rock to one or the other mode of fracturing in either environment.

Nevertheless, there are important differences. A brittle specimen in the laboratory may shatter after parallel, longitudinal extension fractures form, since the pillars of rock between the fractures bow outward against the low-magnitude confining stresses: yield (the end of elastic deformation), fracture (the formation of planar breaks in the sample), and failure (the inability of the sample to support an imposed stress differential) occur nearly simultaneously.

In contrast, the formation of the same type of extension fractures in the laterally confined conditions at depth minimally reduces the ability of a rock mass to support the imposed fracture-parallel compressive stress since the pillars of rock between the fractures remain intact. The rock has fractured, failing in the local sense that the rock has no cohesion or strength across newly formed fracture planes, but fracturing does not mark failure of the rock mass if failure is defined in a broader sense as the point where the rock is no longer able to support an imposed stress differential. The formation of extension fractures accommodates minimal strain, commonly less than one percent in the directions both parallel and normal to the maximum compressive stress (commonly referred to as the "axial" and "lateral" strains, respectively), and Mandl (2005) suggests that this type of fracturing "....would not in any way disturb the (macroscopic) stress field."

For shear fractures, an inflection in the stress-strain curve, the "yield" point, typically marks the end of elastic deformation and precedes the formation of a shear fracture in the laboratory. The formation of the laboratory shear fracture does not lead to immediate and complete failure of the sample since these fractures typically form under more laterally constrained conditions, holding the sample together and allowing friction between the faces of the new shear plane to continue to support some of the axial stress. Yield, fracture, and failure are more widely separated events.

Shear fractures in outcrops and core commonly have limited offsets, measured in centimeters or even millimeters, suggesting that shear fracturing in the subsurface, like extension fracturing, does not always accommodate much strain and may not significantly reduce the maximum applied compressive stress, even though locally the rock has broken and failed along fracture planes. In fact, laboratory tests that use stiff frames which arrest movement as soon as a specimen fails suggest that "violent fracture is not the intrinsic characteristic of rock but due to the rapid release of strain energy from the loading machine" (Abdullah and Amin, 2008). Paterson (1978) points out that the properties and behavior of a laboratory specimen are fundamentally different before and after fractures form, even though the they are plotted as a continuum on the same stress-strain graph, whereas the same is not true of a sheared rock mass in the subsurface.

Griggs and Handin (1960) (Figure 1.2) showed that extension fractures form only in brittle specimens at the farthest-left end of the compression-failure spectrum, which is sometimes considered to be anomalous and as such is not even included in some adaptations of the figure. However, this part of the Griggs and Handin figure depicts extension fracturing in a system of triaxial compression, so it is extremely important as an analogy to the abundant extension fracture systems formed in compressive stress regimes in the subsurface.

Figure 1.2 also shows that the formation of extension fractures in brittle rock does not require much deformation, occurring when the rock has been subjected to less than one percent shortening along the axis of applied stress which is equivalent to shortening a meter of rock by less than one centimeter: extension fracturing does not require large stresses or strains, so it can occur in many geologic environments. These fractures form when a specimen is subjected to a compressive stress parallel to the long axis of the cylindrical sample and simultaneously to lesser-magnitude confining stresses normal to the maximum stress (Figure 1.41).

Gramberg (1965, 1989) was one of the first to argue that the load-parallel extension fractures that form in compression tests in the laboratory are indeed analogous to the widespread sets of subsurface extension fractures oriented normal to the minimum compressive in situ stress. The idea was not universally accepted, in part because it was not clear how the lateral strain required for extension could be accommodated in the subsurface, and in part because it is much easier to break rock in tension. For example, Jaeger et al. (2007) stated that longitudinal splitting fractures must form in tension, and there has been a continuous undercurrent of discussion concerning tension vs. extension in the formation of Mode I fractures at depth. Nevertheless, the interplay between in situ stress and pore pressure, as discussed in Section 1.4.4, largely

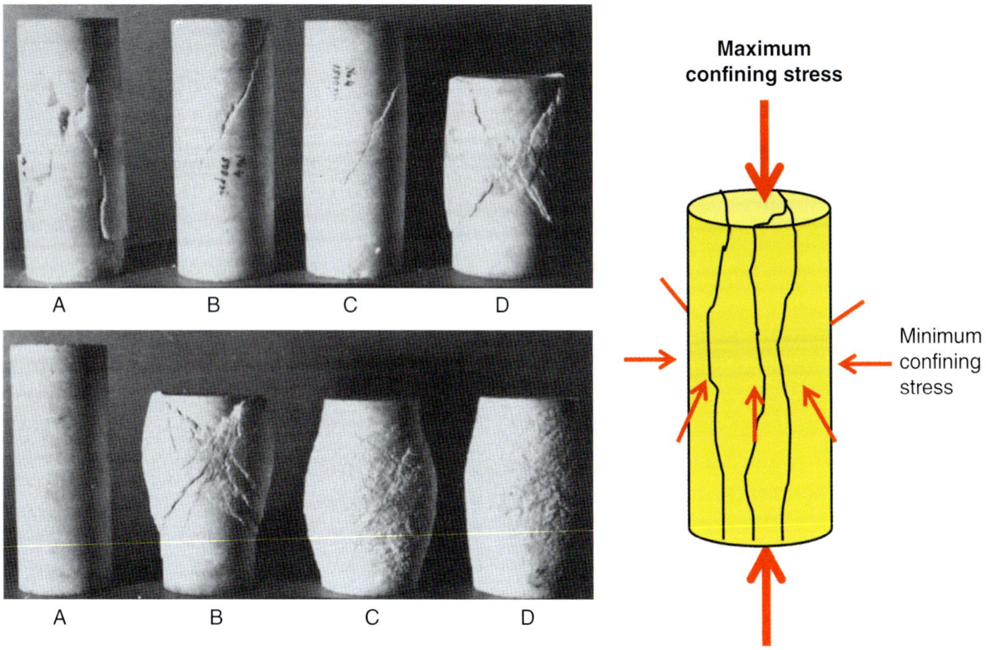

Figure 1.41 The strain-accommodation structures that form in a given lithology are different under different confining stresses. *Left*: laboratory deformation of 2-inch by 5-inch (5.1 cm by 12.7 cm) samples of coarse-grained marble, showing the changes in strain accommodation that occur as samples are deformed under different confining pressures (from Paterson, 1958). *Top*: confining pressures and percent axial strain increase left to right. A: load-parallel extension fractures that formed at >1% strain under atmospheric confining pressure. B, C: single-plane shear fractures that formed at 1% and 2% strain, respectively, under confining pressures of 4 MPa (500 psi) and 9.8 MPa (1,400 psi). D: intersecting conjugate shear planes that formed at 12.5% strain, under a confining stress of 20.6 MPa (3,000 psi) (Paterson noted that the initial dimension of sample D was shorter than the others). *Bottom:* confining pressures increase left to right, but samples B, C, and D were each shortened by 20% strain. A: an undeformed sample. B: intersecting conjugate shear planes that formed under a confining stress of 27.5 MPa (4000 psi). C, D: Unfractured, ductile deformation took place under confining pressures of 45.1 MPa (6500 psi) and 98 MPa (14,200 psi), respectively. Similar left-to-right changes in deformation mode can occur with changing lithology, i.e. with increasing inherent ductilities of the tested lithologies. Note the similar styles of deformation for specimen D in the top row and specimen B in the bottom row: it required a greater percent strain to produce the same type of deformation as the confining stress increased. All samples except C and D in the bottom row show a brittle response to strain. *Right*: a sketch of the load-parallel extension fractures that form in inherently brittle samples and in samples that are strained while under minimal lateral confining stress, equivalent to sample A in the top row left.

negates the objections to Gramberg's model, and given that tension anywhere above the grain scale is rare in the subsurface it is unlikely to be the mechanism responsible for widespread, closely spaced systems of parallel extension fractures.

1.4.4 Interplay Between Developing Fractures and the In Situ Stresses

Both extension and shear fracturing result in an increase in the length of a rock mass in the direction parallel to the minimum compressive stress and a decrease in length parallel to the maximum compressive stress. Most commonly, i.e. where the maximum compressive stress is vertical, lengthening occurs in a horizontal direction parallel to bedding while shortening occurs in the vertical direction. Regardless of the orientation, lengthening would seem to add to the magnitude of the minimum compressive stress as fracturing progresses. Since the rocks are not being pulled apart in tension, any proposed subsurface fracture mechanism must include an explanation of why the minimum stress does not ramp up to arrest the fracturing process.

A stress anisotropy exceeding the strength of a rock is required to fracture it, and such anisotropies can be created in three basic ways although some combination of them is likely to apply to real geologic settings. One way to increase anisotropy is to diminish the minimum stress, common enough in structural settings such as extensional terrains (for example the U.S. Basin and Range province, or the Texas Gulf Coast). The second is to increase the maximum compressive stress, which is typical within and in front of thrust belts and in settings where ongoing sedimentation adds to the weight of the overburden. The third is to decrease the pore pressure such that the anisotropy of the effective stresses increases. The fracture susceptibility of rock can also be enhanced under conditions of increasing the pore pressure; although important, this reduces stress anisotropy. The important effects of pore pressure are examined next, but the increases in anisotropy due to enhanced or diminished stress in one direction are discussed first.

The first scenario, a reduction in the minimum stress, accommodates lengthening of the rock against that minimum stress since the additional length created by fracturing, typically less than one percent, merely acts to sustain the minimum stress at its original magnitude. A decrease in stress iteratively allows and even causes a fracture to form and open, but the extra fracture volume in turn acts to maintain the value of the minimum stress. In this setting, the opening of extension fractures does not require a shouldering aside of the rock that would increase the minimum stress and terminate or even prevent fracturing.

The second setting involves a stress differential enhanced by an increase in the maximum compressive stress. This is a common occurrence within and in front of thrust-belt systems, where the maximum compressive stress increases until a thrust fault forms, at the same time creating a strong stress anisotropy in the strata that will form the thrust sheet as well as the strata filling the adjacent basin. The stress anisotropy results in a basin-scale dilation fabric of natural fractures.

The mechanism that allows lateral extension of the rock in order to accommodate fracturing without also ramping up the minimum compressive stress in a thrust belt is obscure, but vertical extension fractures striking normal to fold and thrust belts are common adjacent to thrust belts around the world (e.g. Bell and Babcock, 1986; Hancock and Bevan, 1987). Part of the explanation can be attributed to the low *effective* minimum compressive stress in the presence of high pore pressure, as shown next (briefly, as pore pressure increases, the effective minimum compressive stress decreases, and lateral compression in a thrust belt can increase pore pressure). The effect of increasing pore pressure with*in* the system is similar to that created by a reduction of the minimum confining stress imposed externally on the strata, allowing fractures to widen by opening against that stress. The problem of opening fractures against the minimum compressive stress in strata in front of a thrust-belt is complicated by the concurrent increases in minimum and intermediate compressive stresses due to Poisson's ratio (as described for the generation of horizontal stresses from the weight of the overburden), but it is not insurmountable.

Extension fractures striking normal to thrust faults with*in* a fold and thrust belt are easier to explain since such systems commonly extend laterally as thrusting progresses, especially where thrusts form an arcuate system (e.g. Crosby, 1969; Cooper, 1992). Lateral extension within the thrust belt reduces the minimum stress and easily accommodates the opening of fractures against a minimum compressive stress.

Stress differentials large enough to break rock, then, can form either by reducing the minimum compressive stress or by increasing the maximum compressive stress. The third and fourth mechanical scenarios that can lead to fracturing rock are controlled by changes in pore pressure. Increasing the pore pressure can decrease the ductility enough to fracture it under the influence of an existing stress differential without changes in the externally applied compressive stress magnitudes. In turn, decreasing pore pressure can increase the stress differential enough to fracture rock of a given ductility. Exploring these scenarios requires the following discussion of the effects of pore pressure.

1.4.5 The Importance of Pore Pressure

1.4.5.1 Introduction

The same rock that becomes both strong and ductile due to the high compressive stresses when it is confined at depth returns to a relatively weak and brittle condition at the same burial depth when the internal pore pressure increases, because pore pressure counteracts and reduces the *effective* compressive confining stresses. And when it is brittle, rock can be broken by small stress differentials, and/or after being subjected to small amounts of strain, typically less than one percent. Reducing the minimum effective compressive stress not only allows fractures to form on geologic time scales, but it also allows extension fractures to open and shear fractures to shear, extending fractured rock in one or more directions even though it is still under compression in all three axes. In contrast, pore pressure adds to the *total* stress acting on a system, controlling the shorter-term aspects of reservoir behavior such as industrial hydraulic fracturing.

1.4.5.2 The Relationship between Pore Pressure and Stress

Pore pressure is an integral component of any system of total and effective subsurface stresses, controlling both natural fracturing and hydraulic stimulation fracturing even though they are totally different mechanical processes. Paterson and Wong (2005) note that the relationship between pore pressure and effective stress is not simple, that "…there is no simple universal definition of effective stress which permits a uniform treatment of all phenomena affected by pore fluid pressure." Nevertheless, it is an important relationship and will be described in detail since it is not widely appreciated.

The *total* stresses confining a reservoir along each of the three axes consist of a combination of the externally imposed stresses (overburden weight, lateral stresses derived from that weight and Poisson's ratio, tectonic stresses, etc.), *plus* a percentage of the pore pressure. In contrast, the *effective* confining stresses consist of the externally imposed stresses *minus* a percentage of the pore pressure. This is an important relationship; both the total and the effective stresses acting on rock change with changes in its internal pore pressure. Since the mechanical properties of a rock that govern its susceptibility to being naturally fractured are in part controlled by the effective confining stresses acting on it, pore pressure is an important control on natural-fracture genesis. Laboratory observations become more relevant to the understanding of natural-fracture mechanics when provisions for variable pore pressures are included in the test apparatus.

The minimum total compressive stress acting on a reservoir can be measured in several ways, the most common being to measure the pressure required to inject a stimulation fracture into the rock and overcome the minimum stress. However, the maximum and intermediate total stresses must be calculated from knowns such as pore pressure, rock density, and its mechanical properties. *Effective* stresses are also calculated rather than measured.

Terzaghi developed the concept of effective stresses in the 1920s (e.g. Terzaghi, 1923) while working with soil stability. Hubbert and Rubey (1959) used the concept to explain the seemingly impossible mechanics of moving large sheets of strata across large distances in fold and thrust belts. More recently, the integral and important relationship between pore pressure and at least one of the three total stresses has been measured (Salz, 1977; Breckles and van Eekelen, 1982; Teufel et al., 1991; Teufel, 1996) and discussed in detail (e.g. Hillis, 2000, 2001, 2003). These references have compiled data from reservoirs in different parts of the world, from the Texas Gulf Coast to Brunei, and the data consistently show that the total minimum stress acting on a formation changes in lockstep with changes in the in situ pore pressure, although the degree of coupling is not consistent.

Salz (1977) published an important figure derived from a comprehensive dataset, the figure definitively showing the changes in the total minimum stress acting on a formation that accompany changes in the magnitude of the in situ pore pressure (Figure 1.42). Salz documented this relationship by comparing the pressures needed to inject stimulation fluids into reservoirs (a measure of the total minimum compressive stress) to the formation pore pressure in the zone of injection, using data from the over-pressured sandstone reservoirs in the Oligocene-age Vicksburg Formation at depths of 10,000–15,000 ft (3000–4600 m) on the Gulf Coast of Texas. Salz normalized both the minimum-stress and the pore-pressure data by depth to get gradients,

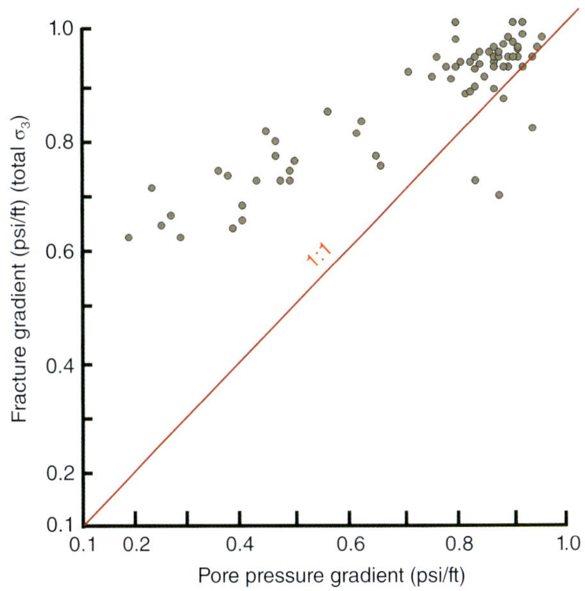

Figure 1.42 A cross plot between measured pore pressures and the associated measured hydraulic fracture pressures, in reservoirs in the Oligocene Vicksburg sandstones of the Texas Gulf Coast. The plot shows that fracture gradient (the pressure needed to inject a hydraulic fracture into a formation and thus a measure of the minimum total in situ compressive stress, with the different data points normalized by depth) varies definitively with the similarly normalized pore pressure gradient in a formation. The plot also shows that both the minimum total stress and the pore pressure converge on the compressive stress of one psi per foot created by the weight of the overburden (adapted from Salz, 1977). The original Salz figure placed the horizontal axis at the level of 0.6 psi/ft fracture gradient, obscuring the convergence.

so that his cross plots show both the stress and the pore-pressure gradients in psi/ft, allowing the data points to be compared regardless of depth.

The Salz data are seminal; they show that the total minimum stress increases with increasing pore pressure in over-pressured reservoirs. Moreover, when Salz tested reservoirs that had been depleted by production (where pore pressures were less than the hydrostatic gradient of about 0.43 psi/ft), he found that the relationship continued at the other end of the trend such that the total minimum stress is proportionately less in reservoirs with diminished pore pressures. The relationship holds in reservoirs ranging from severely depleted reservoirs with a pore pressure gradient of no more than a 0.2 psi/ft through to severely over-pressured reservoirs with pressure gradients of up to 0.95 psi/ft. The data show that it consistently and proportionally requires more pressure to inject a hydraulic stimulation fracture into over-pressured reservoirs than to inject into depleted reservoirs. These data definitively show that the total minimum compressive stress is related to pore pressure, regardless of depth.

Salz calculated correlation coefficients of 0.88 and 0.95 for the relationship between pore pressure and total stress, the difference depending on whether over-pressured or under-pressured reservoirs were measured, but the two coefficients are high and they are not materially different. As pore pressure increases, the total minimum stress increases from 50% of the formation pore pressure to 90% of that pressure, converging at about 1 psi/ft, which is also the lithostatic gradient, i.e. with the maximum compressive stress.

In contrast, the *effective* stresses acting on a formation, which control the strength and brittleness of rock and thus control its natural-fracture susceptibility, *de*crease with increases in pore pressure such that, under geologically produced stress differentials, strata with high internal fluid pressures will fracture naturally more easily than strata with low pressures. This emphasizes an important difference between hydraulic fractures and natural fractures: high pore pressure facilitates natural fracturing, but it inhibits hydraulic fracturing, and the two processes should not be confused.

1.4.5.3 Biot's Coefficient

Significantly, only a percentage of the formation pore pressure adds to the compressive stress to form the total minimum stress, and a percentage of the pore pressure counteracts the confining stresses to create the effective confining stresses (see Appendix 1.A). The salient fact that the stresses commonly change by less than 100% of any change in the pore pressure was explored in a series of papers by Biot in the 1940s and 1950s (e.g. Biot, 1941; Rice and Cleary, 1976), and the percentage is often referred to as the "poroelastic parameter" or "Biot's coefficient." Note that it is a *coefficient* and not a *constant*. The coefficient is often assumed to be 1.0 (100%) since the data needed to calculate this factor are not always available, but where it can be calculated it is commonly less than 1.0. Biot's coefficient is not the same for all formations. It is not constant for different stress and pore-pressure conditions within a given formation. Teufel (1996) compiled and compared data from six oilfields and concluded that "The magnitude of changes in stress state with net changes in pore pressure is dependent on local field conditions and cannot be accurately predicted by the uniaxial strain model…" Nor is it constant for the three compressive stresses acting on a formation, which commonly change by different percentages of any change in pore pressure (Teufel et al., 1993).

Authors differ on whether to apply pore-pressure corrections and Biot's coefficient to an overburden stress since rock is unconfined at the earth's surface. Usually no pore-pressure correction is applied to the *total* overburden stress, which is assumed to be the weight of the rock, whereas all or some of the pore pressure may be subtracted from the weight of the overburden to calculate the *effective* overburden stress. Pore pressure and Biot's coefficient influence a number of other stress-related reservoir parameters, including permeability, deformation, and sonic velocity (see Nermoen et al., 2013).

1.4.5.4 Mohr Diagrams and Pore Pressure

As pore pressure increases, both the maximum and minimum effective stresses plotted on the horizontal axis of a Mohr diagram, as well as the circle defined by the diameter between them, move to the left and therefore closer to the sloping, experimentally determined failure envelope. The points and the circle move left because pore pressure is subtracted from the maximum and minimum compressive stresses to calculate effective stresses.

Extension fractures are sometimes explained in reference to this relationship, i.e. that if pore pressure is elevated to a level exceeding the magnitude of the minimum compressive stress, the effective minimum stress theoretically becomes negative and moves left into the tensile field (Figure 1.43). Although intuitively appealing, this usage does not account for the effect of Biot's coefficient, pore pressure, and the convergence of the maximum and minimum effective stresses as pore pressure increases.

Just as the maximum and minimum *total* stresses increase, converge, and approach the limiting overburden stress as pore pressure increases as shown by the Salz data, so also do the maximum and minimum *effective* stresses *de*crease, converge, and approach zero as pore pressure rises (Figures 1.44, 1.45) (e.g. Addis, 1997; Addis et al., 1996; Hillis, 2000, 2003). Because they converge, pore pressure does not, and theoretically cannot, exceed the minimum compressive stress, and therefore elevated pore pressure should never produce tensile stress conditions or "explode" a rock from within in the subsurface. As the effective stresses converge, the stress differential, which creates preferential fracture strikes, diminishes to zero. Convergence decreases the stress differential available for creating fractures, but the rock simultaneously becomes weaker and more brittle, prone to fracture even under reduced stress differentials.

1.4.5.5 Pore Pressure Makes Rock Weak and Brittle

Elevated pore pressure turns rock at depth into an effectively unconfined and therefore a relatively weak and brittle material that is prone to fracturing, even under the reduced differential stresses created by elevated pore pressure. However, fracturing must occur before the three stresses converge to an isotropic stress state since the formation of a set of systematically oriented fractures requires a stress anisotropy.

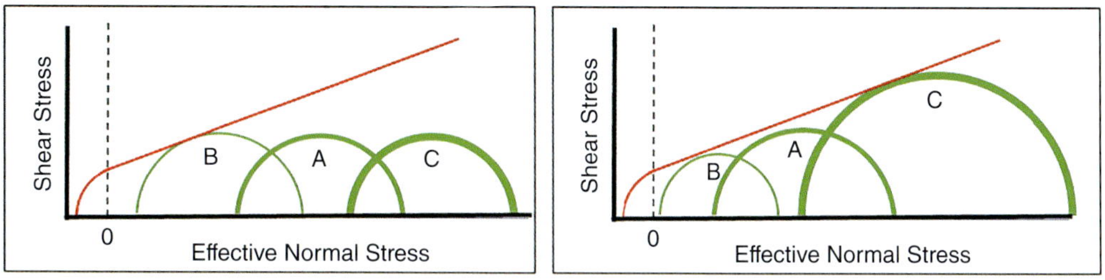

Figure 1.43 *Left*: a depiction of the change in effective stress state due to changes in pore pressure, assuming the effective compressive stresses are the result of subtracting 100% of the pore pressure from the maximum and minimum compressive stresses. The position of the initial-condition stress circle (circle A) moves to the left to become circle B and towards the failure envelope (the red line) by subtracting an amount equal to the increase in pore pressure from both the maximum and minimum compressive stresses as the formation becomes over-pressured. A decrease in pore pressure moves the stress circle to the right in the same manner (circle C). *Right*: a more accurate depiction of the relationship between pore pressure changes and the Mohr stress circles shows that the circles defined by the effective minimum and maximum compressive stresses not only move to the left as pore pressure increases (circle B), but they also decrease in diameter since less than 100% (here 70%, i.e. Biot's coefficient is 0.7) of the pore pressure is subtracted from the minimum and maximum confining stresses each time pore pressure increases (see Appendix 1.A). Failure occurs if stress circle B intersects the failure envelope. Failure can also occur if the stress circle intersects the failure envelope as it moves to the right and expands under conditions of decreasing pore pressure (circle C) (adapted from Hillis, 2003). The "end cap" of the experimentally determined red failure envelope to the left of 0 is a nebulous area, usually included because rock fails so readily in tension, but usually poorly defined. As drawn, the stress circles decrease rapidly and never intersect the failure envelope as pore pressure increases, but that is not always or even usually the case.

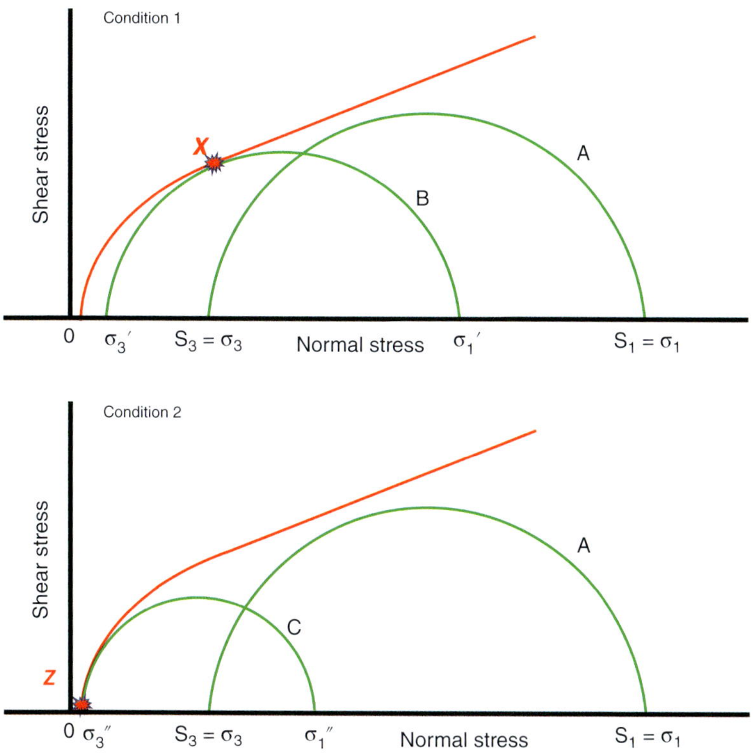

Figure 1.44 The circle diameter defined by the maximum (S_1) and minimum (S_3) stresses (without pore pressure), moves to the left as pore pressure is added to the system. As the circle moves left, it also decreases in diameter (Appendix 1.A). Pore pressure has a greater effect in reducing the maximum effective compressive stress than in reducing the minimum effective stress, and they converge. (The total stress S_1 equals the effective stress σ_1 in the absence of pore pressure). The green circles A, B, and C indicate the stress conditions under different pore pressure conditions (e.g. Hillis, 2001; 2003). Shear fractures form if the green stress circle touches the experimentally determined failure envelope as at X in Condition 1. Opening-mode extension fractures form at Z in Condition 2, when pore pressure reduces the size of the stress circle, pushing the circle to position C. Although tensile stress breaks rock easily in the laboratory and although the failure envelopes are typically drawn to extend to the left of 0 on the normal-stress axis and thus to include true tension, laboratory tests show that none of the confining stresses need to be tensile for extension fracturing to occur.

Some of the early yet still-significant laboratory tests that demonstrate the important relationship between pore pressure and fracture susceptibility were conducted by Robinson (1959), Robinson and Holland (1969), and Murrell (1965). These authors took samples of sandstone, limestone, and shale, and tested them to failure under different confining stresses ranging from 0 to 10,000 psi (0 to 68.9 MPa). As expected, the samples had increasingly higher yield strengths when strained under increasingly higher compressive stresses (Figure 1.46). Robinson then tested his samples to failure under the same lateral confining stresses with different levels of elevated pore pressure within the samples. He found that yield strengths of the samples, as measured by the increasing axial stresses that were required to achieve failure while under a constant lateral confining stress,

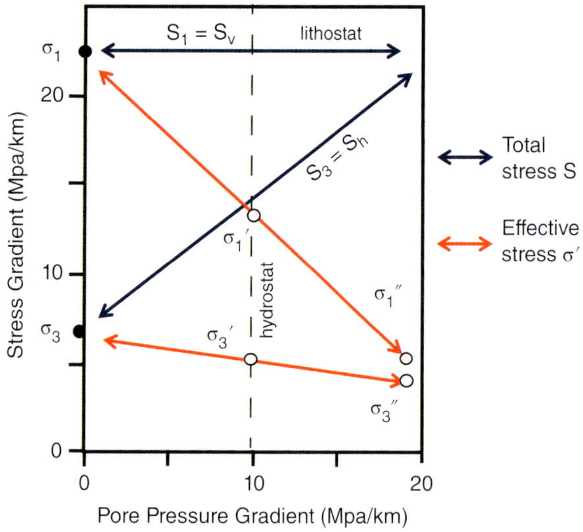

Figure 1.45 The relationships between changing pore pressure and changing total stresses, and between changing pore pressure and changing effective stresses. The blue arrows show total stresses. S_v is the total vertical overburden stress (which in this case is also S_1, the maximum of the three compressive stresses), and S_h is the total minimum horizontal compressive stress (which in this case is also S_3, the minimum of the three compressive stresses). Pore pressure is typically assumed not to contribute to the total overburden stress, which remains constant as the weight of the overburden, since the free surface of the earth is not a confining boundary. The red arrows show the effective stresses. The maximum and minimum *effective* compressive stresses, σ_1 and σ_3, respectively, decrease as pore pressure is added. The prime and double-prime markings indicate two stages of added pore pressure, first to hydrostatic conditions (′) and then to over-pressured conditions (″). The minimum horizontal stress changes by 70% of the pore pressure, but the maximum, overburden stress is assumed to be reduced by the full amount of the pore pressure since that stress is unconfined. The *differential* stresses, both total and effective, (depicted by the vertical distance between the arrows of the same color) increase with decreasing pore pressure, and decrease with increasing pore pressure (modified from Hillis, 2000). The effective stresses do not all change at the same rate relative to changes in pore pressure: the ratio of the change in effective minimum horizontal stress to the change in effective overburden stress is called the "stress path" (Teufel and Rhett, 1991). The differential between the maximum and minimum stresses, and the related potential for shear fracturing, increase more rapidly with pore pressure decreases where the stress path is low than where it approaches 1. Teufel and Rhett documented a stress path of only 0.2 for chalk reservoirs of the North Sea, meaning that the differential stress increased rapidly with reservoir pressure depletion, leading to the formation of new shear fractures in the reservoir during production.

decreased in proportion to the increased pore pressure within the samples.

Robinson's graphs demonstrate several additional important relationships. First, they show that the samples consistently failed as pore pressure *approached* the magnitudes of the lateral confining stress, *before* pore pressure exceeded them (i.e. failure was in the compressive, not tensile field of the Mohr diagram). Second, they show that as pore pressure approached the confining stress the failure mode transitioned from "malleable" (ductile) to "brittle" failure. Robinson's samples were effectively weak and brittle at the highest pore-pressure conditions tested, even under the highest confining-stress conditions. These tests are analogous to geologic systems where rock is susceptible to fracturing at depth in over-pressured formations.

Murrell (1965) and Rhett (2001), among others, did similar experiments and obtained similar results, showing that the strength of sandstones decreases significantly as pore pressure increases under conditions of constant confining stresses. The experiments showed consistent results whether the authors held pore pressure and lateral confining stresses constant while increasing the axial stress until the samples failed, or held the triaxial confining stresses constant and increased pore pressure within the samples until they failed. The former condition would be analogous to settings such as thrust belts where the maximum compressive stress increases due to tectonic compression, the latter would be analogous to settings where the pore pressure increases due to organic maturation within a system of static but asymmetric confining stresses.

In making rock brittle, elevated pore pressure makes rock more susceptible to both shear and extension fracturing. Fracture mode is controlled in part by the intrinsic strength and brittleness of the rock, thus somewhat ductile rock such as limestone or poorly consolidated sandstone is more likely to fracture in shear within the same pore-pressure and confining-stress system that creates extension fracturing in more brittle rock. Qualitative outcrop observations confirm this, suggesting that shear fractures, and trains of en echelon fractures within tabular shear zones, typical of more ductile deformation, are more common in limestones than in sandstones.

Fluids do not support shear stresses, so pore pressure does not directly inhibit or enhance the initiation of shear offset along a fracture plane. However, pore pressure does reduce the component of compressive stress acting normal to an existing shear plane, reducing shear friction and the resistance to shear on an existing shear plane: once a shear plane forms, it takes less energy to offset rock along the existing plane in the presence of high pore pressure than under conditions of low pressure. The U.S. government discovered this accidentally in the 1960s during fluid injection at the Rocky Flats facility in Colorado where they initiated earthquakes by disposing of fluids via a process of injecting them into the subsurface (at pressures below the fracture gradient), slowly elevating the formation pore pressures which

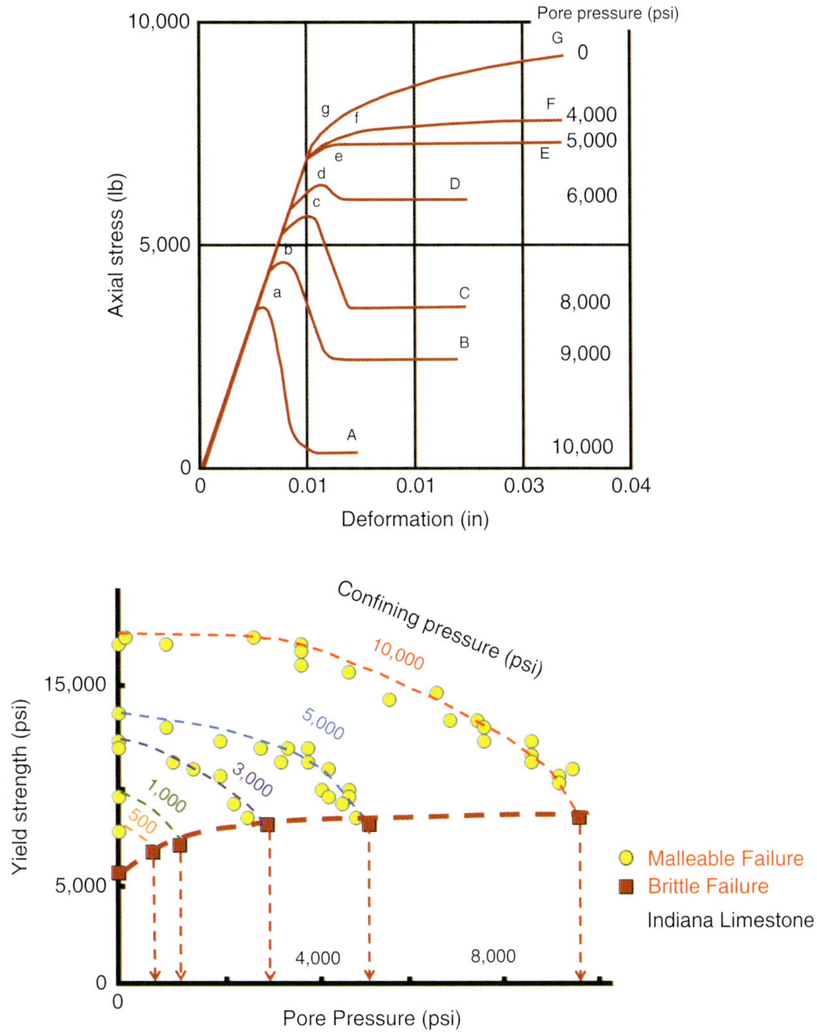

Figure 1.46 The effects of pore pressure on rock strength. *Top*: "Force-Deformation" (i.e. stress-strain) curves showing that samples of Indiana Limestone become weaker, failing at lower-magnitude axial stresses under conditions of increasing pore pressure, *and* showing that the samples become more brittle as indicated by the more abrupt failure (changes in the shape of the stress-strain curve), as pore pressure increases. All samples were tested under 10,000 psi lateral confining stress. The shape of the top curve (G), representing zero pore pressure and 10,000 psi confining pressure, is typical of ductile behavior, whereas the shape of the bottom curve (A), representing pore pressure equal to the confining stress, is typical of brittle failure. *Bottom*: the strength (the vertical axis) of the limestone samples increases as they are subjected to increasing confining pressures (the five curves), but strength then decreases–under unchanged confining pressure–as the pore pressure within the samples (the horizontal axis) is increased. The failure mode also changes from "malleable" to "brittle" as the pore pressure approaches *but before it exceeds* the confining stress. (Both figures adapted from Robinson, 1959.)

reduced the normal stresses on fault planes and allowing them to slip under the existing stress anisotropy (Evans, 1966; Healy, 1968; Ellsworth, 2013). The phenomenon was rediscovered in Oklahoma 50 years later, where injected oilfield waste fluids also triggered earthquakes.

*De*creasing pore pressure can also create fractures, since it increases the anisotropy between the maximum and minimum effective compressive stresses (Figure 1.43). The stress anisotropy becomes large enough to exceed the strength of the rock even as the rock becomes stronger under conditions of increasing effective confining pressures. An example of the real-time shear fracturing under conditions of decreasing pore pressure was documented in the chalk reservoirs of the Ekofisk oilfield in the North Sea, where fractures have formed in the reservoir chalks as pore pressure has been depleted by more than 28 MPa (4000 psi) over several decades of production (Teufel and Rhett, 1991; Teufel et al., 1993). Decreasing pore pressure and the related increase in stress anisotropy can also reactivate shear along faults (e.g. Molina and Zeidouni, 2017).

1.4.5.6 Sources of Pore Pressure

Pore pressure changes are key to making confined subsurface rock susceptible to natural fracturing. Current reservoir conditions are only a snapshot of the geologic history of the strata, so although not all naturally fractured reservoirs are presently over-pressured, any geologic system is likely to have been highly over-pressured at least once during its history. This often occurred at the same time that it was either deeply buried or undergoing tectonic deformation and therefore under conditions of high stress magnitudes and high stress differentials, increasing the probability of fracturing.

Compaction disequilibrium and hydrocarbon generation are two common mechanisms for creating over-pressure conditions in a reservoir (see Law and Spencer, 1998; Swarbrick and Osborne, 1998). Compaction disequilibrium is usually the result of high sedimentation rates, often in rapidly subsiding basins. In this mechanism, deposition and burial exceed the rate at which fluids can escape from the water-charged, under-compacted strata below the accumulating overburden. Similar disequilibrium may also result from lateral compaction within thrust belts although the strata are likely to be more completely lithified in such settings. The second mechanism, hydrocarbon generation, occurs under conditions of increasing temperatures and pressures during burial; Barker (1990) calculated that cracking a barrel of oil in a thermally maturing system generates an average of 3000 cubic ft (85 m^3) of natural gas, and the increased volume should be instrumental in over-pressuring a formation.

Pore pressures generated by these mechanisms could conceivably raise the overburden since the surface of the earth is an unconfined, free surface, but fractures and faults are likely to form first, so pore pressures that approach the magnitude of the overburden are typically relieved by fluid escape from the system, commonly by cyclic "valving" along faults. Most pore pressure-depth profiles show that elevated pore pressures approach and even join the lithostatic gradient, but few reliable measurements exist showing pore pressures in excess of the lithostatic gradient. Several authors (e.g. Moore et al., 1988; Sibson, 1990, 2003; Hunt, 1990; Grauls, 1998) have suggested that an equilibrium is commonly established between the high pore pressures created by ongoing gas generation and episodic pressure releases through faults, and Barker (1990) estimated that a significant percentage of the gas generated at depth has been lost due to such fault valving.

Other over-pressuring mechanisms, typically associated with increasing burial depths, have not been as widely applied to the problem but are viable. Such mechanisms include the thermal expansion of water, the conversion of gypsum to anhydrite, the maturation and dehydration of clay minerals, and hydrocarbon buoyancy (e.g. Fertl, 1976; Swarbrick and Osborne, 1998).

The fact that so many reservoirs are normally pressured and are therefore likely to be connected hydraulically to the surface does not indicate that the reservoirs were never over-pressured. Rather, it indicates that most reservoirs have been sufficiently over-pressured to develop pressure-relief connections to the surface at one or more times during their history, and that the plumbing is still operative. This inference is supported by the hydrocarbon footprints that can be detected at the surface over many reservoirs (e.g. Arp, 1992; Rice et al., 2002; Schumacher, 2012).

In addition to production-related pressure depletion, under-pressured reservoirs and an associated increase in stress anisotropy can be created within strata that were normally pressured but that lost their connection to the surface during continued burial. Under-pressured reservoirs also occur within strata that have been cooled due to a diminished thermal gradient over geologic time.

1.4.5.7 Alternate Theories

The terms "expulsion fracture" and "natural hydraulic fracture" are sometimes applied to fractures, often on a theory-only basis without assessing whether the characteristics of a fracture set support this very specific mode of fracture origin. For many geologists, "natural hydraulic fracture" implies a natural fracture that formed by the same mechanism that is used to hydraulically fracture reservoirs in non-natural, industrial settings, i.e. by an injection into the reservoir of high-pressure fluid from an external source. Lorenz et al. (1991) argued that few natural fracture sets show evidence for this mechanism and that it is not applicable to most natural extension fractures, although other geologic structures such as igneous dikes and some clastic dikes certainly formed in this way.

Fall et al. (2015) used the term "natural hydraulic fracture" in the title of their paper, intending to imply a natural-fracture process that includes a hydraulic, pore-pressure component, but inadvertently suggesting to many readers that the process is analogous to industrial hydraulic injections. The term "natural hydraulic fracturing" has also been used to suggest an implausible fracture-origin mechanism where the pore pressure in a formation is elevated to the point where it exceeds a combination of the tensile strength of the rock and the minimum in situ compressive stress, breaking the rock by the force of the internal fluid pressure. Although intuitively appealing and although it has been "proven" on Mohr diagrams, this mechanism is non-viable because there is no mechanism for concentrating pore pressure in any subset of pores that would preferentially be inflated to become fractures, or for maintaining an elevated fluid pressure exclusively within the growing fractures. As discussed earlier, since pore pressure and the minimum confining stress converge, pore pressure theoretically cannot exceed the minimum confining stress anyway. Empirically, the consistency of orientations of the fractures within most natural fracture sets vitiates the concept of pore pressure exceeding the minimum stress: if the pore pressure in a rock increases to the point where the three effective stresses become equal, as it would need to do if it were to continue to the point where it exceeds the minimum stress, no stress anisotropy would remain to govern a preferential fracture orientation. As pointed out by Gretener (1979), any fractures that *did* form under such conditions should have random, not systematic, orientations.

1.4.6 Summary

Griggs and Handin's (1960) laboratory-generated shear and extension fractures are approximate analogs for natural fractures formed in the subsurface. Extension fractures form when the rock is 1) brittle and susceptible to fracture; 2) minimally confined; and 3) subjected to a stress anisotropy. Shear fractures, still considered to be brittle-deformation structures, form in rock that either has an inherently more brittle-ductile composition and/or that is rendered somewhat more ductile by higher confining stresses. An integral piece of the puzzle, pore pressure, must be included in order to reduce the confining stresses and make rock brittle when it is deeply buried and confined under the weight of the overburden. Deeply buried rock can be brittle and can fracture regardless of its inherent ductility; even inherently ductile mudrock reservoirs can be fractured in the presence of a stress anisotropy and high pore pressure (Figure 1.47).

Rock will fracture in the subsurface:

1. If the stress anisotropy is increased to exceed rock strength by increasing the maximum compressive stress;
2. If the stress anisotropy is increased to exceed rock strength by decreasing the minimum compressive stress;
3. If the stress anisotropy in increased to exceed rock strength by decreasing the pore pressure (favors shear fracturing);
4. If the rock is made weaker and more brittle by increasing the pore pressure to the point where the existing stress anisotropy exceeds the rock strength (favors extension fracturing), even as that anisotropy diminishes with pore-pressure increases.

Most formations have been subjected to conditions of high pore pressure and/or high stress anisotropy, and have therefore been susceptible to fracturing, at one or

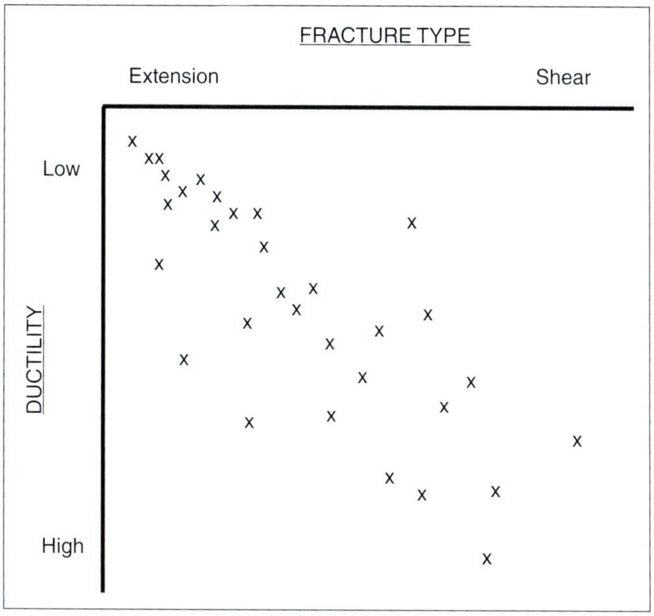

Figure 1.47 *Left*: a conceptual representation of changes in fracture mode (extension vs. shear) as controlled by ductility. Ductility in turn depends initially on the basic lithology of a rock, which itself is a function of composition, diagenesis, cementation, porosity, etc. Shear fractures at the right of the graph are more likely to form in relatively ductile lithologies, and extension fractures at the left of the graph are more likely to form in relatively brittle lithologies. However, rock that is inherently brittle can be made sufficiently ductile to deform in shear at the right side of the graph during burial as the temperature and/or the confining stresses increase. Likewise, intrinsically ductile rock at the right side of the graph can be turned into brittle rock that is prone to extension fracturing by increasing the pore pressure in the formation, increasing the stress anisotropy, or increasing the strain rate. Rock that starts as brittle at the left of the graph but that then becomes more ductile and moves towards the right as it is buried can be moved *back* to the left by increases in pore pressure, strain rate, and stress anisotropy. Hybrid shear fractures form in the field between extension and shear. En echelon veins within wider shear zones form in even more ductile conditions to the right of shear, but not so far right that the rock becomes plastic and accommodates strain without fracturing. The scatter in the conceptual data is intended to reflect geologic heterogeneity. *Right*: an unexpectedly well-developed, irregular fracture system in the clay-rich marine Mancos Shale (Cretaceous, New Mexico; photo courtesy Ryan Hillier). Under the right conditions, any lithology can become ductile and prone to shear, and likewise any rock can be made brittle and prone to fracturing.

more times during their history. The pervasive fracturing that is present in most formations is not difficult to explain.

1.5 Other Fracture Types

1.5.1 Introduction

Many reservoirs are cut by fractures types other than the common shear and extension fractures. Some of these types have the potential to contribute to a reservoir permeability system, others to detract from it, and still others have marginal if any effect, or at least their effects have yet to be measured. The dimensions and characteristics relevant to fluid flow of some of these fracture types are reviewed here, along with discussions of the mechanics of forming them.

1.5.2 Deformation-Band Shear Fractures, Compaction Bands, and Dilation Bands

1.5.2.1 General Characteristics

Deformation-band shear fractures, and the related compaction-band anticracks and dilation-band extension fractures (see Du Bernard et al., 2002; Holcomb et al., 2007; Fossen and Bale, 2007), fit the broad definition of "fracture" in that they are localized, planar zones of strain accommodation in rock where the rock failed under an externally imposed stress. Unlike most fractures, however, these bands were never open breaks in the rock even though porosity is commonly enhanced along dilation bands. Fossen et al. (2007) differentiated the three structures by the degree of shear, compaction, and dilation, and by the degree of cataclasis along the bands. More is known about deformation-band shear fractures since they are the most common form of the three band-type structures.

Shear offset creates deformation-band shear fractures, comminuting component grains of the host rock along narrow shear zones, collapsing the local porosity, reorienting elongate grains, and, if present, converting softer grains along the shear plane into a pseudo-matrix. Permeability in the direction normal to a deformation-band shear can be several orders of magnitude less than that of the host rock (e.g. Antonellini and Aydin, 1994, 1995; Sternlof et al., 2004; Fossen and Bale, 2007; Rath et al., 2011), so where they are well developed this fracture type can have a deleterious effect on reservoir quality. As with other shear fractures, deformation-band shear planes form at angles to the in situ stresses.

Compaction-band anti-crack fractures have similar characteristics but are oriented normal to the maximum compressive stress and form by a reduction in volume and porosity due to compaction within the bands caused by band-normal compression rather than band-parallel shear. For the same reasons of comminuted grains and collapsed porosity, compaction bands also degrade reservoir system permeability. Post-fracturing dissolution and local cementation can also occur along deformation and compaction bands, but these processes are not common since the bands do not typically offer open pathways for mineralizing fluids.

Deformation shear bands can form as well-ordered, systematic, conjugate, shear-fracture pairs, complete with ideal 60° intersection angles and formed under a stress system where the maximum compressive stress bisected the acute intersection angle (e.g. Aydin, 1978; Olsson et al., 2004). They may also form individually, or as irregular splays off the ends of faults where they record local stress fields. In many areas the shear bands are amalgamated into bundles of tens or a few hundreds of sub-parallel planes with significant cumulative offset, with complimentary oblique deformation-band shears forming ladder rungs as secondary Riedel shears within the system (e.g. Davis et al., 2000). A fault in brittle strata where it is expressed as a single discrete shear plane can change into a system of amalgamated deformation-band shears as it extends into an adjacent high-porosity sandstone. Deformation-band shear fractures can form as en echelon segments centimeters or decimeters in extent and which are expressed as non-congruent steps on the rare exposures of deformation-band surfaces. The surfaces are not marked with striations or slickenlines except where shear continued beyond the minimal offset normally associated with shear banding.

1.5.2.2 Dimensions and Distributions

Unsegmented, single deformation-band shear planes can extend for tens or a few hundreds of meters across an outcrop, measurable length being limited only by the size of the outcrop. Fossen et al. (2007) show that there is a linear relationship between displacement and trace length, but the slope of the correlation line changes for different types of bands in different structural settings. These band-type fractures form most readily in poorly cemented, high-porosity rock such as sandstones, bioclastic limestones, and chalks (Nelson, 2001; Rath et al., 2011). Deformation-band shear fractures cut across minor sedimentary bedding planes, commonly extending top to bottom in a reservoir, but the minor amounts of shear associated with individual deformation bands is often absorbed as more pervasive deformation at contacts with more ductile strata.

Where they form as conjugate pairs, normal dip-slip deformation bands can have nearly ideal 60° dip angles, reverse dip-slip bands have 30° dip angles (Figure 1.48), and strike-slip bands are vertical; the system that forms

Figure 1.48 *Left*: thrust-oriented, reverse-dip-slip deformation-band shear fractures (dipping towards each other as indicated by the planes of the geologist's hands) exposed on the upper bedding surface of a high-porosity sandstone of the fluvial Morrison Formation in New Mexico. This is the youngest of three superimposed conjugate pairs in this formation, the three pairs forming sequentially as the horizontal compressive stress ramped up to exceed the weight of the overburden (see Olsson et al., 2004). *Right*: intersecting dip-slip conjugate deformation bands in one bed of the Jurassic eolian Entrada Sandstone of New Mexico. The bed is cut by numerous, closely spaced shear planes and would be a highly compartmentalized reservoir. Pervasive deformation bands did not form in the adjacent, redder sandstone layers, indicating differences in porosity, grain size, or perhaps cementation that created dissimilar mechanical properties in the two layers.

depends on the orientations of the three principle in situ compressive stresses. Changes in the relative magnitudes of the three stresses during shear can lead to the sequential development of all three conjugate orientations within one formation, severely compartmentalizing a reservoir (Olsson et al., 2004).

Deformation bands are easily mistaken for mineralized fractures, but most are not in fact mineralized. The central zone of a shear band which often looks like mineralization is typically a millimeter or less in width and consists of broken and crushed grains and sometimes a pseudomatrix derived from smeared-out softer grains (Figure 1.49). Bundles of deformation bands may be amalgamated to form composite structures centimeters to decimeters in width.

The conjugate pairs that comprise some deformation-band shear fracture systems are more apparent in stereoplots (see Figure 1.26) than in outcrop, since both subsets of the pair may not occur in any given outcrop. Moreover, the ideal conjugate-angle dips and strikes are subject to geologic reality, so the intersection angles may be either more or less than the ideal. The angles also may no longer be ideal if the host strata have been tilted since the deformation bands formed.

The related compaction bands ideally form parallel planes oriented normal to the maximum compressive stress. In Nevada where they were originally described (Hill, 1989), many of the bands are parallel to each other, but some are also oriented at oblique angles to the maximum compressive stress, and these apparently formed as shear-enhanced compaction bands (Liu et al., 2015). Irregular, less systematic deformation bands form in the least well-cemented rock and in structurally more complex settings.

1.5.2.3 Origin

Deformation-band shear fractures form most often in poorly cemented and high-porosity rock, most readily, at least in sandstones, where porosity exceeds 18% (Nelson, 2001). Many deformation bands form at shallow burial depths before the rock becomes well cemented and loses porosity, but bands can form any time that the diagenetic history of the rock creates high-porosity, poorly cemented conditions. Fossen et al. (2007) reported deformation bands in relatively recent glacial deposits, whereas Olsson et al. (2004) recorded a system of deformation bands that formed 100 million years after deposition of the host sandstones. There have also been suggestions that deformation bands can form in poorly consolidated sandy strata around a wellbore due to the near-wellbore stress perturbations associated with production. High-porosity strata can become diagenetically altered after shear bands formed, thus the bands can also be found in what are now low-porosity, well-cemented sandstones.

Offsets along bedding show that individual deformation-band shear fractures commonly accommodate a few millimeters or less displacement, rarely more than a few centimeters. Davis et al. (2000) indicate that a deformation-band shear plane quickly becomes work hardened such that it takes less energy to form successive, parallel shear planes in the adjacent rock than it does to reactivate or extend an existing one.

Figure 1.49 *Left*: a thin section showing the collapsed porosity and grain-size reduction that are typical of deformation-band shear planes (scale bar = 1 mm) (from Fossen et al., 2007). *Right*: an irregular system of deformation-band shear fractures that formed in a poorly cemented, coarse-grained sandstone. The photo illustrates the common multi-stranded nature of such systems, as well as the bands of crushed rock that look like mineralization, and small-scale ladder development along secondary Riedel shears connecting the main shear planes (Vermejo Formation, Raton Basin, Colorado).

Continued strain in a formation that is susceptible to shear banding commonly produces numerous closely spaced, sub-parallel shears.

The two sub-sets of an "X"-shaped conjugate shear pair form sequentially rather than simultaneously (Figure 1.50). This explains the lack of crush zones at the apices of the wedges defined by the acute conjugate angle, crushing that would occur if both legs of the shear pair slipped at the same time.

1.5.3 Faults and Fractures

Faults are large-displacement shear fractures, or perhaps shear fractures are small faults; we find both terms useful even if the threshold for the size distinction is unquantified. The term "microfault," which is used to denote shear fractures when constrained by a lexicon where all shears must be faults, could almost be considered to be an oxymoron. Regardless, the difference between shear fractures and faults is more than just one of size and is not entirely arbitrary since the larger displacements along faults commonly create associated gouge, breccia, clay smear, fracture halos, and antithetic shear fractures.

Both shear fractures and faults can have the three basic Andersonian orientations, i.e. normal dip-slip, reverse dip-slip, and strike-slip, when they form in systems where one of the three compressive stresses is vertical and the other two are horizontal. However, they can be tilted after formation so that their present geometries no longer indicate the stress fields in which they formed.

There are as many or more fault types and variations as there are fracture types, and discussing fault geometries, mechanics, and characteristics would be a book in itself. Faults have been described in numerous publications (see Faulkner et al., 2010), and some of their populations have been described. For example, the relationship between fault offset and fault length has been noted and plotted by Marrett and Allmendinger (1991), and relationships between fault offset and the width of the associated damage zone of gouge, fault breccia, and fractures have been plotted by Robertson (1983) and by Marrett and Allmendinger (1990) (see Figure 3.18). Although the data defining these relationships have considerable scatter, they are linear on log-log plots and provide a starting point for estimating fault lengths and the widths of fracture halos around faults. The relationships can provide preliminary assessments of fracture-enhanced reservoir permeability related to faulting.

Faults may be open, high-permeability pathways within reservoirs (Figures 1.51, 1.52), or they may be occluded by gouge and mineralization. Often enough they change from one condition to the other along their height and length as they cut across different lithologies in a heterogeneous formation. Faults can and do cut across significant lithologic boundaries so they can provide an important degree of hydraulic connectivity within and between reservoirs provided that they are not occluded by mineralization.

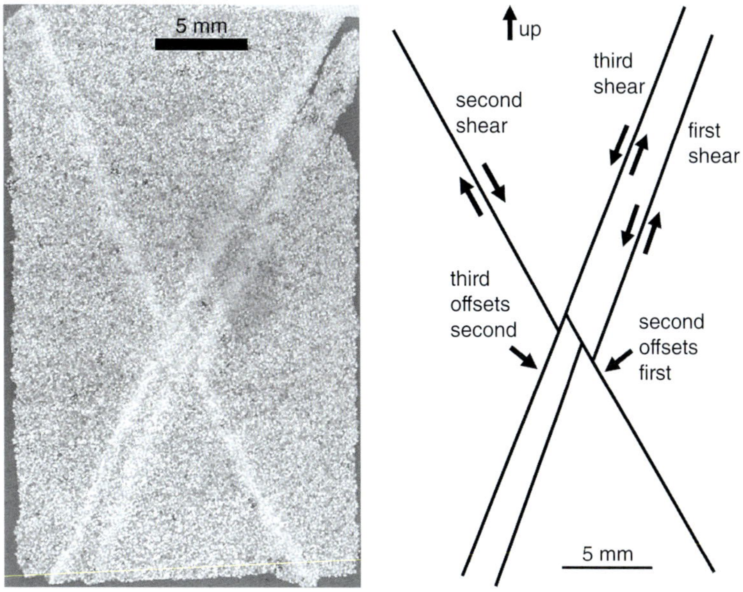

Figure 1.50 *Left*: a thin section cut across a conjugate deformation-band shear pair in sandstone. *Right*: a sketch made from the thin section, highlighting the minor offsets that record sequential rather than simultaneous shear on the three planes, similar to the progression suggested by Ferrill et al. (2000) for conjugate shear fracturing ranging in scale from centimeters to kilometers (from Olsson et, al., 2004).

Figure 1.51 Two views of a normal dip-slip fault system in The Chalk, Cretaceous, southeastern England, showing that a relatively small magnitude of offset, as defined by the horizontal flint layer, can produce enough fault-related, antithetic shear fracturing between the fault segments to enhance local permeability, as suggested by the rusty discoloration along the fractures.

1.5.4 Microfractures

Microfractures were discussed earlier in the context of their origin as early strain-accommodation structures which become more numerous and ultimately coalesce into macrofractures as rock is subjected to increasing strain. Since microfracturing precedes macrofracturing, most macroscopically fractured rock should also contain microfractures and indeed microfractures are reported to be present, even pervasive, in limestones, siltstones, and quartzose sandstones (e.g. Olsson, 1974; Milliken and Laubach, 2000; Anders et al., 2014; Ukar and Laubach, 2016), although they are less common in mudrocks (Loucks and Reed, 2016). Microfractures are scattered within rock but are commonly concentrated adjacent to macrofracture planes: laboratory data suggest that microfracture density increases rapidly within a few grain diameters of a microfracture (Kranz, 1983).

The microfractures that are concentrated around extension fractures may not be entirely equivalent to those associated with shear fractures. Microfractures start to form in random locations in a rock as it is strained, but once an extension fracture starts to propagate, a process zone of microfractures forms in front of and adjacent to the advancing fracture tip due to deformation of the rock around the crack tip as it advances and widens to form a macrofracture. In contrast, the zone of localized microfractures associated with shear fractures

Figure 1.52 Two views of a strike-slip fault in limestones of the Niobrara Formation in Colorado. The multiple, sheared, lenticular layers of calcite mineralization filling the fault are evidence for repeated shear offset and remineralization. Mineralization indicates that the fault has been permeable and capable of transmitting mineralizing fluids to the point where it is now nearly occluded. However, pockets containing centimeter-scale euhedral calcite crystals within the mineralization indicate that scattered open void spaces created by the shear offset still exist within the fault. Slickencrysts and slickenlines record strike-slip shear. The absence of a fracture halo or damage zone, and the width of the mineralized zone, suggest that the normal stress across the fault plane was low during shear.

forms prior to the propagation of shear through the rock, controlling the location of subsequent shear offset.

Not all grain-scale microfractures found in a sandstone are related to the larger-scale macrofractures. Microfractures in second-cycle sedimentary grains may be inherited and unrelated to the macrofractures in the younger rock. Olsson (1974) reported that two unrelated types of microfractures form in limestone when it is strained in the laboratory: "subaxial microfractures" that form parallel or nearly so to the applied stress, and oblique "microfaults" that form and coalesce into shear fractures. Hooker et al. (2009) report on a microfracture system in sandstone where intragranular microfractures comprise about 98% of the population, but where only the transgranular microfractures are genetically related to macrofractures. Some microfractures have no apparent macrofracture equivalents: Lash and Engelder (2005) and Márquez and Mountjoy (1996) have described bitumen-filled horizontal microfractures, oriented parallel to bedding in clay-rich mudrock formations. These authors suggest the microfractures formed in over-pressured strata under conditions of high horizontal compressive stress, and that the horizontal microfracture orientations are largely governed by the fine, horizontally laminated sedimentary fabric of the host rock.

Nevertheless, authors including Friedman (1969) and Moore and Lockner (1995) have noted a parallelism between micro- and macrofractures, suggesting that since microfractures increase in frequency with proximity to a fault they might be used to predict the characteristics and locations of faults. With increasing degrees of structural deformation, microfractures not only coalesce to form macrofractures, they may also change in character. Knipe and Lloyd (1994) describe a deformation progression where, as a minor reverse fault increases in throw, microfractures in the adjacent sandstone change from intragranular extension microfractures to transgranular extension fractures, and finally become "cataclasite-filled shear microfaults" with associated particulate flow in the shear zone.

Many microfractures in sandstone cores are artifacts, created by the elastic and anelastic rebound that occurs after the core is cut from a formation at depth and released from the asymmetric in situ confining stresses (Teufel, 1983; Lin et al., 2006). Distinguishing natural from induced microfractures is not straightforward, and few of the authors who have described microfractures have listed the criteria or characteristics that were used to discriminate between induced microfractures and unmineralized natural microfractures. Anders et al. (2014) suggest that most natural microfractures are mineralized, and mineralized microfractures are unquestionably natural unless they have been filled by a mineral efflorescence derived from the evaporation of formation pore waters, or with devolatilized reservoir oil. The effect of mineralized microfractures on the porosity, permeability, and mechanical properties of a rock, especially where the mineralization is similar to that of the host rock, is probably negligible.

Orientation relative to the in situ stress can also help distinguish natural from induced microfractures. Stress-release microfractures should strike approximately normal to the maximum compressive stress whereas tectonic microfractures should strike parallel to that stress, thus thin sections oriented relative to the induced, stress-controlled petal or centerline fractures in a core may help determine the origin of unmineralized microfractures. S. Brown (personal communication, 2005) suggested another criterion, indicating that stress-relaxation microfractures in laboratory tests most frequently form *between* grains in a sandstone (i.e. intergranular), whereas tectonic microfractures are more commonly found *within* grains (intragranular). Moreover, where a population of relaxation microfractures has been superimposed on rock containing tectonic microfractures, the relaxation cracks viewed in thin section should be more commonly filled with epoxy resin, whereas tectonic microfractures, typically striking 90° to the relaxation fractures, are more likely to be filled with natural cements.

The characteristics and distributions of natural microfractures vary by formation and structural setting. Onasch (1990) found abundant microfractures, locally over 160 per millimeter, in thin sections cut from tightly folded, well-cemented, Silurian quartz sandstones in Appalachian fold belt of Virginia. The microfractures are one to five microns wide and filled with quartz that is marked by trains of fluid inclusions. Onasch inferred that they formed during the final stages of folding, in association with pressure dissolution. Despite the abundance of microfractures, open microfractures that would contribute to porosity and permeability in a reservoir are rare in this setting, forming such a small percentage of the population that Onasch did not feel it necessary to describe them.

In contrast, Liu et al. (2013) described a system of more open microfractures in Cretaceous sandstones in the Kuqa Basin, China, at depths on the order of 20,000 ft (6000 m). The microfracture population includes intragranular microfractures (many of which consist of curved hertzian point-contact fractures radiating away from the point contacts between grains), grain-edge microfractures, and transgranular microfractures. Since the strata have been folded and faulted, related micro-deformation includes crushed grains and microbreccias. (In contrast, Onasch inferred that rapid healing of the Appalachian microfractures prevented grain crushing and brecciation in that setting.)

The microfractures in the Kuqa basin formed in extension, and are either unfilled, or filled/partially filled with calcite, anhydrite, bitumen, or clay. The fractures have lengths on the order of a hundred microns, commensurate with the grain sizes, and widths on the order of half to a few tens of microns. Widths are locally enhanced by dissolution. The microfractures are commonly concentrated within a few millimeters of large fractures and stylolites, but microfracture concentrations also occur in apparently undeformed rock. Transgranular fractures, the most likely to be related to macrofracturing, comprise only 16% of the microfracture population and are concentrated around faults.

1.5.5 Stylolites and Associated Extension Fractures

Stylolites are perhaps the most common type of anti-crack, forming when a rock loses volume along a planar zone and shortens in the dimension normal to the plane due to removal of material by pressure dissolution. Stylolites form most readily in carbonates because of their relatively high solubility, although stylolites can also form in sandstone under conditions of elevated pressures and temperatures (e.g. Dutton and Willis, 1998). In both lithologies, dissolution of the most soluble components of the rock leaves a residue of less soluble materials, commonly clays and organic materials, along an irregular seam in the rock.

Most stylolite planes are parallel to bedding, but stylolite planes can form at any angle including normal to horizontal bedding in stress systems where the maximum compressive stress is other than normal to bedding, as for example within or in front of thrust belts. These are commonly distinguished as "tectonic" stylolites.

Dissolution usually produces three-dimensional, irregularly conical teeth that are outlined by the insoluble residues, but teeth can also form as two-dimensional ridges where bed-normal stylolites cut across thin beds of varying solubility. The walls of well-developed stylolite teeth may be striated, recording shear where rock from one side of the stylolite moved against rock from the other side as the material between them was removed by dissolution.

Benedicto and Schultz (2010) found a relationship between the amplitude of stylolite teeth and the lateral extents of stylolites, suggesting that larger amounts of dissolution and shortening correlate to greater lateral extents of a stylolite (Figure 1.53). Heap et al. (2014), and Toussaint et al. (2018), suggest that stylolites may be too poorly developed in most formations to be true barriers to flow. This is supported by the limited lateral dimensions measured by Benedicto and Schultz (2010).

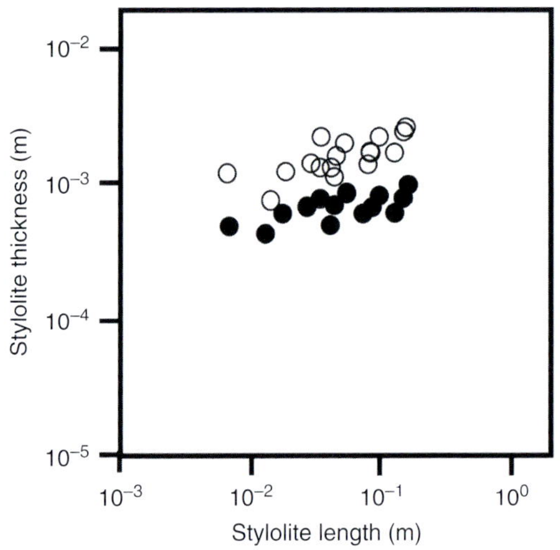

Figure 1.53 Thickness-length correlations for stylolites in limestones, Gubbio, Italy. The measured lateral extents of fractures in this database do not exceed one meter (three ft), supporting a "perforated layer" conceptual model for the minimal effect of stylolites on a reservoir (Toussaint et al., 2018), although stylolites that formed along sedimentary clay seams may be more laterally extensive (adapted from Benedicto and Schultz, 2010).

However, many stylolites form along bedding planes, so the lateral extents of stylolites may be dictated as much by the lateral extent of a sedimentary clay parting as by geomechanics and geochemistry.

Short, relatively wide fractures are associated with some stylolite systems. The fractures are widest where they originate at the teeth of a stylolite and pinch out rapidly to terminate blindly within the adjacent rock (Figure 1.54). Sometimes called "tension gashes," these stylolite-associated extension fractures form most readily where compression normal to the plane of the stylolite is associated with concurrent lateral extension of the rock parallel to the stylolite plane; stylolites without associated fractures probably formed under more uniaxial strain (Nelson, 1981).

A stylolite reduces permeability normal to the stylolite due to the concentration of insolubles along the stylolite seam, but the associated fractures have the potential to enhance permeability in the adjacent rock (Nelson, 2001; Wennberg et al., 2016). Lorenz and Cooper (2018a) provide an example of a well-developed, oil-filled system of stylolite-associated extension fractures in core.

1.5.6 Bed-Parallel Shear Fractures

Bed-parallel shear fractures are common within folded, heterogeneously layered strata, as well as within thicker folded shale formations, where bedding provides a fabric of parallel mechanical weakness planes in the rock. Bed-parallel shear planes are typically slickenlined,

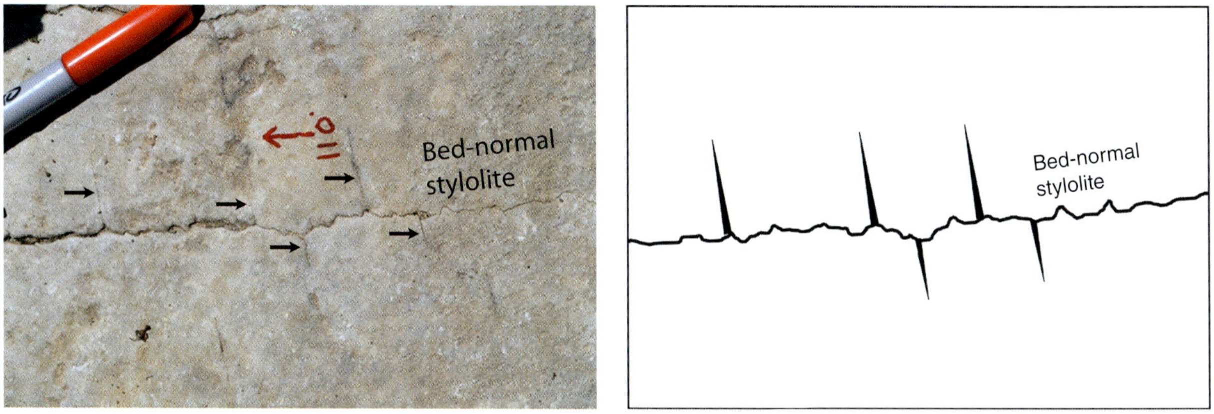

Figure 1.54 Map-view photo (*left*) and sketch (*right*) of a horizontal bedding plane cut by a vertical, "tectonic," bed-normal stylolite, with five associated extension fractures (black arrows) in the Eocene Pila Spi limestone in northern Iraq. The strike of the vertical stylolite (110°–290°) is normal to the maximum compressive stress, which was horizontal, parallel to bedding, and had a trend of 20°–200°.

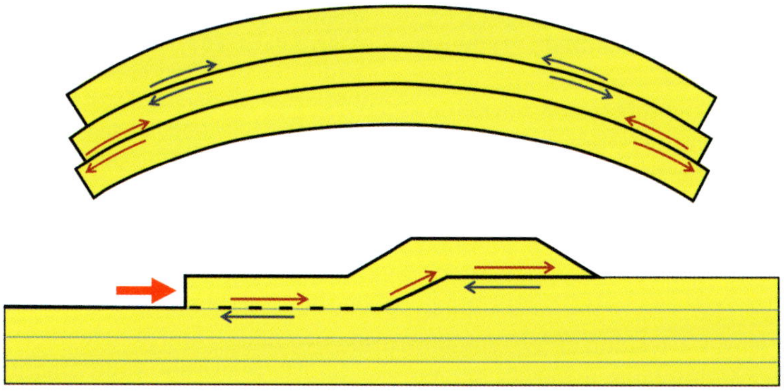

Figure 1.55 *Top*: layered strata are prone to bed-parallel shear during folding, particularly if the layering consists of relatively thin alternating units of strong and weak rock or if the interfaces between layers are weak. Bed-parallel shear can also develop in folded thicker, more homogeneous shale formations since bedding planes in shales are relatively weak. Folding in thick carbonates or sandstones is likely to be accommodated by other mechanisms. *Bottom*: shear can form along bedding planes where the layered strata are part of a thrust belt, both along the main bed-parallel thrust planes and within the formation where it is folded at the bends in the ramp as a thrust cuts up-section.

and the slicks often strike parallel to the dip azimuth of the tilted strata and normal to the local fold axis, reflecting shear in the sense of an arched deck of cards (Figure 1.55). The outer layers of the fold slide towards the crest of the fold relative to those of the inner layers in order to accommodate the strain imposed by curvature. Some sheared bedding surfaces are marked by superimposed, oblique sets of slickenlines that record multiple shear events and directions during the development of more complex folds. Many sheared bedding planes in muddy strata have a reflective, polished surface due to the alignment of clay platelets on the shear face.

Bed-parallel shear is also characteristic of thrust belts, where thrust faults follow stratigraphic layering, periodically ramping up-section at shallow angles to bedding. Bed-parallel shear can follow the thrusts along bedding, and additional bed-parallel shear can occur in the folds where the thrust sheet migrates over bends in the ramps.

Bed-parallel shear typically has little or no signature to differentiate it from bedding in an image log, and it can also be easy to overlook as just another break in the core when logging core unless the ends of core pieces are examined (Figure 1.56).

Bed-parallel shear occurs in folded strata because the weak interfaces between the layers control the location and orientation of strain accommodation, so geomechanics and stress axes have less influence on fracture orientation. Although they are shears, bedding-plane shears typically do not form as conjugate pairs and do not follow Anderson's (1951) ideal shear orientations (Figure 1.57). Nevertheless, we have logged bed-parallel shear planes in gently dipping strata where the shears form one sub-set of a pair of conjugate shears, the complimentary sub-set consisting of irregular, bedding-oblique shear planes, with the sub-sets forming a nearly ideal, thrust-oriented conjugate-shear pair. This geometry suggests that the

Figure 1.56 Two views of a bed-parallel shear fracture in the butt section of a 4-inch (10-cm) diameter core. *Left*: the minimal expression of the shear plane on the slab surface of the core (just below the double-headed arrow) could easily be dismissed as a bedding-plane break in the core. *Right*: the polished face of the same fracture as seen on the end of the core is marked by a pattern of slickenlines parallel to the dip azimuth (parallel to the red arrow).

Figure 1.57 *Left*: the steeply dipping sandstones and limestones of the Pennsylvanian Tensleep Formation on the forelimb of Beer Mug Anticline in Wyoming were folded over the tip of an underlying thrust fault (see Cooper, 2013). Folding created several zones of significant bed-parallel shear between the relatively thick lithologic units. The white arrow in the photo on the left shows the location of the bed-parallel shear zone shown in the photo on the right. *Right*: the bed-parallel shear zone is filled with a thick deposit of layered and multiply sheared white calcite that is up to six inches (15 cm) thick between a thick gray limestone on the left, stratigraphically below the shear plane, and a thick eolian sandstone on the right, stratigraphically above the shear.

host strata were folded by a high-magnitude horizontal compressive stress, and that a crumpled rug rather than a drape fold is the best model for folding.

The spacings of bed-parallel shear planes, measured normal to bedding, depend in part on the amount of strain imposed on the strata, and in part on the sedimentary and mechanical composition of the folded formation. Spacings are closer and offsets are less within thinner-bedded strata, whereas spacings are wider and offsets are greater in thicker-bedded formations.

Bed-parallel shear can accommodate most of the strain in some formations and on some folds (e.g. Stephenson et al., 2007), rendering deformation by other structures such as bed-normal fracturing or pervasive ductile strain unnecessary.

1.5.7 Beef-Filled Fractures

A unique type of fracturing occurs in many marine mudrock formations, typically but not always oriented parallel to bedding and commonly in the clay-rich, deeper-marine deposits. These fractures are filled with a characteristic fibrous mineralization (Figure 1.58) that reminded early observers of the muscle fibers in a beefsteak, so the mineralization has traditionally been called "beef" (see Gale et al., 2014).

Of all fracture types, only beef-filled fractures are defined almost entirely by their mineralization. Other fracture types may be barren or they may be filled with any of a variety of minerals displaying any of a variety of crystal habits, regardless of the mechanical origin of the fracture. In contrast, beef-filled fractures are invariably filled with diagnostic fibrous crystals of calcite or less commonly gypsum or quartz. If the fragile mineralization is broken up and lost, common in a core slab, the rock displays no evidence for the fractures.

The fibrous, prismatic crystals that fill these fractures are usually interpreted to have grown concurrently with fracture opening, so that increasing width kept pace with mineral precipitation and open void was never present between the fracture walls (see Hooker et al., 2019). This contrasts with the mineralization in other fracture types which is often precipitated within open fracture voids during geochemical events unrelated to the mechanics of fracturing. Moreover, mineralization in most fracture types grows from the fracture face inward into the fracture aperture, whereas the "antitaxial" mineral-fiber growth in beef-filled fractures typically grew in the opposite direction, i.e. by the addition of material at the contact between mineralization and the fracture face.

The prismatic fibers of the beef are oriented normal to the fracture walls, and a medial line may be present near the middle of the band of mineralization, separating two layers of crystals. The medial line is often marked by bits of host-rock debris. The host-rock faces of beef-filled fractures are unornamented, mineralization is weakly cemented to the faces, and some of the fractures are associated with indications of soft-sediment deformation.

The fact that these fractures are defined by their mineralization rather than their mechanics highlights the current discussion concerning their origin. Ukar et al. (2017) called them "bed-parallel fibrous calcite veins," and often they are denoted merely by reference to the filling material (i.e. "layers of beef" in the rock). Most of the initial studies concluded that these are early-diagenetic, shallow-burial structures. Marshall (1982), Stonley (1983), and Al-Aasm et al. (1993) cited petrographic textures, supported by oxygen-isotope data, to suggest that the fractures formed by displacive growth of the fibrous calcite during diagenesis at burial

Figure 1.58 *Left*: a bedding-parallel, beef-filled, calcite-mineralized fracture in 2.5-inch (6.4-cm) diameter core cut from a deep-marine mudrock in Texas shows the common medial line and the bed-normal fibrous calcite crystals, pencil point for scale. *Right*: a beef-filled fracture in the same core, with an unusual vertical orientation. The 3.5-mm wide fracture is filed with calcite having the same fibrous crystal habit and medial line found the more common horizontal beef-filled fractures; pencil point for scale.

depths of tens to a few hundreds of meters, within poorly lithified sediment.

More recent studies have used isotopes and hydrocarbon-bearing fluid inclusions in the calcite mineralization to suggest that the fractures originated during deeper burial and hydrocarbon generation (e.g. Cobbold et al., 2014; Rodrigues et al., 2009; Ukar et al., 2017; Weger et al., 2019). Fluid inclusions in calcite and gypsum can stretch and deform and they are less reliable than inclusions in stiffer minerals such as quartz, and calcite and gypsum can be unstable and susceptible to alteration at higher temperatures and pressures; this evidence should be used cautiously. Cobbold and Rodrigues (2007) have suggested that "seepage forces" created by vertically migrating fluids in overpressured basins can reduce the effective vertical stress (σ_V) in deeply buried mudrocks, moving σ_V into tension to form these horizontal fractures. The concept of seepage forces, the viscous, frictional drag between a fluid and its host medium caused by fluid flow, is common in soil mechanics where flow rates are typically greater than geologic rates. Hooker et al. (2019) cite compositional and isotopic evidence to suggest that the fracture-filling beef mineralization is derived from local fluids in a relatively closed geochemical system, whereas Weger et al. (2019) use thermometric reconstructions based on isotopic data to suggest that the mineralizing fluids were derived from deeper, hotter sources.

Regardless of their origin, the typical bed-parallel orientation of beef-filled fractures offers potential pathways for connecting high-angle fracture types to form fracture-controlled permeability networks in a reservoir. Beef-filled fractures up to three feet (a meter) in thickness and up to 0.6 miles (1 km) in lateral extent (Figure 1.59) have been reported from the Vaca Muerta Formation in Argentina (Rodrigues et al., 2009; Weger et al., 2019), providing potential horizontal streaks of enhanced permeability. We are unaware of published measurements of restored-state permeabilities along these universally occluded fractures, but if the fractures are well developed and interconnected, fracture-parallel permeability would not have to be more than microdarcy in scale to impact production in extremely low-permeability mudrock reservoirs.

1.5.8 Ptygmatically Folded Fractures

Ptygmatically folded extension fractures are a class of early-diagenetic structures (e.g. Bishop et al., 2006; Hooker et al., 2019) that are typically short, occluded, and strata-bound. They commonly occur as sets of parallel-striking fractures that are closely spaced and well developed but limited to specific beds in muddy and calcareous marine formations (Figure 1.60). Less often

Figure 1.59 White, laterally extensive layers of beef form resistant ledges in a stream bed where mudrocks of the Vaca Muerta Formation crop out in Nuquén Province, Argentina. (Photo courtesy of Luciano Monti and Yacimientos Petrolíferos Fiscales [YPF] S.A.)

they are scattered and isolated in these rocks. Rarely, a system will consist of superimposed, oblique-striking and therefore intersecting fracture sets in the same bed, and some sets have non-systematic strikes.

If these fracture systems are well developed and have higher permeabilities within the mineralization than that of the host matrix rock, they may create horizontal permeability streaks in a reservoir, but in general they probably have minimal impact. As of this writing we are unaware of published measurements of the in situ permeability of these structures, although Landry et al. (2016) calculate that there could be permeability within the mineralization of other apparently occluded fractures in mudrocks based on SEM images showing intercrystalline void spaces.

These structures are not well exposed in outcrop since they are hosted by mudrocks that are prone to weathering, obscuring such small features with their sub-millimeter widths and centimeter-scale heights. Little is known about the lateral extent of such systems, but even the vertically shortest fractures typically show no indication terminating laterally within the diameters of four-inch cores, and the fracture systems probably extend laterally as far as the host depositional/diagenetic environment. Although their heights are typically measured only in centimeters or even millimeters, we have logged some of these early-formed, ptygmatically folded fractures that extend vertically for up to five feet (1.5 m) along cores in both mudrock and limestone.

Where these fractures are well developed, they have spacings on the order of millimeters to a few centimeters, and they are typically restricted to specific beds that are commonly only a few centimeters thick. There may be

Figure 1.60 Two views of short, ptygmatically folded vertical extension fractures. *Left*: three fractures are exposed on the slab face where it is cut by a break in the core along a bedding plane. *Right*: the same three fractures as exposed on that bedding plane in the equivalent butt of the core, showing that although short, they are parallel and laterally extensive. They are crenulated only in the vertical dimension, reflecting shortening during compaction of the host strata. The obvious doublet exhibited by the fracture on the right in the right-hand photo is the result of overlap of the mineral-filled fracture plane during compaction and vertical shortening.

as many as ten such intensely fractured beds per foot of vertical foot of core within intervals ten or more feet thick. In aggregate, such well-developed systems have the potential to influence reservoir permeability even though the individual fractures are short, narrow, and typically occluded. Minerals that fill these fractures include calcite, quartz, gypsum, pyrite, barite, and bitumen, and rarely they are filled by more than one mineralizing agent. Pyrite is only dominant locally, but it is more common in these fractures than in other fracture types.

Most of the fractures are vertical despite compaction and shortening along the vertical axis, but subsequent lateral, soft-sediment deformation has locally altered dips and further contorted the structures (Figure 1.61). Vertical shortening mechanisms include ptygmatic folding as well as shear overlap of the mineralization that filled the fracture widths, and some of the mineralized fractures formed stiff struts that were pushed upward and downward into the adjacent semi-lithified layers as the host layer compacted. Where fractures cut more than one lithology, the degree of contortion of the fracture plane and its mineralization varies with lithology, being more deformed in the muddier strata that had a larger degree of compaction. Rare exposures of the fracture faces show a crenulated, corduroy texture that, unlike the fractography of most fractures, records post fracturing compaction and ptygmatic folding rather than the origin of the fracture (Lorenz and Cooper, 2018a).

Their characteristics indicate that ptygmatically folded fractures formed in extension as unfolded, vertical planes. Folding and other shortening mechanisms indicate that they formed and were mineralized within under-compacted, poorly lithified sediment, and that the mineralized planes were folded during later compaction of the host strata. The mechanics of forming planar, parallel, systematic fractures in poorly lithified sediment have yet to be fully explained, but high pore pressures, as from microbial generation of gas and decomposing organic material in the sediment, might have made even the semi-lithified sediment into a relatively brittle material capable of fracturing.

The strikes of one set of ptygmatically folded extension fractures in core we logged could be oriented and were found to be parallel to the associated paleo-shoreline. It is possible that the stress anisotropies that fractured the poorly consolidated strata were related to the reduced minimum compressive stress caused by down-slope gravitational sliding during deposition and burial.

1.5.9 Alteration of Fracture Systems by Dissolution

Finally, dissolution can affect any type of fracture system by widening apertures, extending lengths, and interconnecting fracture planes. Dissolution can remove existing fracture mineralization and make fracture walls rough and irregular (Figure 1.62), and the altered fracture walls can in turn be remineralized.

Fractures affected by dissolution may be so altered that they resemble vuggy porosity and are barely recognizable as fractures. The dimensions of these structures no longer have systematic distributions, making them difficult to quantify and even more difficult to predict from limited subsurface samples. Fractures offer the initial pathways for dissolution fluids, but even where mineralized, the relatively pure calcite that often fills fractures is typically more soluble than the host rock. Large fissures, filled with

Figure 1.61 *Left*: an early-formed vertical extension fracture was filled with calcite and then shortened along its vertical axis by ptygmatic folding as the poorly lithified strata compacted. Compaction and fracture shortening were greatest within the muddy intervals. The bottom of the fracture was wide and too stiff to fold so it was punched downward into the underlying semi-lithified layers as a stiff strut during compaction. *Right*: ptygmatically folded fractures that were deformed both vertically and laterally during compaction.

Figure 1.62 *Left*: a bed-normal dissolution slot that followed a short, strata-bound extension fracture in tilted strata. The tilted strata were cut by a vertical core, which is being held so that bedding and fracture are in their pre-fold positions. Mesozoic marine shale-siltstone sequence, Iraq. The original fracture terminated at the overlying and underlying shales, limiting the height of the subsequent dissolution slot. The slot has been incompletely remineralized with crystalline, oil-stained calcite. *Right*: core containing a vertical extension fracture that has been subject to dissolution, resulting in an irregular fracture slot. From the Cambrian-Ordovician Arbuckle Dolomite, Kansas. The dissolution surfaces have been partially remineralized with scattered dolomite crystals. Both cores are 4-inches (10-cm) diameter.

exotic materials, may even develop from dissolution slots that followed an existing fracture system.

Dissolution fluids have several possible sources, including meteoric waters derived from an overlying subaerial exposure surface, hydrothermal waters from deep sources, and sulfuric acids associated with hydrocarbon migration (e.g. Hill, 2000). Dissolution can lead to cavernous porosity where elongate cavern passageways are roughly linear and aligned with a precursor fracture system (Figure 1.63). Reservoirs developed in such systems consist of large voids, filled with hydrocarbons if the operator is lucky, and marked

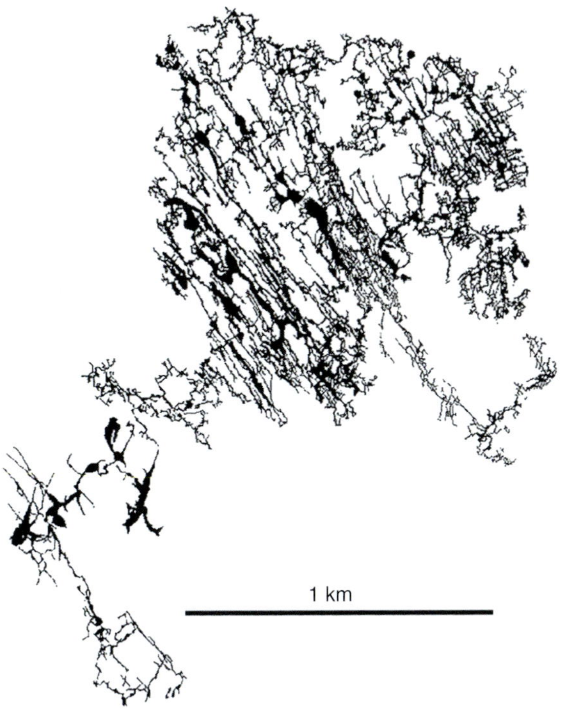

Figure 1.63 A rectilinear pattern of dissolution along a fracture system is recorded by the systematic cave passageways of Wind Cave, South Dakota (modified from Horrocks and Szukalski, 2002). Dissolution produced 136 linear miles (217 linear km) of passageways within an area less than a square mile (2.6 km^2) in extent.

by bit drops when drilling and by immediate and often unexpected inter-well hydraulic communication in specific directions. However, dissolution voids may also be secondarily filled with locally derived cave breccias and washed-in exotic materials (e.g. Loucks, 1999). When lithified, these cave-filling conglomerates and breccias can in turn be fractured. Stylolites and insoluble dissolution residues are commonly associated with such dissolution processes and deposits.

APPENDIX 1.A The Relationship Between Pore Pressure and the In Situ Effective Stresses

Introduction

The relationship between pore pressure and stress in a reservoir is subtle and not always fully appreciated, but it is supremely important because it controls the susceptibility of rock to both natural fracturing and industrial hydraulic fracturing. This Appendix gives the equations (Rice and Cleary, 1976; N. Warpinski, personal communication, January 2019) that are commonly used to calculate the in situ *effective* compressive stresses (i.e. the in situ stresses as reduced by pore pressure) in a reservoir, which control natural fracturing. Putting plausible numbers into the equations demonstrates the stress changes that occur in a reservoir at 10,000 ft depth as a function of pore pressure. For further reading on effective stress see Robin (1973) and Berryman (1992).

To calculate the effective stresses in a reservoir one needs to know the in situ pore pressure, the weight of the overburden, Poisson's ratio for the reservoir rock, and Biot's coefficient. The numbers used here for pore pressure are based on hydrostatic gradients and common over-pressure conditions, the numbers used for the overburden weight are based on the typical lithostatic gradient, the assigned Poisson's ratio is a common value for sedimentary rock, and the value used for Biot's coefficient is chosen from observed values from various reservoirs. These numbers, when inserted into the equations, demonstrate the reduction and convergence of the three effective stresses as pore pressure increases.

All stresses discussed here are compressive. The pore pressures discussed here are the pervasive pressures within a reservoir that change relatively slowly over the course of years during production or over the course millions of years during geologic time; they are distinct from the rapid and very localized fluid-pressure increases associated with injected hydraulic stimulations. These calculations focus on the maximum and minimum stresses and the difference between them. The role of the intermediate stress is less well understood; Mogi (1967) reported that rock strength increases and fracture shear angle decreases in laboratory tests as the intermediate stress increases, but Paterson and Wong (2005) suggest that the intermediate stress may have more effect on fracture propagation than on fracture initiation.

A note on terms: S is used here to designate a stress without the effects of pore pressure. Many authors use the symbol σ for that stress (for example, Griggs and Handin, 1960, in Figure 1.2), but others use σ to designate an *effective* stress, i.e. the stress S minus pore pressure, and that convention is used here. Some authors have used the term S' to designate an effective stress, but in this text the "prime" symbols (i.e. S', σ") are used to designate different stages in the development of an effective stress (see Figures 1.44, 1.45). S has also been used to indicate a *total* stress, i.e. the stress S *plus* a percentage of the pore pressure (see Figure 1.45), but total stresses are not discussed in this appendix. The subscripts V, H, and h designate stresses in the vertical, maximum-horizontal, and minimum-horizontal directions. By convention, h is also used to indicate the thickness of the overburden; it is only used as such once here, in the first equation. The subscripts 1, 2, and 3 are used to denote the maximum, intermediate, and minimum compressive stresses, respectively.

Vertical Stress

The vertical stress magnitude in a reservoir (S_V) is a function of the thickness (h) and density (ρ) of the overlying rock as it is affected by the pull of gravity (g), and is given by the formula:

$$S_V = \rho gh \qquad (A.1)$$

Precise estimates of the weight of the overburden are commonly made using the actual densities and thicknesses of the different overlying rock layers as derived from sonic logs, but for simplicity, the oilfield rule of thumb based on the average density of sedimentary rock is used here, i.e. the weight of the overburden (the vertical stress, S_V) increases at one pound per square inch per foot of depth (1 psi/ft, or 22.6 MPa/km). Thus, conveniently, the overburden stress at a depth of 10,000 ft is close to 10,000 psi. S_V is typically the largest of the three stresses imposed on the rock, i.e. $S_V = S_1$, although S_V may be the intermediate (S_2) or even the minimum (S_3) stress axis, in strike-slip and thrust-belt tectonic settings, respectively.

Horizontal Stress

In a basin where $S_V = S_1$, the two horizontal stresses at depth (the maximum and intermediate horizontal stresses, S_H and S_h) are created by the inability of confined rock to expand laterally under the weight of the overburden. The degree to which the rock would expand if unconfined, and thus the lateral stress created by that confinement, is governed by the weight of the overburden (S_V) and Poisson's ratio (υ) in the equation:

$$S_h = S_V\,(\upsilon/[1-\upsilon]) \qquad (A.2)$$

Given a υ of 0.25, a common value for sedimentary rock, then by Equation 2 the horizontal stress S_h in the reservoir at 10,000 ft depth is calculated as follows:

$$S_h = 10{,}000\ \text{psi}\ (0.25/[1-0.25])$$

$$S_h = 10{,}000\ \text{psi}\ (0.33)$$

$$S_h = 3300\ \text{psi}$$

In the absence of tectonics, the two horizontal stresses should be equal, i.e. $S_H = S_h = S_2 = S_3$. With the imposition of a tectonic stress component, one of the horizontal stresses commonly increases to become the maximum horizontal stress, $S_H = S_2$, leaving $S_h = S_3$ as the minimum horizontal stress. For simplicity in this example calculation, the horizontal stresses are assumed to be equal and both are designated S_h.*

Effective Vertical Stress

The pressure of any fluid in a formation, the pore pressure, designated Pp, acts to partially support and counteract the vertical and horizontal stresses imposed on a reservoir by the weight of the overburden. The remaining stresses are the *effective* stresses (σ). (In contrast, pore pressure *adds* to the *total* stresses acting on a formation, governing the pumping pressure required to inject a hydraulic fracture into a reservoir as demonstrated by Salz, 1977; see Figure 1.42.)

The weight of a column of fresh water increases by about 0.43 psi per foot, so if the formation at 10,000 ft depth is hydraulically connected to the surface, the pore pressure in the formation will be 4300 psi and the formation is said to be "normally pressured." The hydraulic, pore-pressure gradient can be as high as 0.5 psi per foot for some brines, and can be increased up to the 1 psi/ft overburden gradient by mechanisms such as organic maturation. The vertical and horizontal stresses are reduced by a percentage of the pore pressure, the percentage being Biot's coefficient, α (see the discussion of α in the text), to produce effective vertical and horizontal stresses (σ_V and σ_h respectively). The effective vertical stress is given by:

$$\sigma_V = Sv - (\alpha Pp) \qquad (A.3)$$

For a reservoir at 10,000 ft depth where the vertical stress is the weight of the overburden, i.e. 10,000 psi, and given an α of 0.8, one can use Equation A.3 to calculate the effective vertical stress, σ_V, of 6560 psi for a normally pressured reservoir. If the hydraulic connection to the surface is cut off and the formation becomes over-pressured, the effective vertical stress is proportionately less at successively greater pore pressures, as shown by Table A.1. (If, as some authors suggest, Biot's

* The two horizontal stresses at depth may not be equal even in basins where $S_1 = S_v$ since few basins are entirely tectonically inactive and since stress anisotropies can remain locked into rock long after the tectonics that created them have become inactive. For example, in the Piceance Basin of Colorado, measurements show that the stresses in the fluvial sandstones have a horizontal stress anisotropy of 600–800 psi (4.1–5.5 MPa), but this anisotropy must be residual since the enclosing mudrocks are isotropically stressed and therefore cannot be transmitting an active stress anisotropy to the sandstones. The stress anisotropy in the sandstones is a remnant of the stress that was imposed by basin-margin thrusting during the Laramide Orogeny at least 40 million years ago (Lorenz and Finley 1991). The mudrocks were anisotropically stressed at that same time, but mudrocks are visco-elastic and the stresses reverted to an isotropic state in the 40 million years since the basin-margin thrusting stopped (see Warpinski, 1989).

Table A.1 Calculated effective vertical stresses in a reservoir at 10,000 ft depth for six pore-pressure conditions, where Biot's coefficient is 0.8, using Equation A.3. Stresses and pressures given in psi.

Sv (given)	α (given)	Pp (given)	σv (calculated)
10,000	0.8	4300	6560
10,000	0.8	5400	5680
10,000	0.8	6500	4800
10,000	0.8	7600	3920
10,000	0.8	8700	3040
10,000	0.8	9800	2160

Table A.2 Calculated effective horizontal stresses in a reservoir at 10,000 ft depth for six pore-pressure conditions, where Biot's coefficient is 0.8 and Poisson's ratio is 0.25, using Equation A.4. Stresses and pressures given in psi.

σv (calculated)	ν (given)	ν/1−ν	σh (calculated)
6560	0.25	0.33	2165
5680	0.25	0.33	1874
4800	0.25	0.33	1584
3920	0.25	0.33	1294
3040	0.25	0.33	1003
2160	0.25	0.33	712

poroelastic coefficient does not apply to the overburden stress, then all of the pore pressure is subtracted from the weight of the overburden and the effective overburden stress for a normally pressured reservoir is reduced to 5700 psi. In some reservoirs, a different coefficient can apply to each of the three stresses [e.g. Teufel et al., 1993]. These considerations would change the following calculations, but the trends and results would be the same.)

Effective Horizontal Stress

The effective horizontal stress (σ_h) at depth is a function of the effective vertical stress (σ_V) and the Poisson's ratio of the reservoir rock (ν), as given in the equation:

$$\sigma_h = \sigma_V[\nu/(1-\nu)] \tag{A.4}$$

Table A.2 shows that by this formula, the effective horizontal stress in the reservoir at 10,000 ft also decreases with mounting pore pressure because horizontal stress is a function of the decreasing vertical effective stress. Conversely, both the effective vertical and the effective horizontal stresses would *in*crease if the pore pressure *de*creases, as for example during production.

Stress Differential

As pore pressure increases in the reservoir, not only are the maximum and minimum effective stresses reduced but they also converge as the differential between them ($\sigma_V - \sigma_h$) decreases (Table A.3; Figure A.1). The stress differential is a primary control on the ability of the

Table A.3 Calculated differential between the maximum (σ_V, from Table A.1) and minimum (σ_h, from Table A.2) effective stresses in a reservoir at 10,000 ft depth for six pore-pressure conditions, where Biot's coefficient is 0.8 and Poisson's ratio is 0.25. Stresses and pressures given in psi.

Pp (given)	σv	σh	σv – σh
4300	6560	2165	4395
5400	5680	1874	3806
6500	4800	1584	3216
7600	3920	1294	2626
8700	3040	1003	2037
9800	2160	712	1448

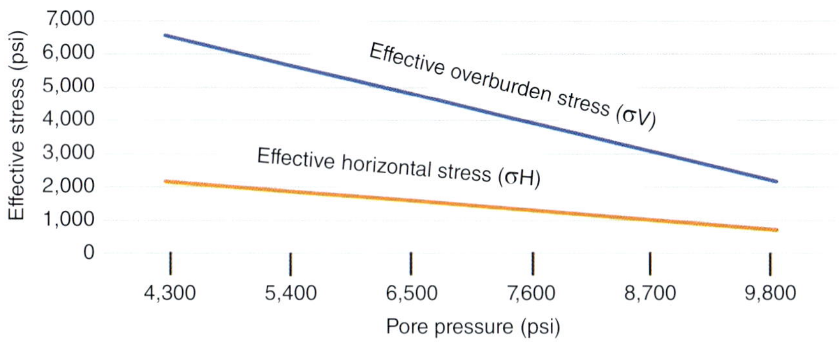

Figure A.1 Convergence of the effective maximum and minimum effective stresses as pore pressure increases. This plot is derived from the calculated effective stresses found in the last columns of Table A.1 and Table A.2. It compares the changes in the vertical and horizontal effective stresses with increasing pore pressure, and shows the associated diminishing stress differential (the vertical distance between the two lines, equivalent to the last column of Table A.3), for a reservoir at 10,000 ft.

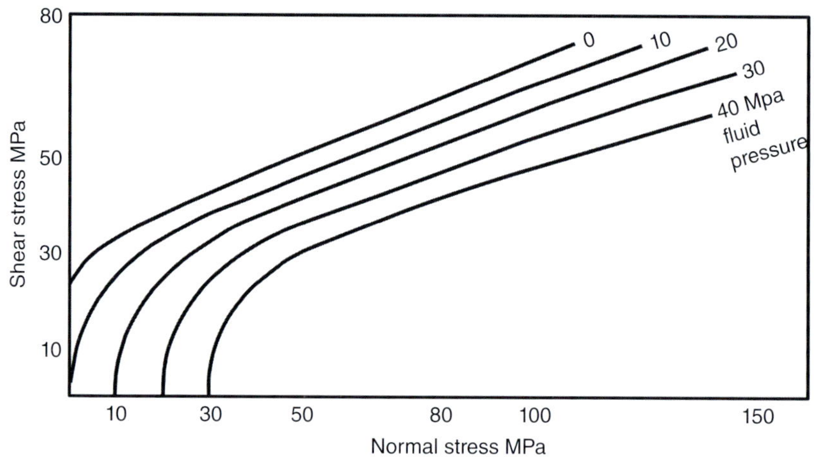

Figure A.2 Not only do the stress circles on a Mohr diagram decrease in size and move to the left with increasing fluid pressure, but the rock becomes weaker and more brittle, so the experimentally determined failure envelope moves to the right to meet the left-moving stress circle. (From Bayly, 1983, his figure 6.29).

in situ stresses to fracture rock, so a decrease in the stress differential decreases the ability of the stress system to create natural fractures. However, the same reduction in effective confining stresses both increases the brittleness of rock and decreases its strength, both factors increasing the fracture susceptibility of rock more rapidly than the decrease in stress differential diminishes the ability of the stress system to create fracturing (see Figure 1.46 and discussion in the text). In reference to a Mohr diagram, the stress circle gets smaller, and the position of the failure envelope shifts to the right (Figure A.2). Despite the competing trends, increasing pore pressure increases the probability of natural fracturing. Conversely, decreasing pore pressure causes an increase in the stress differential, also increasing the potential of the system to fracture rock.

PART 2

Measuring and Analyzing Fractures in Reservoirs

2.1 Introduction

This is the applied, nuts and bolts part of the book, where we describe the specific techniques that can be used in measuring fractures, the advantages and limitations of those techniques, and some of the methods that can be used to analyze data derived from fracture measurements. The material presented here has benefitted from and, we hope, complements the descriptions of techniques published by Nelson et al. (1987), Kulander et al, (1990), Lorenz and Hill (1992), Skopec (1994), Lockman et al. (1997), and Nelson (2001, 2020), but it is based primarily on our own experience.

As shown in the previous chapter, fracture systems can be multifaceted, and each fracture set within a system is typically complex in its own right, being composed of fractures of non-uniform sizes. Fractures are not typically distributed uniformly throughout a reservoir, but neither are their distributions random. More importantly, different fracture types and even different members of a fracture set have different effects on reservoir permeability. Fracture models are easy to construct from datasets composed of basic dip and strike data, but dip and strike alone do not capture all the fracture characteristics that control permeability in a reservoir. More valid models can be built from detailed, comprehensive fracture measurements and from in-depth analyses of fracture populations.

Makel (2007) suggests that five characteristics are required to model fluid flow through a fracture set: density, orientation, dimensions (height and length), connectivity, and aperture. Aguilera (2003) adds water saturation, and stress magnitude/orientation to that list, and some models have other requirements, but these are the basic input data. Some of these parameters can be measured directly from outcrops, cores, and image logs, but in fact, most datasets derived from hands-on measurements are incomplete because data collection is limited to essentially two-dimensional outcrops and one-dimensional cores whereas a fracture model is three-dimensional. In addition, some of the characteristics that are needed for modeling must be reconstructed indirectly from the dimensions that can be measured directly, and some of the dimensions that can be measured directly can be misleading. Most fracture datasets do not lend themselves to direct, prescriptive inputs for fracture modeling, so direct measurements of subsurface fracture data are often best used to constrain rather than to explicitly build models; subsurface fracture datasets may be more applicable to building continuous rather than discrete fracture models (e.g. Ouenes, 2000, 2019; Jenkins et al., 2009).

Hands-on fracture data represent reality but a limited piece of reality, so fracture measurements must be collected and analyzed carefully and in a knowledgeable manner. Fracture data must also be fleshed out with data from other sources including geophysical and engineering measurements. This chapter, then, describes the types of data that should be collected during a fracture study, how to collect the data, and some of the analyses that can be used to address common industry questions. We will also present several sanitized case studies of fracture data-collection and data-analysis projects based on work we have performed for clients.

We describe fracture measurements and analyses in detail in this chapter, but as one astute participant on a recent field trip observed: "Who cares? We're just going to drill and frac the reservoir anyway." That observation is fair, especially for smaller companies that cannot devote time, effort, and funds to coring or logging a reservoir and analyzing the data. Understandably, small operators often prefer to drill and complete a well on an empirical basis: "It worked last time, and if it works again great. If it doesn't work, plug and abandon and move on to the next well." The ultimate empirical approach would be to drill randomly and blindly, but no one does that; all companies put as much knowledge into drilling a prospect as possible, and fracture characteristics can be part of that forethought. Greater efficiency can be achieved in both exploration and production if one makes the effort to understand a reservoir first, using knowledgeable data-collection techniques and analyses

Applied Concepts in Fractured Reservoirs, First Edition. John C. Lorenz and Scott P. Cooper.

to squeeze as much information as possible out of the available data sources.

An analogy for the cost of operating in a data vacuum, for incorrectly assuming that geology provides a homogeneous reservoir media and running engineering calculations without understanding the complexity of the medium, is provided by the application of nuclear energy to hydrocarbon reservoirs in the late 1960s and early 1970s (e.g. Lorenz, 2001). These nuclear tests were carried out in both the U.S. and the former Soviet Union and were surprisingly ineffective despite the impressive amounts of energy applied to the reservoirs in the attempt to stimulate them. It wasn't obvious why the tests were unsuccessful because few data were collected from the reservoirs prior to the attempted stimulations, thus it was not recognized that the reservoirs contain extensive natural-fracture systems. In turn it could not be recognized that the nuclear blasts destroyed those fractures which, undamaged, create millidarcy-scale permeability and allow economic production rates from microdarcy-scale matrix rock.

2.1.1 Reasons to Take Core

Cores and image logs have shown that natural fractures are pervasive in the subsurface, even in some of the deepest holes drilled, and that they play an important role in many if not most hydrocarbon reservoirs. The most direct way to assess natural fractures in a reservoir is by cutting core from it. Since it is not always obvious to management exactly what advantage core provides when assessing a reservoir permeability system, the following is a list of the benefits of taking core, running image logs, and collecting and analyzing the acquired fracture data in detail.

- Core provides a means of calibrating fracture signatures on image logs, which commonly image significantly fewer fractures than are present in a formation, and do not accurately capture fracture apertures since apertures are irregular.
- Once an image log is calibrated across a cored interval, the rest of the image log, as well as image logs in adjacent wells, become significantly more valuable.
- Cores allow direct measurement of reservoir matrix permeabilities, and sometimes of fracture permeabilities in the lab. Permeability (k) in three directions can be measured from core (k_V, k_H, and k_h).
- Cores allow direct measurement of geomechanical rock properties. (Logs can measure secondary properties such as sonic travel time that are assumed to be related to the geomechanical properties.)
- The observable interactions between natural fractures and centerline induced fractures in a core provide direct evidence for predicting interactions between stimulations and natural fractures.
- The induced petal and centerline fractures in core record the in-situ stress orientations.
- Cores show the distribution of natural fractures relative to lithology.
- All fractures are not equal: cores show the range of fracture sizes and types, allowing a determination of which fracture populations control reservoir permeability.
- Fracture surfaces can only be observed in cores. These surfaces distinguish between shear fractures, which form intersecting permeability networks, and extension fractures, which may not.
- Core allows direct measurements of matrix permeabilities, which can be compared to reservoir system permeabilities to determine the contribution of fractures to the permeability system.

Taking core is one in a series of integrated steps that can be taken to understand the fracture and stress system in a reservoir (see Appendix 2.A). All or none of the steps can be taken, depending on budget and what level of understanding of the fracture-stress system is needed. The larger the reservoir and the greater the potentially recoverable reserves, the greater the financial reward is for improving the efficiency of hydrocarbon recovery from a reservoir. A company with a large target has more incentive to underwrite the cost of fracture and stress characterizations, whereas the cost-benefit ratio of such a study may be prohibitive for smaller plays.

There are no magic-bullet shortcuts in fracture analysis: the magic is that if you know what you're doing you can extract significant information from the small samples of a reservoir provided by outcrops or core, and from the indirect geophysical data provided by image logs, CT scans, and seismic data. Understanding a fracture system requires an up-front investment in time, expertise, and data collection: properly logging a core for fractures is labor-intensive.

Fracture measurements in U.S. cores are a wonderful mixture of metric and English units. Fracture depths and heights in U.S. cores are typically measured in feet and tenths of feet (not inches) while at the same time fracture widths are typically measured in millimeters. International fracture measurements are more uniformly made in metric units. Fracture data are useful regardless of the units used.

The different types of fracture data described and discussed in this section are interrelated and mutually supporting. The goal is quantitative fracture modeling, although semi-quantitative data collection and analysis techniques can and must be used to approach that goal.

2.1.2 Analyses

Once fracture data are collected, they need to be interpreted, analyzed, and then applied, otherwise they are not useful and there is no point in collecting the data. Many fracture studies get shelved because there is no practical way to use the collected data other than as a general "the observed fractures probably explain the behavior of the reservoir." On the other hand, the data from some fracture reports get plugged directly into discrete fracture models to mechanically populate the entire volume of a modeled reservoir without an appreciation for the limitations of the data or a real understanding of the fracture system.

Discrete fracture data can be plugged into a model, but that input should be tempered with the general insights gained from the data, i.e. a conceptual fracture model should be developed before a quantitative fracture model is built. For example, determinations of the logged fracture types, extension vs. shear, provide insights into whether or not fracture-controlled permeability is likely to extend vertically across bedding within a reservoir even if that piece of information is missing from the dataset. Likewise, knowledge of the fracture types and their strikes relative to each other and to the in situ maximum horizontal compressive stress affords insights into the orientation and aspect ratio of the fracture-controlled drainage ellipse in a reservoir, as well as into the nature of the interaction between natural fractures and hydraulic stimulation fractures.

2.1.3 Fracture Data Sources

Several sources of fracture data are described here, but the emphasis is on core. Core provides an actual sample of the subsurface rock and the fractures it contains, whereas other types of subsurface fracture data are measurements of specific properties of a fracture (i.e. its electrical conductivity, the contrast in density between the fracture and the host, the sonic anomaly provided by a fracture, etc.). Image logs and seismic surveys provide more data over larger volumes of a formation than does core, but both tools measure properties of the rock that are assumed to be directly related to fracture characteristics. Fracture assessments using data from these remote technologies also tend make the unwarranted assumption that all fractures are equal in size and effect. Both image logs and seismic surveys become more valuable if they can be calibrated with core.

CT scans (Computed [X-ray] Tomography) provide data from within a core and thus are exceptionally useful three-dimensional characterizations, but the fracture images provided by CT scans are not a substitute for hands-on fracture data. CT scans, as with image logs, should be compared directly to the core to assess and calibrate the fracture CT signatures, which become more valuable when such comparisons are made.

Fracture data reduced to computer screens seem to be clean and unambiguous, and they are therefore easy to analyze, but raw fracture data, particularly from a core, are often messy and less than definitive. Initial fracture logs of a core commonly need to be refined, assigning fractures to different sets, throwing out spurious data, etc., to make sense of them during an analysis.

2.1.4 Quantitative vs. Semi-Quantitative Data

Many cored fractures are well-exposed, discrete structures that are easy to measure when logging core, and for these structures we can measure exact widths, heights, strikes, etc. However, some of the loggable data are semi-quantitative; i.e. although we record discrete numbers for the various fracture parameters, those numbers are often best estimates based on context, experience, and locally on the resemblance of an obscure fracture to similar, better-exposed fractures in other parts of a core. Data sets are often incomplete; for example, a fracture may terminate within the core and so be exposed on only one side of the core such that its strike cannot be determined, but its height, width, and mineralization can still be measured and it can still be included in a fracture-intensity analysis. Likewise, many of the measurable fracture heights in a core are minima because the full fracture height is unknown: it exits the sides of the core upward and/or downward before terminating, or the termination is lost in a zone of missing core or rubble. Such minima are still useful, providing lower boundaries on fracture heights.

2.1.5 Timing of a Fracture Study

For maximum efficiency, fracture data should be collected and assessed early in the development of a play (e.g. Nelson, 2001, 2020). This involves upfront expense, but opportunities to collect important fracture data become fewer as a reservoir-development drilling and production program moves forward. An image log cannot be run in a cased well, nor can a drilled well be cored. If fracture data are not gathered until later in a field-development program, existing wells cannot be repositioned and horizontal wellbore azimuths cannot be changed to take advantage of newly gained knowledge of fracture systems such as the orientation of a fracture-controlled drainage ellipse. It is expensive, but as Nelson (2001) has pointed out, failure to acquire and analyze fracture data early in the development of a field can lead to an inability to acquire the data later when the need for it is recognized.

2.1.6 Need for Experience

Core data comprises a miniscule sampling of a reservoir, so it is imperative to take the time to gather a full, informed data suite from this expensive sample. One must extract as much information as possible from it, and then analyze the data in an informed way. Fractures are easy to measure for dip and strike, but if fracture geometries are indiscriminately analyzed without distinguishing between natural and induced fractures or between extension and shear fractures, at best the subsequent analyses do not reach their full potential, and at worst they will be misleading. Data collection can include some 20 measurements and observations for each cored fracture, and fracture studies that do not make maximum use of the cored data are incomplete.

2.1.7 Other Data Sources

Outcrop studies and engineering tests are also valuable sources of fracture data, providing insights into the three-dimensional distributions and effects of the fractures captured by a core. Outcrop and engineering data should be integrated with the core and image-log data so one may build conceptual models of subsurface fracture systems, and ultimately build quantitative fracture/fluid-flow models. Examples of the engineering data that provide insights on fracture systems at depth are provided in Part 3.

2.2 Planning a Core Program for Fracture Analysis

2.2.1 Introduction

A plan for acquiring fracture data from core starts with decisions on core diameter, core length, core intervals, whether to orient the core, whether to run an image log over the cored interval, whether to run CT scans on the core, and what tests to run on the core. The basic details of core retrieval and processing, and the schedule for marking, plugging, sampling, logging, and slabbing are also important parts of a fracture-data acquisition plan, and simple questions of how a core will be cut to length for boxing or for transport to the lab impact the ability to recover fracture information from the core, since fracture data can be and in fact commonly are lost with each step in the core processing. For an extensive discussion of coring operations see Ashena and Thonhauser (2018).

Coring is an expensive operation, so it is important to be able to explain to management how the fracture data will be preserved and even enhanced by the various components of the plan. Many of these steps are carried out by service-company technicians with a less than perfect appreciation for processing a core for fracture data, thus close monitoring of the processes can help prevent data loss. Considerations include the following:

2.2.2 Core Diameter and Length

Decisions on core diameter and length are often driven by logistical and financial limitations, but larger core diameters and longer cores enhance the probability of capturing a representative fracture population from a reservoir. The probability of capturing a vertical fracture with a vertical, four-inch diameter core is rather low (Lorenz, 1992; see Figure 2.34) so a reservoir may contain significant fractures even if no fractures are present in a short, small-diameter core. In addition, the larger samples of a mineralized fracture provided by larger-diameter core are stronger and therefore they have a better chance of remaining intact during coring operations, both reducing the chance of the core jamming in the barrel before a full run has been cut and improving the chance of obtaining valid fracture-aperture measurements. Successful long core runs can often be cut from fractured formations with stacked 30-foot core barrels.

2.2.3 Substituting Sidewall Core Samples

Sidewall cores are sometimes considered as an alternative to a full core. Typical sidewall cores are an inch in diameter and less than two inches long and have a random azimuth in a wellbore, thus the probability of capturing a fracture with a sidewall core is low in anything but the most heavily fractured formations. If a sidewall core does capture a natural fracture the small-diameter plug is likely to break during coring, reducing the chances of successful retrieval of the most important plugs. Sidewall cores provide little information on fracture heights or distributions; they are better than no core but only if they capture a fracture. Sidewall cores are unlikely to meet the objective of characterizing a fracture population.

2.2.4 Orienting a Core

Orientated cores provide valuable fracture-strike information, allowing for a relatively easy determination of whether a population of natural fractures is composed of parallel or intersecting fractures, which in turn helps answer questions of the degree of fracture-controlled permeability anisotropy in a reservoir. It also helps determine fracture orientation(s) relative to the in situ stresses and thus affords insights into their likely behavior during the stress changes that occur in a reservoir during production. The core-orientation techniques presently in use are based on inertial systems and are much more reliable than the older systems based on a magnetic compass and a 35-mm film camera, but they are expensive and they are not failure-proof.

The local stresses imposed on a core by the scribe knives, necessary for orientating a core, can split apart natural fractures and break up a core, and some companies prefer not to orient cores to avoid this risk, but we have not found it to be a major problem. In contrast, scribe knives have sometimes been run on unoriented core to help keep core pieces oriented during processing, and Nelson (2020) suggests that running a scribe shoe can even help prevent core from jamming in a core barrel. If a core is to be oriented, the coring service company may offer a choice in the scribe-knife configuration, and if so, an asymmetric geometry is best, as described in Section 2.6, "Oriented Core."

2.2.5 Drilling Parameters

The drilling parameters for cutting a core, including the weight on bit, revolutions per minute, mud weight, and pumping pressure, are usually set by the coring company based on experience in the formation or lithology to be cored. In areas where there is no previous experience, it may take a few core runs to determine the optimum drilling parameters for successfully cutting cores. The petal fractures that form in a core from the bit weight hammering on the bottom of the hole (Lorenz et al., 1990) are less likely to form if the brake on the drilling rig's draw-works operates smoothly and frequently, which results in smaller, more frequent adjustments of the bit weight and less impact on the formation. However, petal fractures provide valuable core-orientation information, so while it is best to minimize them it is not necessary to eliminate them entirely.

Centerline fractures typically originate from petal fractures, propagating as small hydraulic fractures due to the mud weight augmented by the pressure of the mud pumps jetting against the bottom of the hole and into the fracture (e.g. Lorenz and Cooper, 2018a). Centerline fractures can split a core for tens of feet down its axis, making slabbing and various whole-core measurements difficult, and reducing the mud weight and/or the pumping pressure can reduce the probability of forming centerline fractures. Nevertheless, as with petal fractures, centerline fractures provide a useful orientation reference.

2.2.6 Trip Time for Core Recovery

Core from some formations comes out of a core barrel resembling a stack of discs or poker chips separated by closely spaced induced fractures cutting across the core. Core discing can form in several ways, one of which is inferred to be the expansion of pore fluids within the core as it is pulled out of the hole. Where disc fractures form in this way, the probability of creating them increases as the rock matrix permeability decreases, as the reservoir pressure increases, in formations with weak bedding planes, and as the rate of recovery as the core is tripped out of the hole increases, since these factors increase the pressure gradient between the core interior and the core surface during recovery, allowing high internal pressures to expand and split the rock. The susceptibility of a core to damage during a trip out of the hole depends on the change in volume of the pore fluid due to the changes in temperature and pressure during the trip, and the diffusivity of the matrix rock, i.e. the ability of the pore pressure in the core to equilibrate to the changing surrounding condtions. Ashena et al. (2019) calculate that water-bearing cores are minimally susceptible to damage by pore-fluid expansion, whereas oil-bearing cores are most susceptible.

The only factor that can be controlled is the rate at which the core is pulled out of a hole, and slower rates allow pore pressure to equilibrate more slowly with the mud pressure in the hole, minimizing the probability of this type of core discing. It will not, however prevent other types of discing such as discs that form in shales during dehydration or discs that form in brittle rock due to shear forces at the interface between the bit and the formation.

Coring companies commonly have enough experience in a basin to know whether discing is a problem for a specific formation, and they can make recommendations for core-recovery rates. A visit to the core warehouse to assess discing in other cores from the formation may also be in order.

2.2.7 Collecting Data on Site

Since artificial fractures are commonly created in a core due to changing drilling conditions while the core is being cut, a fracture analyst should try to acquire copies of the coring report, which lists the drilling and coring parameters such as weight on bit, rates of penetration, core cut vs. core recovered, mud weights, pumping pressures, depths of connections, and bit revolutions per minute. These parameters are often changed in the middle of a core run hoping to improve penetration rates or to overcome incipient jamming, and induced fractures can occur at the depths where the parameters change. The coring records can show where to expect spinoffs in the core and can provide insights as to why a formation did or did not core easily. The records may offer reasons for abundant induced fractures in a core – information that might be used to reduce them in the next core run and information that would help differentiate natural from induced fractures at specific depths in a core. The closer a fracture analyst is to the coring and processing operations the better, since important information is lost if the analyst waits to log the core until it is cut to length, cleaned, marked, slabbed, and laid out on tables in the laboratory.

2.2.8 Running an Image Log

It is often easier fund an image log than a core since logging requires less rig time than coring, and since a log provides large sets of easily analyzed data over long intervals of the wellbore. Image logs are indispensable to the industry and come with robust data-analysis software. Image logs are the current industry standard for fracture analyses, and other than core they provide more, and more reliable, information than any other fracture-assessment technology. An image log commonly undercounts a fracture population, so it can be made more valuable by calibrating it with core; core provides a reality check on the image data.

2.2.9 Back-to-Back Cores

Sequential, back-to-back core runs are common when coring thick formations. The core plan should include contingencies for the possibility that a core run comes up short, since when a core run recovers fewer feet of core than the driller's log suggests were cut, the missing core may have dropped out the bottom of the core barrel and been left in the wellbore. The next core run may have to be shortened by the amount of the missing core because lost core at the bottom of the hole can be caught by the next core run and if so, cutting a full 30-ft (9 m) core will overfill the barrel and damage the core.

If cored, poorly-mineralized natural fractures are oblique to the axis of a wellbore (Figure 2.1), the probability that shear along the fractures will jam the core in the barrel increases as more core is cut since each foot of core adds about 15 lb/7 kg to the core at the bottom of the barrel, increasing the shear stress acting on the fracture planes.

In contrast, fracture planes in a horizontal core may part, allowing the detached core sections to slide up into the empty core barrel since there is no overlying core weight to press the sections together as there is in a vertical core. If the fractures are oblique to the core axis, the core segments will be heavier on the side of the core with projecting wedges of rock, and they will roll around the core axis to put that wedge on the bottom-side of the core barrel. The pieces will then line up point-to-point in the barrel, filling the core barrel before the anticipated 30 feet (9 m) of core has been cut.

Figure 2.1 Poorly mineralized fractures that are oblique to the core axis may slip and wedge the core within the confined core barrel, jamming the core and terminating a core run, especially if there is a significant length of core above the fractures.

2.2.10 On-Site Processing

Service companies typically cut a core into manageable lengths while they are still in the inner core barrel on site, for ease of transport back to the laboratory. The lengths of the sections and the nature of the cuts between them should be specified beforehand; the sections should be as long as possible to optimize core continuity, and the cuts should be made to facilitate reconstruction of that continuity.

Core used to be broken to length with a hammer, leaving ends that could be easily locked together, but more recent practice has been to cut the core with a masonry saw. Saws make more precise cuts, but they may not preserve the core continuity since unlike hammer breaks, one saw cut looks much like another. Some service companies cut a core partway through and break the remaining segment, leaving locking pieces of the core to preserve continuity. Other companies cut a core at an angle of less than 90° to the core axis, which can also help maintain core continuity.

2.2.11 CT Scans

Service companies typically offer the option to scan a core with CT imagery, and a decision needs to be made early whether to do so or not. The circumferential view of core in a CT scan is analogous to the cylindrical, wellbore-surface view of fractures provided by image logs, but CT scans are more versatile, also offering cross-section views of fractures within the interior of the core. CT scans help decipher fracture strikes where they are difficult to see on the rough outer surface of an unslabbed core, and they offer insights into fracture distributions relative to subtle lithology changes that may not be obvious visually. CT scans also provide important information on core continuity since they are records of the core before it has been extruded from the inner barrel, before rubble zones fall apart, and before the core is sampled (see Section 2.7, "Using CT Scans"). CT scans enhance but they do not replace a hands-on core examination.

2.2.12 Removing Core from the Barrel

Pre-drill decisions need to be made regarding coring equipment and the techniques for core retrieval and processing. Service companies provide various options for core barrels, for core barrel liners, and for removing core from a core barrel. The old system of catching core pieces as they are dropped onto the rig floor from the bottom of a dangling core barrel is, thankfully, almost a thing of the past. Most coring service companies now use inner core barrels, and once removed from the outer core barrel at the rig site, the inner barrel is typically cut into short sections for transport to the laboratory where the core is removed from the sections.

Cores should be removed cautiously from these sections of the inner barrel to minimize the distress to the core and the potential for creating artificial fractures. This is easier to recommend than to do given the proclivity for fragmented core to wedge within a core barrel and the determination of technicians with hammers to remove it. Technicians often resort to ramming a jammed core from the core-barrel liner with shovel handles, hammers, and iron bars, especially at 2 a.m. The development of split aluminum inner barrels of different types has simplified the problem of core removal, minimizing the opportunities for creating induced fractures and maximizing core continuity, but this technology is not universally available.

2.2.13 Core-Jam Prevention Measures

Core from some formations tends to jam in a core barrel more frequently than others. A jam during coring is usually recognized by loss of torque and reduced penetration rates since the weight of the core bit and its teeth are no longer being applied to the bottom of the hole, the bit being held off the bottom by a core stub that cannot be pushed upward into the barrel due to the jam. At the same time, the mud-pump pressure falls off because the bit no longer seals against the bottom of the hole to maintain pressure in the pumping system.

Service companies have developed several techniques for dealing with core jamming: in some systems, weak sections that are intended to part below a jam are built into the inner barrel, allowing the section of the inner barrel containing the jam to move upward within the outer barrel despite the jam. Decisions need to be made as to whether a formation is heavily fractured or otherwise prone to jamming, and whether such preventative measures should be taken.

2.2.14 Maximizing and Documenting Core Continuity

Pre-coring discussions should include a method for maintaining the core-orientation continuity around its longitudinal axis as the retrieved core is cut to lengths, and for reconstructing that continuity as the core is removed from sections of the inner core barrel. Ideally this is done by laying out the core end to end and locking the pieces together in the laboratory as they are removed from the inner barrel, marking each continuous interval with a green Master Orientation Line (MOL) at the same time that depths and the red-black, uphole orientation line pair are drawn on the core surface, before sampling and slabbing.

A Master Orientation Line or MOL indicates the rotational core continuity for each continuously lockable core section and must include either a written log or specific core markings denoting the locations of each re-set in the line where the core *cannot* be locked together such as spin-offs and rubble zones. Some technicians are unfamiliar and consider the MOL to be an arbitrary line to be marked on the core, running it indiscriminately across unlockable discontinuities and even across the plastic bags containing handfuls of core rubble. The details of marking MOL lines and their uses are discussed in Section 2.3.6, "Making and Using a Master Orientation Line."

2.2.15 Slabbing Protocol

Some companies slab consecutive core pieces without regard to the orientation of the sequential slab planes relative to each other, degrading the ability to determine relative fracture orientations. Other companies provide a reference mark on the core for the slabbing technician to follow, often relative to a MOL, or relative to core-orientation scribe lines if the core is oriented, and such slabs are usually consistent along continuous core intervals, which is usually desirable for fracture work. Some technicians are asked to cut the core so that the slab face provides a photogenic surface, so they try to avoid fractures that would break up a slab; others have been specifically instructed to cut slabs normal to obvious fractures.

Without a look at the core and the fracture system before the core is slabbed, it is difficult to decide on the optimum orientation for a slab plane. In some cases, a random slabbing orientation has in fact been useful, providing multiple views of a complex fracture system in a core. However, a consistent slab-plane orientation that cuts normal to the dominant fracture strike is usually more desirable for fracture logging. Horizontal cores should be slabbed in a vertical orientation, normal to bedding, but that cannot always be determined prior to slabbing (see Appendix 2.C).

The length of core that can be slabbed is limited by the size of the slabbing saw, but for fracture work longer core pieces are best. Six-inch slab lengths, used for the convenience of the slabbing technician or to accommodate the limitations of a slab saw, are marginally suitable for fracture work.

2.2.16 Scheduling Fracture Logging and other Core Processes

Pre-core planning should include a discussion of when the core will be logged for fractures relative to other planned core analyses. Large fractures should be visible regardless of slabbing, and cutting across them degrades orientation data, suggesting that a core should be logged before it is slabbed. In contrast, small fractures may not be visible unless they are exposed on a slab surface, in which case fracture logging should be done after slabbing. Data on fracture orientations and dimensions are inevitably lost with each core-processing and sampling step, so the earlier a core can be logged the better, preferably immediately after the core has been recovered and marked, and before it has been sampled. However, there are often competing and time-sensitive needs for core sampling and the relative priority of a fracture study must be established during planning. A fracture study needs a clean core surface so that subtle fracture characteristics can be logged, but porosity and permeability samples need to be taken before a core is thoroughly washed.

Ideally a core would be logged before slabbing and again after slabbing, but time and money rarely allow this. If samples must be taken from a core before the fractures are logged, and if a core-long MOL has not been marked, a short, temporary MOL line can be drawn on the core surface parallel to the core axis and extending across the sample and onto the core above and below the sample. This at least preserves the relative orientations of the core segments on either side of the sample.

We offer a core-processing protocol in Appendix 2.B, which starts at the point the core has been recovered from the rig site and is laid out on tables in the laboratory, ready to be marked for depth and orientation.

2.3 Logging Core for Fractures

The following section lists best practices to be used when logging core. Use of these practices can optimize both the quality and quantity of collected fracture data. Some of these are simplistic and seemingly obvious ("wash the core so you can *see* the fractures"), yet in fact they are *not* common practice.

2.3.1 Wash the Core!

A core that is cleaned sufficiently for depth markings is still often coated with enough drilling mud to obscure fractures, and you can't log what you can't see. Service companies are not expected to clean all the drilling mud off a core because the rock will be nicely exposed once it is slabbed. However, much of a fracture is exposed on the still-muddy outer core surfaces, so a fracture-logger's toolkit should contain a toothbrush, a fingernail brush, and even a floor scrub brush to clean enough of the remaining mud off a core in critical areas to be able to log fractures. The commonly available water in a spray bottle is often insufficient for removing drilling mud, so having water in a five-gallon bucket on the floor in front of the logger can be useful. Scrub the fractures clean but beware of removing depth markings. If porosity-permeability plugs will be taken after logging it may be necessary to dry-scrub the core. Natural mineralization is rarely removed by scrubbing a fracture face.

Cores cut with oil-based drilling mud or saturated with reservoir oil are messy, sticky, and difficult to handle and mark. Often the fracture analyst is logging these cores while wearing gloves, maybe with a glove on one hand to manipulate the core but leaving the other ungloved and clean to take notes, operate a computer, and take photographs. Soapy water may be needed in scrubbing such core to expose fractures. Washing the core involves removing it from the core boxes, but the analyst needs to do this anyway to examine all core surfaces for fractures since there is no guarantee that a cored fracture will intersect whatever surface of the core is exposed as it lies in a core box. Washing core is labor-intensive, but you cannot log what you cannot see.

Figure 2.2 The typical tightly packed layout of core boxes on tables leaves no room for working with the core, so the analyst often needs to rearrange boxes to open up a usable workspace on the table. This analyst has removed boxes from the table where he is working, temporarily piling the boxes on top of other boxes for later reference. The workspace migrates as the core boxes are logged and moved from one side of the space to the other. Plastic core bags retard evaporation of core fluids but are an impediment to working with the core. If the core is fresh, the bags should be unsealed just before logging and resealed after logging.

2.3.2 Use all the Core and Remove it from the Core Boxes

If a core has been slabbed, both the slabs and the butts must be examined for fractures. Sedimentologists and stratigraphers usually need to look only at faces of the core slabs because lithology, ripple marks, and unconformities cut across the core and will be exposed on a slab face regardless of its orientation. However, slabs comprise less than a third of a core's volume, and there is no guarantee that a slab has been cut at the best angle to reveal a cored fracture or that a fracture will even intersect the slab face.

For fracture work, slabs and butts must often be reassembled on a layout bench to get the full three-dimensional configuration of a system of natural and induced fractures. In some cases, however, a core is too fragmented to be removed from the box, and all core continuity would be lost if the myriad pieces were to be taken out for examination. A careful poke and prod through the pieces as they sit in the box can often provide enough information for a fracture analysis.

2.3.3 Laying Out Intervals of Core for Fracture Logging

Technicians usually lay out whole-core or core butt boxes like piano keys on the layout tables, with the tops of the core in each box away from the viewer. The boxes are typically pushed together to minimize the required table space (Figure 2.2), and the slab and butt boxes are usually laid out on separate tables. This layout pattern is not optimal for logging fractures.

Since core must be removed from the boxes for fracture logging, and since the analyst must move the core boxes around to do so, it is best to lay out boxes with a few inches between them. It is also best to leave an open space, wider than the length of a core box, on the table next to the first box to be logged so that the box can be rotated sideways to be parallel to the edge of the table. This allows the analyst to remove the core from the box with a minimum of leaning over and related stress on the back muscles. As each core box is logged it gets moved from the unlogged to the logged side of the open space.

Likewise, it is best to lay out slab boxes close to the equivalent-depth butt boxes, i.e. with a slab box interspersed with the equivalent butt boxes, so that the analyst does not need to carry core boxes or individual core pieces from one end of the room to the other to compare slabs to butts. Such a layout requires an early and clear understanding between logger and the lab personnel regarding the desired layout arrangements. A sketch and a box of donuts can be helpful if the technicians are to be persuaded to accommodate the needs of the logger. Regardless, a fracture logger can expect to be moving core boxes while logging, so handling core includes safety considerations since dropped core can damage toes.

Figure 2.3 Piecing core together on a layout table exposes all core surfaces and allows measurement of full fracture heights and geometries.

Piecing a core together for fracture analysis is an integral and necessary part of the logging process: it provides information on the three-dimensional configuration of fractures as well as the relative orientations of fractures in different sections of core. If the workflow allows, core layout and assembly can be done on long core racks after the core is marked and before it is boxed (Figure 2.3).

Short intervals of core a few feet long can be laid out on flat tables, being careful to prevent them from rolling off the table with chocks of some sort, even if they are just marking pens. Longer core sections should be laid out on racks, often built into the table surface, but other expedients such as the channeled styrofoam beds of some types of core boxes can also be used. If no racks are available, the lids of the large, flat core slab box can be turned upside-down and chocked at a tilted angle and the core pieces placed so that they rest in the angle between the side and the top of the box. This setup holds the pieces in place while exposing the core for view and photography, and allows the core pieces to be rotated as necessary. The brown color of the cardboard also provides a relatively neutral background for photography, depending on lighting and the color of the core.

2.3.4 Core-Logging Toolkit

A basic fracture-logging toolkit includes the following:

- *Camera and spare batteries* Most fractures cannot be characterized by a single photograph, so take lots of photographs from different angles. Photographs add significantly to reports, and in conjunction with notes and measurements are excellent memory aids.
- *Digital microscope-camera* such as a Carson Digital Microscope or a Dinolite™, for documenting details of fracture surfaces, widths, and apertures at the millimeter scale.
- *Laptop computer* for running the digital microscope, viewing CT scans, image logs, and music while logging.
- *Hand lens.*
- *Hydrochloric acid.*
- *Flashlight* for the oblique lighting that is necessary to highlight fracture-surface fractography. Since the objective is to make shadows, single-bulb lights, or lights with a linear alignment of multiple bulbs are best; flashlights with bulbs arrayed on a circular surface throw light from too many directions to provide good shadows.
- *Scrub brushes* of various sizes.
- *Logging forms and pencils.*
- *A circular protractor*, with the interior hole $^{1}/_{16}$ inch (1.6 mm) larger than the diameter of the core to be logged, for measuring fracture strikes.
- *A carpenter's protractor*, for measuring dip and strike angles.
- *Athletic or adhesive tape*, for covering fingertips rubbed raw after picking up core for a week.
- *Rubber gloves* for picking up cores from an oil-filled reservoir or that were drilled with oil-based mud.
- *Marking pens* of different colors for writing depths and notes on the core to show in the photographs. Do not write on the slab faces, but most companies will allow writing on the backs of the slabs and the core butts. Slab faces can sometimes be annotated with a pencil since that the graphite is easily removed after the photo is taken.
- *Tape measure*, marked in tenths of a foot for US work, metric for international work, for measuring fracture depths and heights.
- *Millimeter measure* for smaller measurements such as fracture widths.

A convenient way to organize these tools during logging is to use the cardboard top of a slab box as a mobile desk that can be slid along on top of the core boxes as fracture logging progresses (Figure 2.4). The direction of logging is largely irrelevant for fracture work, but we find it convenient to log a core from the top downward, in the

Figure 2.4 Core-analysis tools in a traveling desk made of a core-box lid. Protractors, cameras, hand lenses, scrub brushes of different types, and core-marker pens are among the important items. Silver-colored sharpies for marking dark-colored rock.

direction that the core was cut, whereas stratigraphers typically log core from the bottom up, in the direction of sediment accumulation.

2.3.5 Recording Data

Before starting a fracture log, one should first record basic information about the well and the core, including:

- Date the core was logged.
- Well name, location, and date it was drilled.
- Depths of the cored interval(s) and how much core was recovered from each run.
- General condition of the core (i.e. heavily sampled or rubblized vs. complete, lockable or not lockable, etc.).
- Core diameter.
- How much of the core volume was available for study (i.e. whole core, butts, slabs, or both).
- Whether the service company's coring report was available.
- For oriented core: scribe configuration, scribe quality, and whether the associated orientation survey is available.
- Depths of the formation tops.
- For cores cut from deviated wells, the deviation angle (measured from the vertical, so a well with a 90° deviation is horizontal) and the azimuth of that deviation.

These pieces of information help determine the reliability of the fracture data that will subsequently be logged from the core, and they bear on calculations of the fracture intensities and distributions for each of the cored formations.

One way to log a core for fractures is to have two clipboards, with two forms. The first form is a graphic log, sketching fractures onto a paper representation of the core, ten feet or two meters per column, usually two columns per page. This log serves primarily as notes to accompany the quantitative data that are collected on the second form, where up to 20 pieces of information are collected for each fracture. The collected data can include the following, wherever possible, for each fracture:

1. Depth of the fracture (we log this as the top of the fracture, some conventions use the vertical midpoint of the fracture).
2. The number of fractures at this depth if they are part of the same set.
3. Fracture type (i.e. shear, extension, etc.).
4. Fracture mineralization type (i.e. calcite, clay, etc.).
5. Fracture height, i.e. its extent along the core axis (this number will mean something different for vertical, inclined, or horizontal fractures in a vertical core, and of course the measurable vertical fracture heights in a horizontal four-inch diameter core will all be four inches or less).
6. Fracture width (total width, from one host-rock wall to the other).
7. The estimated remnant fracture porosity within the fracture width, i.e. percentage of unmineralized void space within any mineralization that fills the fracture.

8. The spacing measured directly between this fracture and any adjacent parallel fracture, measured normal to the fracture planes.
9. Dip angle of the fracture, using the core axis from a nominally vertical well as the reference for vertical. If the core is from a deviated well, the high side of the core must be known before true dips can be measured.
10. The acute intersection angle between the fracture being logged and any nearby natural fractures.
11. The acute intersection angle between the fracture and any nearby petal and centerline fractures
    a) Notations of whether the natural fracture strikes clockwise or counterclockwise from the induced fracture.
12. For oriented core, the strike of the fracture relative to the Principal Scribe Line
    a) The true fracture strike if the orientation survey is available and considered to be reliable.
13. The lithology that hosts the fracture.
14. The thickness of the bed that hosts the fracture.
15. The location and type of the upper vertical termination of the fracture (i.e. does it terminate at a change in lithology, does it extend across bedding, is it lost in a rubble zone or missing in a removed sample, is it unknown because the fracture exits the side of the core before terminating?).
16. The same information for the lower vertical fracture termination.
17. Notations of whether the fracture crosses internal bedding planes in the host bed, or cuts across multiple lithologies.
18. The nature of the fracture surface, i.e. fractographic markings if any (slickenlines, a plume, etc.), evidence for dissolution. If no fractography, notations as to whether it is rough, irregular, smooth, planar, etc.
19. Miscellaneous notes.

This list can be modified for deviated and oriented core. All of the desired information cannot be measured for many fractures, and the fracture logger may choose to record best estimates for some of the missing data. For example, the width of a broken fracture can be estimated based on the widths of similar fractures in the core or from the size of mineral crystals on the fracture face. On the other hand, some important data blocks will be unfillable, such as strike where a fracture is exposed on only one side of a core. The reason for an unmeasured strike can be recorded in the miscellaneous notes section.

Logging a core accurately for fractures is a labor-intensive process, but the reward is a quantitative, defensible, and analyzable database. Fracture logging should be as empirical and objective as possible, but there is still some artistry to it since fracture categories overlap and cores can throw enigmas at the analyst. It is not uncommon to start work on a core by tentatively assigning the first logged fracture to one category and then either confirming that assignation or amending it as similar but better-exposed fractures are logged or found in more definitive settings.

2.3.6 Making and Using a Master Orientation Line

Because part of the analysis of cored fractures depends on the ability to compare the orientations of fractures to each other, it is often useful to mark an orientation reference on a core. This is usually done by drawing a straight, green Master Orientation Line (MOL) along each lockable, continuous interval the core, a technique that is useful for vertical, deviated, and horizontal cores alike. After removal from the core barrel, a core is laid out horizontally on a core table and locked together wherever possible, given a preliminary cleaning, marked with a red-black line pair indicating the uphole orientation ("red on the right looking uphole"), and labeled with depths. The green MOL is then marked on the core surface.

Two key features make this orientation line useful: it must be as straight as possible, and there must be permanent notations of where the line is interrupted at unlocked core breaks. A straight edge should be used when marking a MOL, and a pipe level can even be used to mark the exact top of a core in several places, sometimes with a chalk line snapped between the markings as the core is laid out on the table or rack. A hand-drawn MOL is not acceptable: measurements with an accuracy of a few degrees will be made relative to it, so it must provide a consistent and straight reference on the side of the core.

The green MOL is continuous *only* as long as the core can be locked together end-to-end. At each point where the ability to lock the core is lost, due to rubble zones, missing samples, spin-offs, etc., there must be a notation, either on the core (commonly an arrowhead at the end of the green line) or in a written log that stays with the core, indicating that rotational continuity has been lost. When continuity is lost, the strikes of fractures measured above the discontinuity cannot be compared to fracture strikes below it. Notations of MOL discontinuities are sometimes forgotten or omitted by technicians, mismarked, or lost over time, rendering a MOL useless.

Within each continuous interval, fracture strikes can be measured relative to the MOL in the same way they are measured relative to the Principal Scribe Line in an oriented core, and although they cannot be related to north, they at least can be compared to each other. If a sample is later removed from a continuous interval, the green MOL still indicates that the continuity of the interval was previously uninterrupted, preserving the

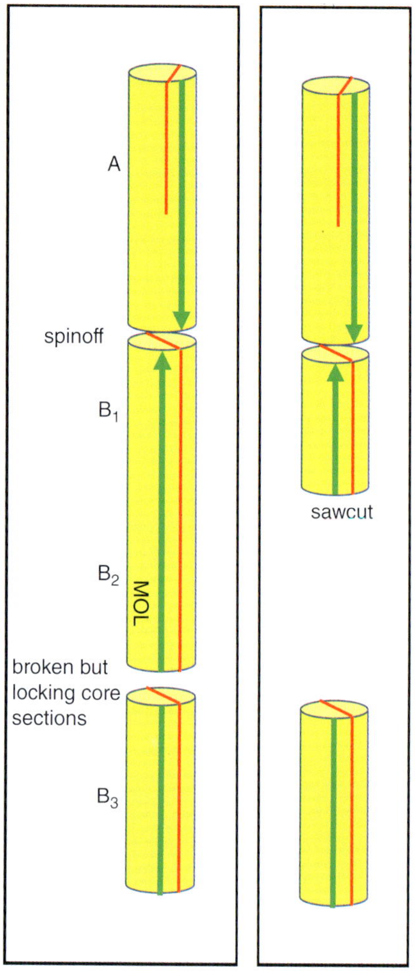

Figure 2.5 *Left*: a Master Orientation Line (the green "MOL") marked on a vertical core containing two vertical natural fractures (red). The strike of the fracture in core piece A cannot be related to the strike of the fracture in core piece B because they are separated by a spinoff which breaks the continuity of the core; they may or may not be part of the same fracture set. Arrowheads on the MOL permanently record the spinoff as a continuity break. The *absence* of arrowheads at the breaks between the B_1 and B_2 core segments indicates that they were locked together when the core was first processed, and the MOL records the relative orientations of the core segments even after the removal of core section B_2. *Right*: the remaining fracture sections in core pieces B_1 and B_3 after sampling piece B_2 can be determined to be parallel, and probably part of the same tall fracture, since they both strike 25° counterclockwise from the MOL.

ability to compare fracture strikes above and below the sample (Figure 2.5).

If a core is oriented, the groove marking the Principal Scribe Line (see Section 2.6, "Oriented Core") is usually set at the top side of the core as it lies on the layout bench, and the MOL is started coincident with that scribe groove. Since orientation grooves usually rotate slowly around the core circumference with depth, the Principal Scribe Line typically diverges from the MOL

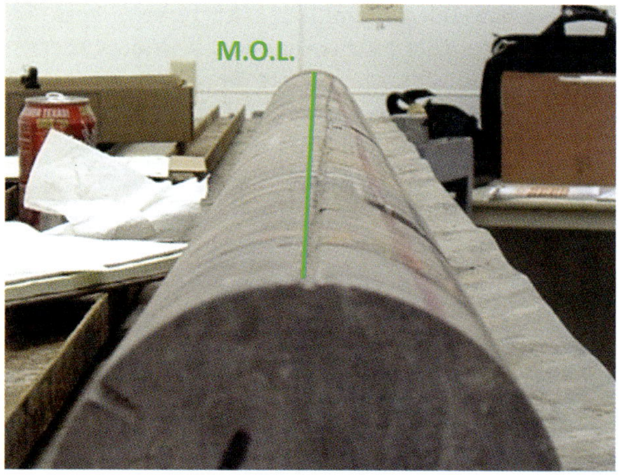

Figure 2.6 Drawing a green Master Orientation Line on a core surface for continuous core intervals provides a reference against which fracture strikes can be measured. This vertical core is viewed looking downhole. The locations of core discontinuities such as spinoffs, rubble zones, or general misfits of the core ends must be noted if a MOL is to be useful. The groove scratched on the core by the Principal Scribe Knife rotates clockwise with depth, as discussed in Section 2.6, "Oriented Core".

(Figure 2.6). Measurements of the amount and rate of rotation, and comparisons to the core orientation report, provide checks on the quality and validity of an orientation survey.

2.3.7 Differentiating Natural from Induced Fractures

Before logging too many fractures in a core it is important to decide which of the fractures are natural and which are "induced," created by the coring and handling processes. Most cores contain induced fractures of several types, and each type has a range of forms, so lists and flow charts for the characteristics that distinguish natural from induced fractures can be good for specific cores but are they not universal. Some types of induced fractures offer useful orientation references, and if these fractures can be recognized they can be used to determine the relative strikes of associated natural fractures and sometimes their true strikes. Regardless, it is imperative to distinguish natural from induced fractures in a core since induced fractures do not contribute to reservoir permeability and should not be included in that part of a fracture database that will be used to model permeability.

The reader is referred to our companion volume (Lorenz and Cooper, 2018a) for detailed descriptions of the induced fracture types in cores, and for discussions of how to distinguish between natural and induced fractures. The range of distinguishing characteristics that define induced fractures becomes apparent with enough core experience, but those characteristics are difficult to tabulate. Petal and centerline fractures (Figure 2.7) are the most useful types of induced fractures because they consistently strike parallel to the maximum in situ horizontal compressive stress. We do not log induced fractures individually or enter them into a database, but we do sketch them on the graphic logs while logging a core. It is important to log any measurable intersection angles between the petal/centerline orientation references and natural fractures. Other common induced fractures include disc fractures that cut across the long axis of many cores, conchoidal impact fractures, and barbed core-bending fractures.

2.4 Taking, Measuring and Analyzing Fracture Data

In the context of this volume, the main reason for collecting fracture data is to assess the influence of natural fractures on reservoir permeability, since the contribution of fracture volume to bulk reservoir porosity is rarely significant, as discussed in Part 3. With enough core capturing enough natural fractures, and with proper analysis of the fractures in the core, one can develop a good conceptual model of fracture distributions and effects, and one can even build a quantitative database of fracture characteristics. Fracture measurements can, of course, be used for a variety of other purposes such as reconstructing a structural history or calculating percent strain across a structure.

For assessments of permeability and fluid flow, two main questions can be addressed with core fracture data:

1. What is the permeability of the individual fractures? Fracture apertures that are occluded by mineralization may not enhance reservoir permeability regardless of how numerous they are. Fractures with open apertures but with faces that are lined with slickenlines or clay may inhibit the flow of fluid from the matrix into the fracture apertures.
2. How well interconnected are the individual fractures, i.e. do they form a fracture-permeability network or are they dead-end conduits of little consequence even if individually they are high-permeability slots? If they are connected, do they form three-dimensional networks that provide reasonably radial drainage, or are they only connected end-to-end, creating highly anisotropic drainage?

In the following sections, we will describe how and what to measure, and try to show how those specific measurements can be used to answer these two questions and to reconstruct the effects of a fracture system on reservoir permeability.

Each piece of fracture information should be collected, to the extent possible, for each cored fracture. Here we describe those pieces of information more fully and describe the potential utility of each type of data. Although they are collected and are important individually, the pieces have more value when assessed collectively as a system.

2.4.1 Fracture Type

Determining fracture type is an important first step in fracture analysis. Simplistically, there are two basic fracture types, shear and extension, but there are other fracture types such as bed-parallel shear planes, deformation shear bands, and ptygmatic extension fractures. Each has a typical configuration and creates a unique fracture-permeability fabric in a reservoir, so it is important to identify fracture type. Shear fracture surfaces are usually marked, in approximate order of increasing offset, by small steps, lineations, slickenlines, slickencrysts, gouge, and/or fault breccia. In contrast, extension fractures may be marked by plume structure, arrest lines, and/or twist hackle. Either fracture type may lack fractography if the faces have been covered by mineralization or etched by dissolution. Dip-slip shear fractures should offset bedding, but the offset may be only on the order of a millimeter, and the offset of strike-slip shears is parallel to bedding, so the apparent absence of offset does not preclude an interpretation of shear.

The basic distinction between shear and extension fracturing provides initial insights into whether the natural-fracture permeability system is likely to be strata-bound (extension), or to cut across minor bedding discontinuities (shear). Moreover, a single set of shear fractures is likely to form an intersecting network of conjugate pairs (Figure 2.8), whereas a single set of extension fractures forms a set of parallel fractures that may be connected end-to-end but that has little potential for interconnectivity across fracture strike. Assuming the fracture widths are unmineralized and open, the interconnected pair of a conjugate shear-fracture set is likely to have good lateral interconnectivity within a reservoir, whereas the poorly connected parallel fractures of a single set of extension fractures will enhance permeability only in the direction parallel to fracture strike.

If the stress orientations have not changed since fracturing (not a given but assumed for the following example), shear fractures will strike oblique to the in situ stress axes and are therefore "critically stressed" and susceptible to shear during production. In contrast, extension fractures are susceptible to a narrowing of the fracture apertures but not to shear. Orientation relative

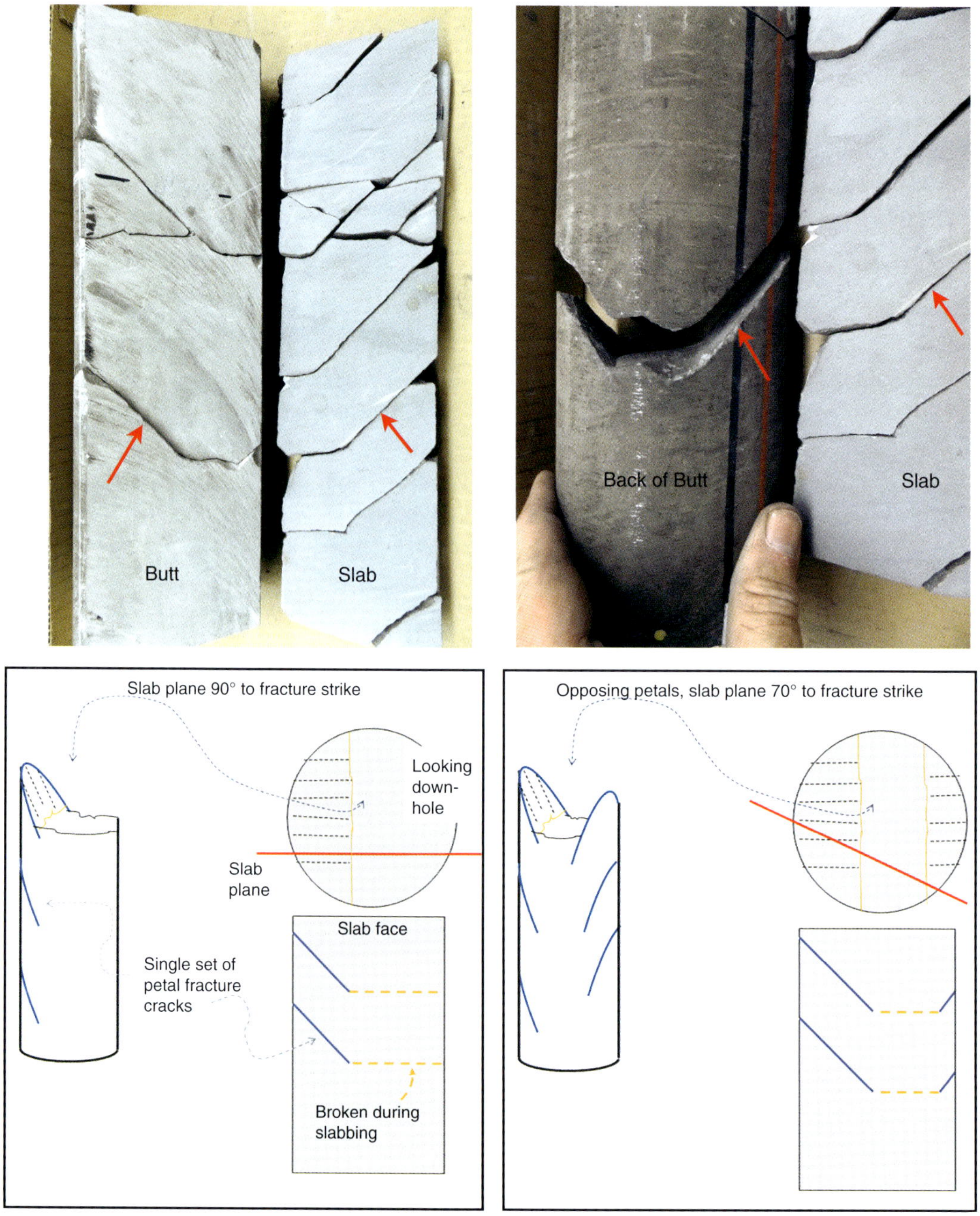

Figure 2.7 Petal fractures, induced by the weight of the bit on the formation during coring, have numerous forms, and their expression on a slab face varies with the angle between the strike of the fracture and the plane of the slab. The opposing surfaces exposed by slabbing a core (*photos, top*) can show oblique and therefore puzzling fracture geometries. However, the true form of the fracture, essential for type recognition, becomes apparent when the fracture is assessed in three dimensions *(sketches, below)* instead of limiting the view to the two-dimensional slab plane.

to the in situ stresses also bears on the likely interactions between hydraulic stimulation fractures and natural fractures since the strikes of stimulations are dictated by the present-day in situ stress orientations and anisotropy. Open natural fractures oriented normal or oblique to the maximum compressive stress are more likely to cause a stimulation fracture to branch, lose pressure, and drop proppant than are fractures that strike parallel to that stress.

Fractures can also have compound histories, and the fractography of their surfaces does not always reflect the full story of their origin or the exact nature of the fracture network (Figure 2.9). The basic geometries of fractures in core, i.e. their dips and strikes, can and should

Figure 2.8 *Top*: a system of two intersecting sets of extension fractures breaks a limestone bed into nearly equidimensional blocks that have tumbled down slope from the fractured bed in the background. Systems consisting of two sets of intersecting open fractures extending top to bottom of a bed are not rare, but neither are they universal. The two fracture sets formed at different times within different stress regimes, and the most recent maximum horizontal compressive stress orientation is commonly parallel to the planes of the youngest fracture set. *Bottom*: two shear-fracture sets form a conjugate pair. These two intersecting sets formed at the same time, when the maximum and minimum in situ horizontal compressive stresses were oblique to the fracture planes. A single stress system formed these intersecting shear fractures, whereas two sequential stress systems formed the intersecting extension-fracture sets.

be used to help determine fracture type, but fractography should be used where possible to support that determination. Fracture identification is a critical part of a fracture assessment, and this is easier to do with core, where the diagnostic fracture surfaces can be examined, than with remote techniques such as image logs or CT scans which provide only cross sections and the basic geometries of fracture planes.

2.4.2 Fracture Depths: Intensity and Density

Based on frequency alone, a core that captures many fractures suggests a well-developed and probably interconnected reservoir fracture system, whereas core that retrieved relatively few fractures suggests a less-important fracture system, although apparent fracture frequency also depends on the geometry of the wellbore relative to the fracture planes. It is useful to quantify "well-developed," "few," and "less-important," but this is not a trivial exercise. Lateral fracture spacing is one quantitative measure of fracture development, but this parameter is difficult to obtain from vertical core where the absence of fractures in the small sampling of a reservoir does not preclude the presence of important vertical fractures. Spacing is more readily obtainable from deviated core if the wellbore azimuth is oriented at a reasonably high angle to fracture strike.

Figure 2.9 A set of closely spaced fractures with a compound history on a fold: the parallel extension fractures, trending parallel to the axis of folding (upper right to lower left, parallel to the red line), were reactivated in bed-normal shear as the fold tightened, forming steps on the bedding plane on the vertical limb of the fold (red arrow, in front of the geologist). The fracture faces record shear, so limited core samples cut from this part of the fold might misleadingly suggest a network of conjugate, shear fractures even though the fractures are in fact parallel, reflecting the inherited extension-fracture pattern. A more widely spaced set of extension fractures (green arrow) strikes normal to the fold axes.

Fracture development is commonly quantified in terms of "density" and "intensity," the definitions of which change from author to author and whether one is measuring in one, two, or three dimensions. Here we follow an established geological usage as described by Doe and Dershowitz (2008) and Rohrbaugh et al. (2002), where "density" is taken as a basic measurement of the simple number of fractures per length, area, or volume of a sample. "Intensity" is the same as density for a linear sample but as the sample becomes planar or volumetric it incorporates other dimensions to become fracture length per surface area, and fracture area per volume. The definitions of "intensity" and "density" are typically reversed in non-geological applications, and in fact some geologists have used that convention.

Variations of these measures of fracture development can be obtained from core, becoming complicated with changes in the orientation of fracture planes relative to the core axis. As a basic measure, the number of fractures per foot of core is an easy number to come up with. This fracture *density*, the number of fractures in a linear sample, it is readily applicable to vertical fractures captured by horizontal cores, which are analogous to fractures measured in outcrop by scanlines. The definition gets fuzzy when applied to a count of vertical fractures in a vertical core since although that also fits the technical definition of "density," the geometry and thus the probability of fracture intersection are different, thus the measured density number has a different significance.

A simple number of fractures per foot of core does not account for differences in fracture type or whether the fractures are tall, short, wide, narrow, open, closed, all the same, or variable, but it provides a number that at a basic level can be used to compare fracture development in cores cut from the same stratigraphic intervals in adjacent wells. We have found that, if there are no other measures to go by, the presence of one or more vertical extension fractures per 10 ft (3 m) of vertical core typically indicates that fracturing is developed well enough to influence reservoir permeability. The meaning of this number changes if the fracture population consists of another fracture type or if fracture planes are oblique to the core axis.

The next level of assessing the degree of fracture development prevents populations of numerous but short and inconsequential fractures from skewing measurements, by including fracture length or height in the assessment. Fracture *intensity* is usually taken to be a measure of the degree of fracture development calculated as the ratio of cumulative fracture height to core length, sometimes restricted to the height of fractures per foot of fracture-prone lithology in the core. A ratio of greater than about 1 ft (0.3 m) of fracture height per 10 ft (3 m) of core again seems to indicate a fracture system that is sufficiently well developed to influence reservoir permeability, but this observation is also restricted to vertical core and vertical fractures.

The significance of the calculated fracture intensity and density numbers changes if fracture planes are inclined rather than parallel to a wellbore such that they have a higher probability of being intersected by the wellbore. For example, for fracture sets with a given spacing, a vertical wellbore has a greater chance of intersecting inclined or horizontal fractures than it does of intersecting vertical fractures, so a vertical core containing the same number of vertical and horizontal fractures suggests that the vertical fractures are much better developed than the horizontal fractures. Likewise, a horizontal wellbore has a high probability of intersecting vertical fractures that strike normal to the wellbore azimuth, and a low probability of intersecting both horizontal fractures and those vertical fractures that strike parallel to the wellbore azimuth. Add inclined fractures and deviated wellbores to the mix and it becomes difficult to calculate meaningful fracture densities or intensities for comparing fracture development from well to well.

Nevertheless, density and intensity measurements are often the only available means of comparing degrees of fracture development in adjacent wellbores. Intervals of well-developed fractures in a core can still point to better targets for perforations in reservoirs if the limitations of the data are recognized and the fracture-development measures are used in a semi-quantitative manner. The degree of fracture development in a core or image log is still important since it provides direct measures of fracture porosity and spacing, and can offer indirect insights into the degree to which fractures are likely to be interconnected into a permeability network.

Figure 2.10 Measuring fracture dip angle with a carpenter's protractor, using the outer core surface as a reference for vertical. If the core is horizontal and the high side of the core can be determined, the same method can be used to measure both strike and dip. This protractor offers several scales and the correct one must be used, in this case the inside arc with 90° on the left side nearest the core (red arrow). Alternatively, the angle can be mathematically calculated from the other measurement arcs. This fracture has a dip angle of 62°.

2.4.3 Fracture Dip Angles

2.4.3.1 Measuring Dip Angles

Dip angles are easily picked on an image log and can be easy to measure in core, using the core axis as a reference and a carpenter's protractor. If the fracture is broken open along the fracture plane, the protractor can be placed directly on the fracture surface, measuring the angle between the core axis as represented by the outer core surface and the fracture plane (Figure 2.10). Dip angles for fractures exposed on a core slab face, however, are only apparent dips unless the fracture strikes normal to the slab plane, and this must be taken into account by visually tracing the fracture around the outer core surface and measuring the true dip of the fracture plane in three dimensions. The dip angles of fractures in intact core are not always easy to measure, requiring visualization and a little inventiveness. Dip angles are often the only easily obtainable measure of a fracture where a vertical core consists of rubble, since all that is required is a core fragment displaying both a section of the fracture plane and a section of the vertical outer core surface.

Most "vertical" wells are not exactly vertical so cores may in fact be slightly inclined, making the core axis a less than perfect reference for true fracture dip angles, but this is usually inconsequential. For larger deviations, it is useful to use the wellbore-deviation survey that is run in each well, and if a well is intentionally deviated the deviation angle and the deviation azimuth through the cored interval are necessary for determining fracture dip angles in the core. The high side of a deviated core must also be known since, for example, a vertical fracture cut normal to the fracture plane by core cut from a wellbore deviated at 45° from the vertical will appear to be horizontal if the core is rotated 180° around its axis from its in situ position. The problem is relatively constrained if the deviated core has been cut through strata where bedding is visible and arguably horizontal. However, the problem is almost unsolvable where the formation is massive or where the strata have been tilted, unless the core is oriented or an image log is available.

Service companies commonly use a goniometer to easily determine fracture dip angles when they are processing a core, which is sensible if the technician understands fractures, properly orients the core, and makes the important distinctions between natural and induced fractures. Goniometers are less commonly available when an analyst is working with archived core, and trotting off to another room to measure each fracture with a goniometer can be inconvenient and time consuming, so a basic protractor is still useful.

2.4.3.2 Using Dip Angles

Dip angle histograms (Figure 2.11) can be used to compare fracture data obtained from unoriented cores to fracture data obtained from image logs. Strikes cannot be compared since the core is unoriented, but dip-angle histograms can show whether the two datasets are compatible, whether the image log is picking up signatures from all fracture sets, and whether the image log is distinguishing properly between natural and induced fractures. Dip-angle histograms can help determinations of whether horizontal fractures in core can or cannot be distinguished from bedding in the image log. Histograms can also show whether image-log interpretations are biased towards the more-easily-picked inclined fractures, i.e. whether imaged natural vertical fractures have been misidentified as induced fractures. This should not be a game of "gotcha" however; petrophysicists should be included in a core assessment so that they can properly pick and analyze fractures in the associated image log – it is a team effort.

Dip-angle histograms can be used in conjunction with stereonets in determining fracture type and in inferring the degree of fracture interconnectivity. If most or all fractures are vertical, the fractures may be either high-angle extension fractures or strike-slip shear fractures, whereas fracture sets with predominantly 60° or 30° dip angles and opposing dip azimuths are likely to be interconnected conjugate dip-slip and reverse dip-slip shear fractures, respectively.

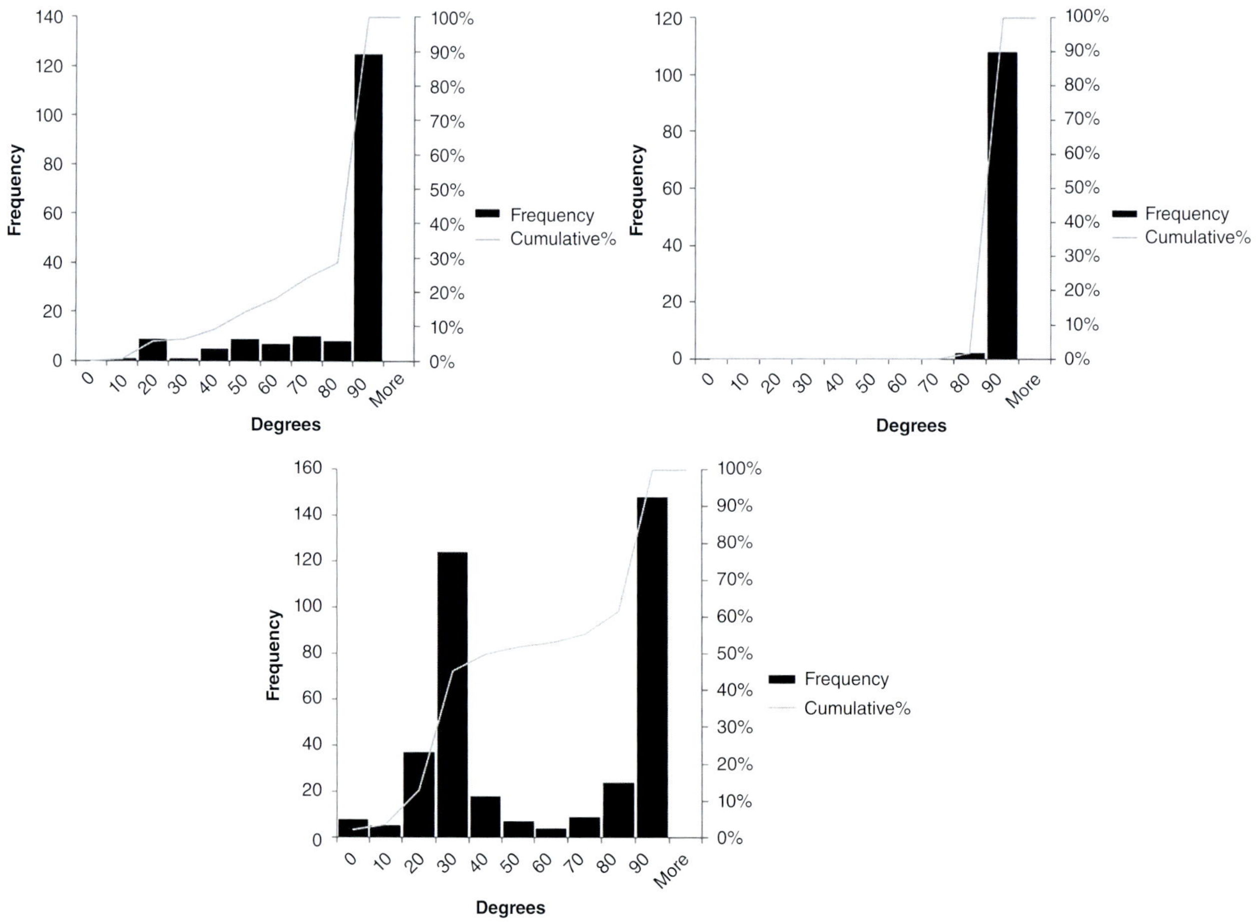

Figure 2.11 Examples of fracture dip-angle histograms derived from core data. *Top left*: a compound fracture system dominated by vertical extension fractures (n = 175). *Top right*: a system composed entirely of vertical or near-vertical extension fractures (n = 110). *Bottom*: a compound fracture system dominated by two fracture sets: intermediate-angle shear fractures and high-angle extension fractures (n = 384). (Unpublished core data).

Interpretations of fracture type based on geometries should be supported with other indicators such as the fractography of the fracture surfaces.

2.4.4 Fracture Distributions

Natural fractures are rarely uniformly distributed along the length of a core. Fractures may be concentrated at certain depths associated with structure, and different fracture types often have affinities for specific host lithologies (Figure 2.12). Care must be taken to differentiate between truly non-uniform distributions created by geomechanical/lithologic variations, and apparently non-uniform distributions created by sampling issues. A core may be short, or the fractures are poorly developed such that a statistically valid, representative sample of the fracture population was not recovered, and seemingly random fracture distributions may be solely the result of a limited sampling of the fracture set. A representative fracture population is only likely to be cored if numerous fractures are present in a formation and if the wellbore is oriented at an angle that has a good probability of intersecting those fractures.

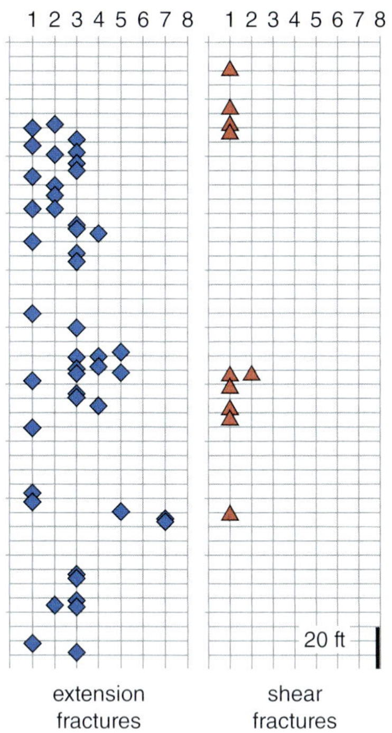

Figure 2.12 Example of variable fracture depth-distributions by type in one core that cut through a limestone-marl sequence. Extension fractures are more common in the limestones whereas shear fractures are localized within marley intervals. The inclined planes of shear fractures have a high probability of being intersected by a vertical well, thus their absence from certain intervals of core through the formation is real. In contrast, the concentrations of vertical extension fractures may more apparent than real since the vertical core may just have missed fractures in the intervals where they appear to be absent. (Unpublished core data.)

An analyst has no control over the make-up of a cored fracture population; a more robust sample cannot be obtained by cutting additional core. Thus it is imperative to understand and make the most of the fractures that are captured, and to respect the limitations of some datasets. Fracture depth-frequency charts for vertical cores provide a basic visualization for fracture distributions in a formation, and can be constructed plotting the locations of individual fractures in a core by depth, by binning the data and plotting histograms showing the number of fractures by foot in a core (Figure 2.13), or by plotting a running average of the number of fractures in a moving interval along the core. Fracture depth-frequency distributions by fracture type are also useful when more than one type occurs in a core, since some fracture types have more effect on reservoir permeability than others.

Fracture distributions should also be assessed by lithology (Figure 2.14). A basic lithologic log of a core is indispensable when logging and analyzing cored fractures, but a detailed sedimentology log usually provides more detail than can be accommodated by relatively gross-scale fracture-distribution plots. The gamma-ray logs that are run on most cores cut since the 1990s are essential for plotting fracture data by depth and for correlating a core fracture log to an image log, but the gamma profile does not always reflect the mechanical properties that control fracture development so a lithologic log is still indispensable.

Fractures captured by cores from horizontal and deviated wells offer another type of fracture-distribution information. A horizontal core may have been cut from a single bed, so all the fractures are from one host lithology. This usually provides excellent data on fracture spacing within that bed but very little information regarding variations in fracture distribution by lithology in the adjacent beds. Core from inclined wells that intersect multiple stratigraphic units can provide information on both vertical and lateral fracture distributions and should be considered for pilot holes cored to characterize natural fractures.

Properly interpreted, fracture-distribution assessments provide information as to which units in a reservoir are likely to contain the best-developed fracture systems. Fracture distributions can also document relationships between fracturing and structures such as folds and faults, where fracture frequency commonly increases near a fold hinge or in proximity to a fault. Distribution data from core can be used for quantitative

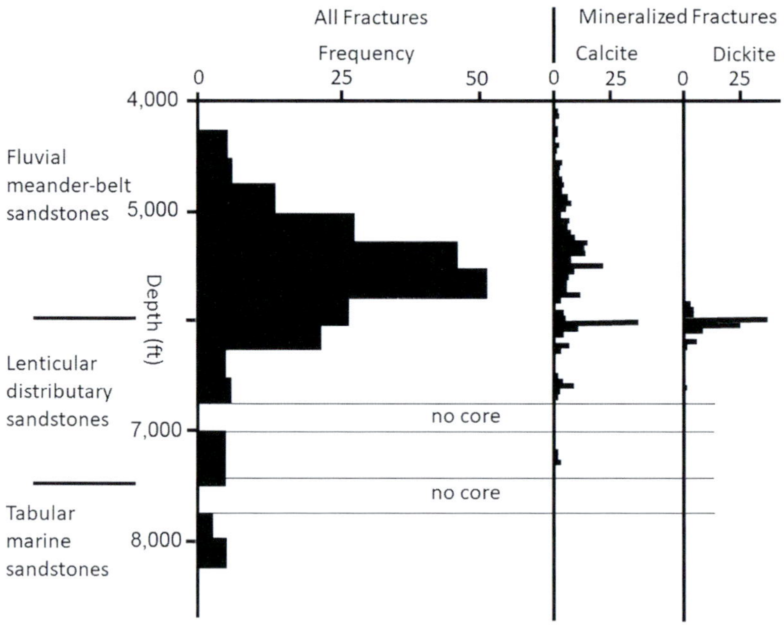

Figure 2.13 An example of fracture distribution, plotting fracture frequency by depth in cores cut from vertical wells in an extensively cored formation. *Left*: this histogram uses 250-ft (76 m) depth bins to plot the distribution of vertical extension fractures (n = 220). *Middle and right*: these two histograms plot all fractures in the same cores, not just extension fractures, by mineralization type, and use 100-ft (30 m) depth bins (calcite n = 208; dickite n = 72). This figure shows two important features: 1) Fracture frequency *appears* to be influenced by depositional environment: fractures are more common in the cores cut from the meander-belt sandstones, with frequency falling off lower in the section where the core penetrated more lenticular distributary-channel sandstones and more laterally extensive, more homogeneous marine strata. In fact, however, the continuous core in the meander-belt interval cut numerous thin overbank sandstones that contain closely spaced fractures, increasing the fracture count, whereas such beds are absent in the cored marine intervals. Horizontal and deviated cores at this site show that the fracture spacings are similar in the reservoir-scale distributary-channel and marine sandstones. 2) A local enhancement in fracture frequency occurs near the depth of 6,000 ft. This anomaly is associated with a concentration of dickite-mineralized fractures having a unique, irregular morphology. Dickite is a high-temperature polymorph of kaolinite, and both mineralization and fracture irregularity suggest a local influx of hydrothermal fluids along a structural anomaly (adapted from Finley and Lorenz, 1988; Lorenz and Hill, 1992).

Figure 2.14 Example of the percentage distribution of three fracture types by lithology in a cored, heterogeneous formation. Extension fractures (*left*) are restricted to the chalk facies, compaction fractures (*right*) are restricted to marls, and shear fractures (*middle*) occur in both lithologies (unpublished data.)

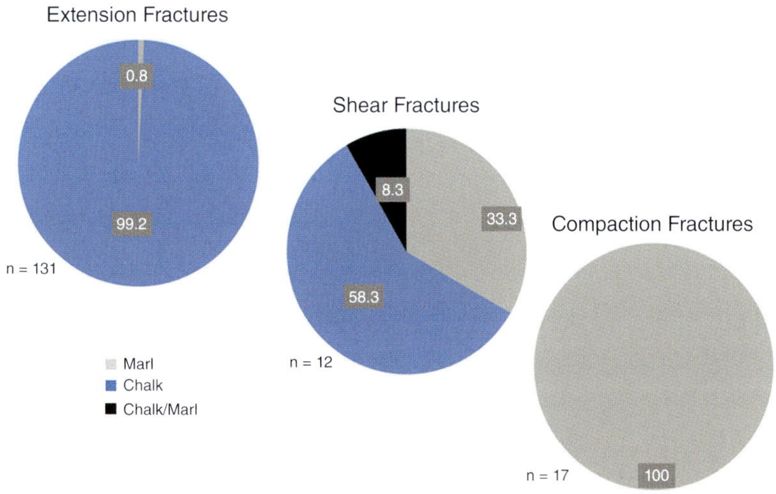

modeling if the cored fracture population is sufficiently robust to support it, but even if the sampled population is relatively small the distributions can be integrated into conceptual fracture models.

2.4.5 Fracture Heights and Terminations

The heights of vertical fractures captured by a vertical core are often definitive and easy to measure, but horizontal cores offer only limited height information since

Figure 2.15 *Left*: an example of a known vertical fracture termination, where a calcite-mineralized fracture in a marine mudstone narrows as it approaches the contact with an overlying limestone and terminates against that contact. *Right*: examples of truncated fracture heights, where the fractures either exit the side of the core (red arrows) or are lost in a missing piece of core (black arrow), before terminating. The vertical line to the right of the number "12" is an orientation groove scratched onto the core surface.

they cannot sample vertical heights greater than the core diameter. In between these extremes, where a fracture is oblique to a core axis, a greater percentage of a fracture's height may be captured, but the total height is still unknown (Figure 2.15). Fracture terminations can also be lost in rubble zones and removed with core samples.

Even truncated fracture-height data are useful in providing minimum dimensions. Such incomplete height databases can be fleshed out qualitatively by combining them with notations of fracture termination types and locations. For example, if most of the observed terminations are at bedding planes, then the inference can be made that even though few of the full fracture heights were measured they are probably equivalent to bed thickness and the fractures are strata-bound, limiting the vertical natural-fracture interconnectivity.

Heterogeneous reservoirs are likely to have extension-fracture heights that are limited by internal bedding contacts, where heights are significantly less than gross reservoir thickness (Figure 2.16). In contrast, fractures that extend across bedding or that formed in homogeneous rock are more likely to be tall and enhance vertical permeability. Strata-bound fractures that consistently terminate at the contact between reservoir rock and interbedded non-reservoir lithologies (Figures 2.17, 2.18) provide good internal vertical permeability but little hydraulic communication between reservoirs.

Some extension-fracture sets, especially in homogeneous lithologies, terminate "blindly," i.e. they terminate for no apparent geomechanical reason within a homogeneous lithology. Fracture sets with numerous blind terminations typically have heights that are less than the thickness of the host bed, suggesting they were not subjected to enough stress during fracturing to drive the fractures vertically to bedding boundaries. This also suggests that the fractures are relatively circular in shape since the same limited stress would not have driven the fractures horizontally any farther than it propagated

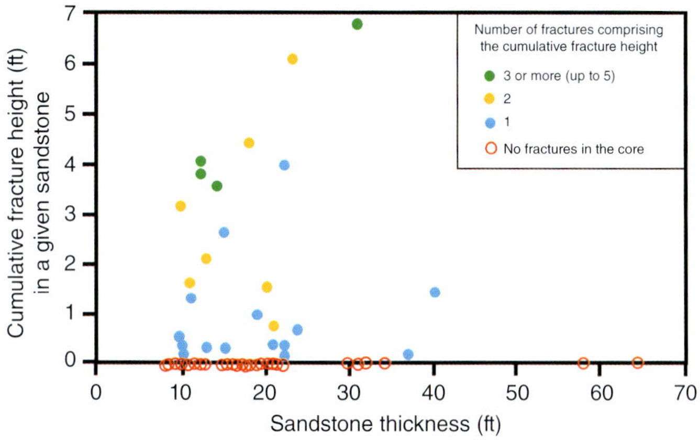

Figure 2.16 *Left*: chart showing the cumulative heights for vertical extension fractures in each sandstone cut by vertical core from heterogeneous fluvial reservoirs of the Mesaverde Group in Colorado. Data from sandstones less than 10 ft (3 m) thick are not included. The factures are relatively short, and even adding together the heights of the cored fractures in each sandstone are never equal the total bed thickness, indicating that most terminate at internal bedding planes or are artificially truncated. Core from about half of the sandstones contain no fractures, but the abundance of fractures in the other sandstones shows that fractures are ubiquitous and closely spaced even in beds where the 4-inch (10 cm) diameter core missed them. Core from a deviated nearby well confirms this (from Lorenz and Hill, 1992). *Right*: a two-dimensional outcrop illustration of this type of fracture distribution in heterogeneous deposits; the fluvial Mesaverde Formation in Colorado.

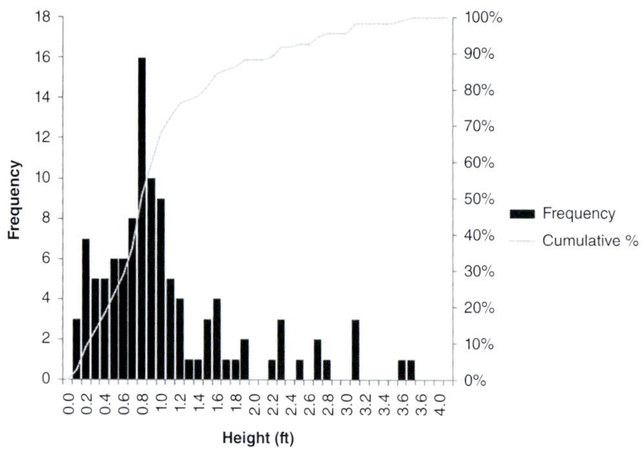

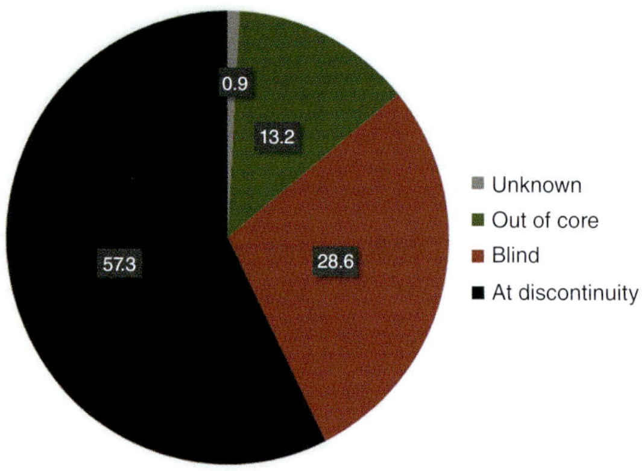

Figure 2.17 Example of vertical-fracture height data from a vertical core, documenting bedding-bound fractures. *Left*: histogram of measurable heights (n = 110). *Right*: pie chart showing that over half of the fractures terminate at a bedding discontinuity and nearly a third are blind, i.e. terminating within a homogeneous lithology. This suggests that even the fractures that have unknown terminations do not extend beyond the reservoir unit and that vertical fracture permeability is limited by bedding (n = 220, unpublished data).

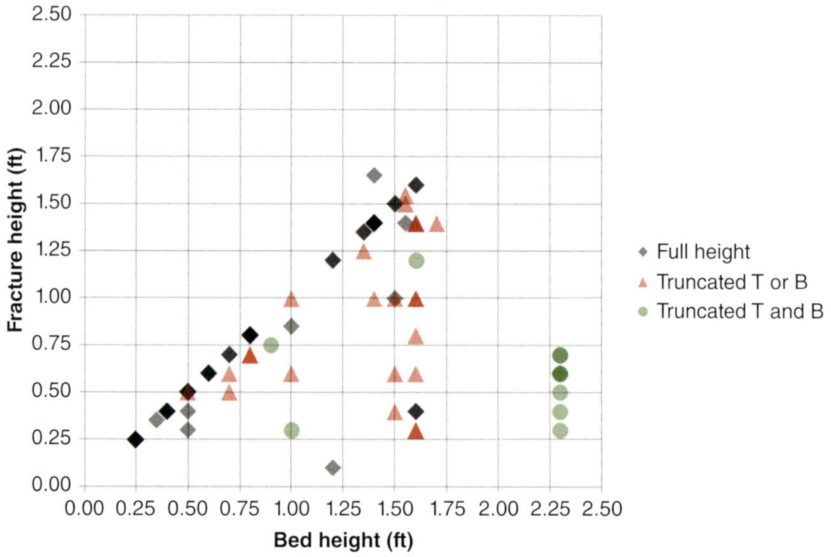

Figure 2.18 Example of a dataset for vertical fracture heights captured by a vertical core. The full heights of 51% of this population of 85 extension fractures could be measured in the core (gray and black data points). The other measured fracture heights were truncated, top (T), bottom (B), or both (T and B), as indicated by the color of the data point. The population of fractures with only one truncation is not obviously distinct from the population of fractures truncated at both ends. The dually truncated fractures might be expected to be shorter, but that is negated by the increased probability of taller fractures to be doubly truncated. Darker points indicate overlapping measurements (unpublished data).

them vertically, providing qualitative information on the fracture lengths.

Shear fractures behave differently than extension fractures, being less constrained by the mechanical stratigraphy of a formation and more likely than extension fractures to cross bedding. Dip-slip normal and reverse shear fractures are likely to exit the sides of a vertical core before terminating since they are typically inclined. Strike-slip shears are usually vertical and parallel to the axis of a vertical core, so the relationship between their vertical terminations and bedding is captured more often. Shear fracture heights and terminations are only captured infrequently by core (Figure 2.19), thus it is important to differentiate fracture type when developing conceptual models of fracture heights and fracture-permeability systems. Conceptual models can in turn be used to refine and enhance an incomplete but quantitative core-fracture database.

Fracture heights and terminations can be used to estimate the vertical connectivity of a fracture system within a reservoir, complementing the previously discussed fracture-type and fracture-distribution data. If fracture surface markings cannot be observed because the fractures are tightly cemented, the dip and termination data can be used to help constrain fracture type.

Records of the locations and types of cored fracture terminations thus provide a means of assessing probable fracture heights in the reservoir from the incomplete fracture heights captured by core. Extension-fracture heights, lengths, widths, and spacings are all initially dictated by the amount of strain imposed onto a formation, but the control on height changes, becoming limited

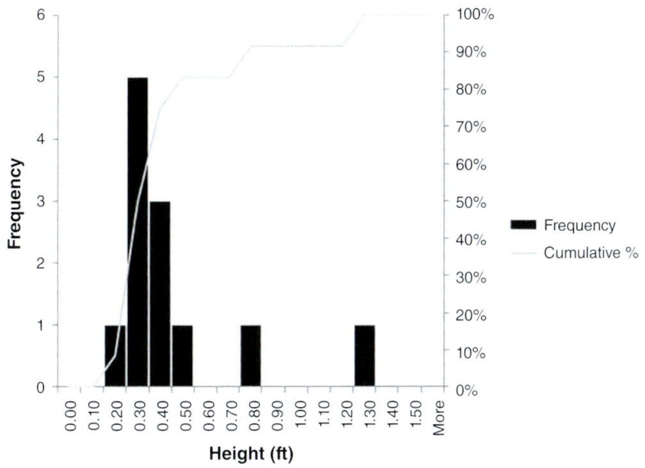

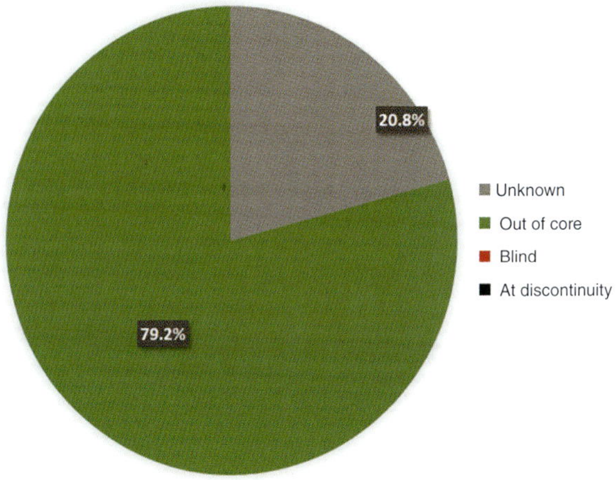

Figure 2.19 *Left*: histogram showing the truncated, measurable heights for a population of inclined shear fractures, where the captured height depends almost entirely on the core diameter and the dip angle of the fracture plane (n = 12). *Right*: pie chart showing the percentage of truncation types for this fracture set, where all fractures either exit the side of the core or are "unknown" and lost in missing samples (n = 24, unpublished data).

by bedding once the fractures propagate to mechanical boundaries.

2.4.6 Fracture Widths, Apertures, and Mineralization

Fracture width, the distance between the opposing fracture surfaces in the host rock regardless of mineralization, is relatively easy to measure for extension fractures by using a physical scale appropriate to the size of the fracture, although a hand lens and even a microscope may be necessary for measuring narrow fractures. Quantitative measurements are desirable, but a fracture logger should not agonize over width and aperture measurements since widths can vary along the plane of a fracture. The widths of narrow fractures can be measured using a reference scale such as that provided by Ortega et al. (2006) or by estimating width in reference to the half-millimeter-diameter of the mechanical-pencil lead being used to record data.

Exact fracture width can only be measured where the plane of exposure cuts across the fracture plane at an angle close to 90°, although the apparent fracture widths measured along planes that cut oblique to the fracture between 60° and 89° overestimate width by less than 15%. A fracture exposed on a cylindrical core surface has a variable apparent width depending on where it is examined. The apparent fracture widths exposed on randomly cut slab planes need to be measured carefully, so, as always, core pieces must be picked up and examined for the best exposure of width. Precision is desirable, but an imprecise estimate is better than no estimate, which would leave the parameter totally unconstrained.

Estimates of fracture widths, apertures, and remnant fracture porosities must often be made where the rock is broken open along a fracture plane or where one face of the fracture is missing. In such cases width and remnant porosity within a mineralized fracture aperture can be estimated based on the mineralization characteristics, assuming that the missing opposing face is similar, or based on the characteristics of similar fractures in intact core. However, the character of the remaining mineralized face must be considered carefully since there are potential pitfalls, as illustrated by Figure 2.20.

Mineralization of some kind is present in most cored natural fractures, narrowing or even occluding the fracture apertures. Incomplete mineralization may leave an open but irregular medial aperture within the fracture width, or it may leave only tortuous pathways between fracture-bridging pillars and crystals of mineralization. In both cases the irregular open part of the fracture width, the aperture, can be expressed as a percentage of the total fracture width. As used here, this percentage is the "remnant fracture porosity" within a fracture's width.

This percentage is most easily estimated semi-quantitatively by comparisons to charts such as that presented by Compton (1985) for estimating percentages of heavy minerals in thin sections (Figure 2.21). More precision could be obtained from thin section point counts or from laboratory measurements, but these require an investment of time and money. Estimates, which with experience can be reasonably accurate, are typically sufficient.

Fracture widths can be enhanced by dissolution, but more commonly they are reduced by mineralization, and it is the open, remnant fracture aperture left within a mineralized fracture that determines the ability of fluids to flow along the plane of the fracture. Many populations of fracture widths have log-normal distributions,

Figure 2.20 If only one face of a mineralized fracture is available for measurement, the fracture may appear to be incompletely mineralized, but the possibility that part of the occluding mineralization has adhered to the opposing, missing fracture face should be considered.

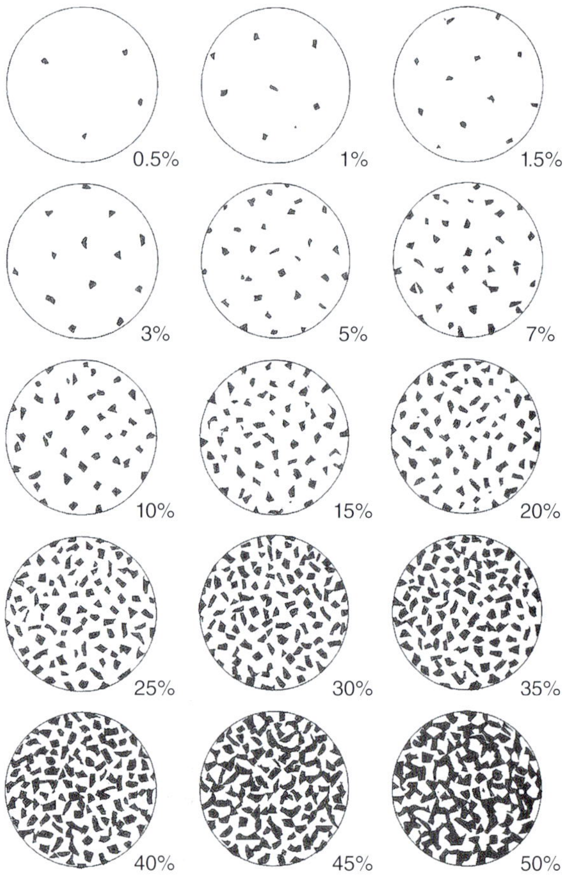

Figure 2.21 A visual reference for estimating remnant fracture porosity (from Compton, 1985). This chart was developed for petrographic applications, specifically for estimating the percentage of heavy minerals in a thin section, but it can be used in other applications.

with numerous narrow fractures and increasingly fewer fractures as the widths increase (Figures 2.22, 2.23). In contrast, remnant fracture apertures are controlled by the geochemistry of the system and are less likely to be log-normally distributed. The two datasets shown in Figures 2.22 and 2.23 come from similar lengths of vertical core cut from the same well and the same formation, but from two beds separated vertically by nearly 600 ft (183 m).

Cross plots of fracture widths vs. their remnant apertures rarely show linear correlations, and when they do they commonly show that wider fractures that have lower percentages of remnant porosity. Nevertheless, even a small percentage of open void space in a wide fracture can provide a significant permeability enhancement over matrix permeabilities, and these cross plots typically show that many of the small fractures have important remnant apertures. Since they are numerous, even small open fractures can contribute an important flow capacity to a reservoir plumbing system.

The ranges and averages of the fracture widths in the two beds shown in Figures 2.22 and 2.23 are identical, suggesting that they accommodated the same amount of strain, as would be expected. However, the fracture sets are dissimilar in their ranges for the remnant-porosity values and in the distributions of those values. The wider range of values as well as the higher average shown in Figure 2.22 suggest that the natural-fracture permeability system in this unit should be more effective, even though the same amount of core captured fewer fractures ($n = 23$) in this bed than in the bed shown in Figure 2.23 ($n = 34$).

The widths of shear fractures are commonly more irregular than the widths of extension fractures, and the widths of any fracture type in a carbonate can be more irregular than the widths of fractures in less-soluble lithologies (Figures 2.24, 2.25). Carbonate systems are more chemically reactive than clastic lithologies, so in addition to the mechanical fracture opening, fracture apertures in carbonates can undergo multiple events of widening by dissolution and narrowing by mineralization. The widths and remnant porosities of fractures

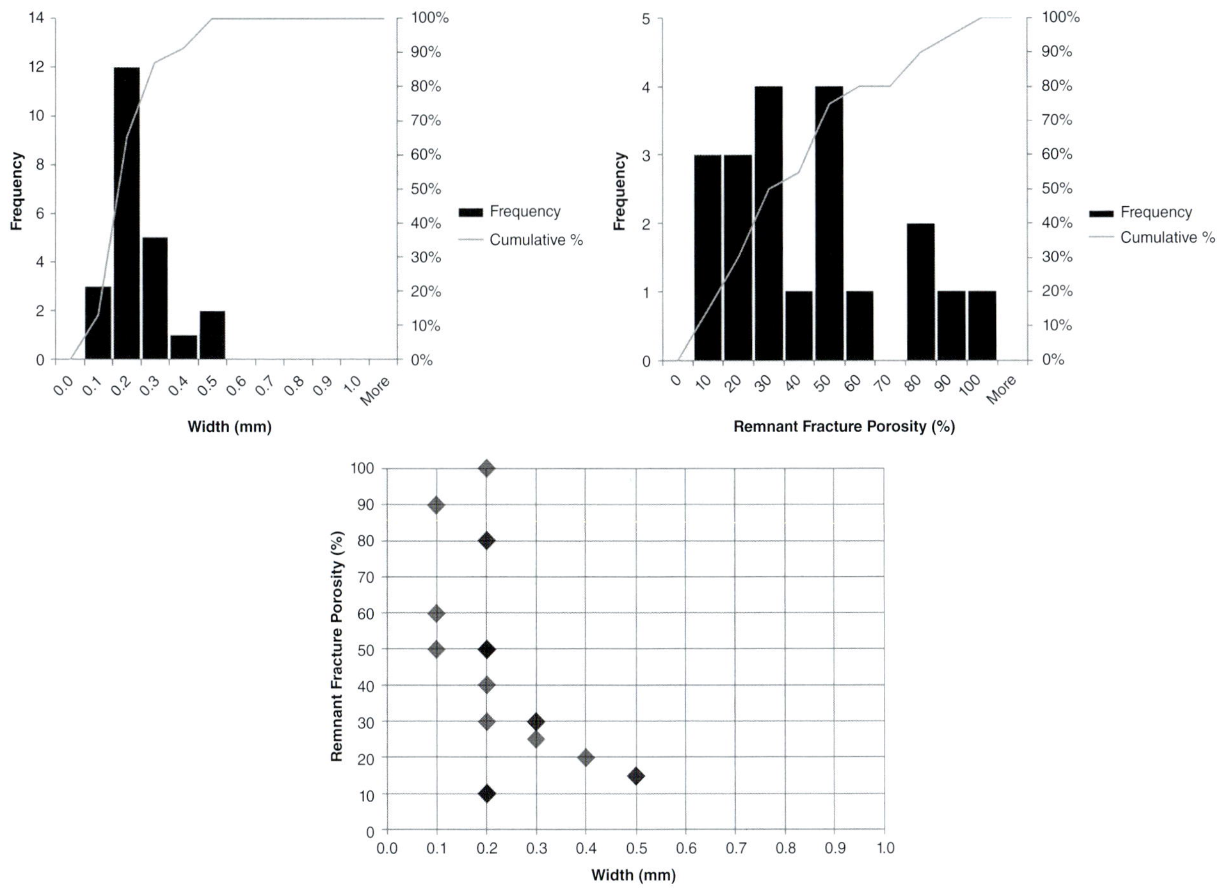

Figure 2.22 *Top left*: a histogram of the widths of bed-normal extension fractures sampled by 150 ft of four-inch diameter vertical core in a calcareous shale (n = 23; min 0.10, max 0.5, ave 0.24 mm). Nelson (2020) suggests that a bimodal fracture-width histogram can indicate the presence of two fracture sets where information on strike or other criteria that would distinguish the sets are lacking. *Top right*: a histogram of the remnant fracture porosities in these fractures (n = 20; min 10%, max 100%, ave 42.3%). Fracture widths are log-normally distributed and are controlled by mechanics; fracture apertures are irregularly distributed and controlled by geochemistry. *Bottom*: a cross plot shows a possible correlation between decreasing remnant fracture porosity and increasing width. (Unpublished data.)

in carbonates may have irregular distributions on histograms and cross plots due to this post-fracturing chemical activity, and what may have originally been a log-normal width distribution can be modified during diagenesis.

The width and remnant-aperture populations of shear fractures (Figure 2.26) can have distributions that are similar to the log-normal distributions of extension fractures. The offset asperities along a shear fracture create an irregular width that is difficult to capture with a single measurement, but these apertures are commonly characterized this way for expediency.

Fracture apertures contribute minimally to the bulk porosity of most reservoirs, and the width-cubed relationship between width and permeability is tenuous at best since fracture apertures rarely consist of slots with parallel walls. Nevertheless, aperture is still the primary control on fracture permeability, and the degree to which fractures enhance reservoir permeability should be assessed semi-quantitatively by measuring fracture widths and remnant apertures where possible. More quantitative fracture permeabilities can be measured in the laboratory, either by testing fractured whole-core samples (prior to slabbing) or by plugging fractures with the fracture planes oriented parallel to the axis of the plug. Only the narrower, more tightly cemented fractures hold together during plugging with a rotary plugging bit (less aggressive water-jet plugging techniques have recently become available), meaning that the tests may indicate only a minimum in the degree of fracture-related system permeability enhancement. The fractured plugs should be tested under restored-state stress and saturation conditions. The sensitivity of fracture permeability to changes in stress can also be measured by testing the plugs under different confining pressures. Control samples of the unfractured rock adjacent to the fracture plugs should be tested at the same time.

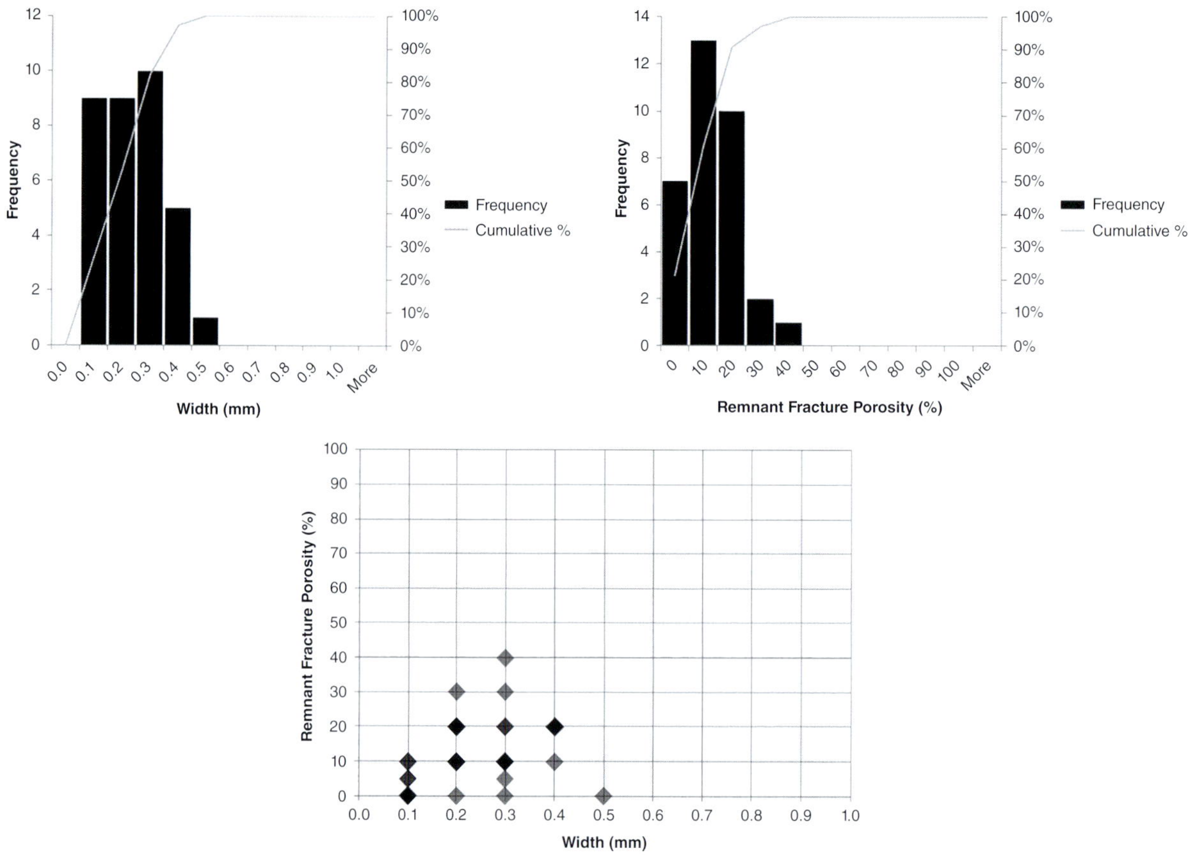

Figure 2.23 *Top left*: a histogram of the widths of bed-normal extension fractures sampled by 153 ft of four-inch vertical core in a similar calcareous shale cut from a bed deeper in the same well as that shown by Figure 2.22 (n = 34; min 0.10, max 0.5, ave 0.24 mm). *Top right*: a histogram of the remnant fracture porosities in the same fractures (n = 33; min 0%, max 40%, ave 12.6%). *Bottom*: a cross plot comparing fracture widths to the remnant fracture porosities for this fracture set. (Unpublished data.) Both fracture widths and the remnant porosities are approximately log-normally distributed.

Thin sections are another way of assessing fracture apertures and their potential effect on permeability. A suite of thin sections cut specifically for fracture study can provide information on remnant void space within the mineralization of smaller fractures. The orientation of the thin sections relative to the core axis and the fracture planes should be recorded when the rock chip is cut for each thin section.

The ability of fluid to flow along a mineralized fracture may be greater than, equal to, or less than that of flow through the matrix. An absence of mineralization or incomplete mineralization will contribute to the first condition, and this is commonly the default conceptual fracture model, but that is not necessarily the most common one. Nevertheless, laboratory tests of fracture permeability suggest that reservoir fluids, especially natural gas under high pressure, can flow at rates significant to production even along narrow fractures that appear to be occluded by calcite (e.g. Lorenz et al., 1989; 2005).

If mineralized fractures have a lower fracture-parallel permeability than that of the matrix rock, then the fractures form baffles and barriers to flow, and flow is channeled within the matrix between and parallel to the fractures. Although these effects can be estimated from core parameters such as fracture width and mineralization, their actual effects can only be measured by laboratory permeability tests and engineering reservoir assessments.

Fracture surfaces also commonly provide important information on their potential effects on reservoir plumbing. The ability of fluid to flow across a fracture face and into the fracture aperture may be equal to flow through the matrix, or flow may be inhibited. The unaltered faces of unmineralized extension fractures offer no resistance to fluid flow across the plane of a fracture, and as with flow along a fracture, even a degree of mineralization may not entirely inhibit fluid flow across a fracture. However, the slickenlines and slickensides created on many shear fracture faces are metamorphic-style textures that severely reduce porosity, permeability, and fluid flow. Thin sections cut normal to a fracture face can provide insights into the potential for permeability

Figure 2.24 *Top*: a histogram of remnant fracture porosity data in carbonates may show irregular patterns due to the chemical activity of these lithologies and multiple dissolution/precipitation events. This histogram shows the remnant fracture porosities of 140 vertical extension fractures measured in 1,600 ft (490 m) of vertical core cut from the Arbuckle Dolomite in Kansas (average remnant fracture porosity is 54%, unpublished data). *Lower left*: a chemically modified extension fracture in the Eocene Pila Spi Formation, northern Iraq, shows the irregular, vuggy aperture created by dissolution along the fracture plane, finger for scale. *Lower right*: dissolution slots that followed vertical extension fractures in the Pennsylvanian Madera Limestone, New Mexico. The slots are orders of magnitude wider than the original fractures, and, like those fractures they are confined to a limestone that is underlain and overlain by muddier layers. A black pocketknife at the bottom of the left slot shows the scale.

reduction across a fracture face, but they must be cut carefully since the fracture face, at the edge of a thin section chip, can be lost during preparation.

Some fracture faces display an embayed, irregular texture that records etching and dissolution of the fracture faces. This may not affect flow from the matrix into the fracture aperture, but dissolution is commonly followed by secondary fracture-face mineralization. The faces of shear fractures, especially in carbonates, may also be coated with a layer of slickenlined clay (Figure 2.27). The clay is the residue of minor pressure-related dissolution along the fracture walls during shear, much like the insoluble residue found along stylolites. Such fractures typically have widths that are completely occluded with clay and no remnant fracture porosity within the fracture width. The clay filling should significantly inhibit flow both along and across such fractures.

Widths, apertures, and mineralization are significant parameters that must be recorded to assess the effectiveness of fractures as conduits for fluid flow and to calculate fracture volumetrics (see sections 2.13, "Case Studies in Estimating Fracture Effectiveness from Core Studies," and 3.6, "Fracture Volumetrics"). The measurable fracture parameters provide a quantitative, basic, database, but that database must be assessed with a qualitative understanding of the fracture system.

2.4.7 Fracture Spacings

Fracture spacings are a critical parameter, defining the "shape factor" or "sigma" that describes block size between fractures in an engineering model. Spacing, the distance between fracture planes measured in the direction normal to those planes, is closely related to "intensity," the number of fractures per sample length

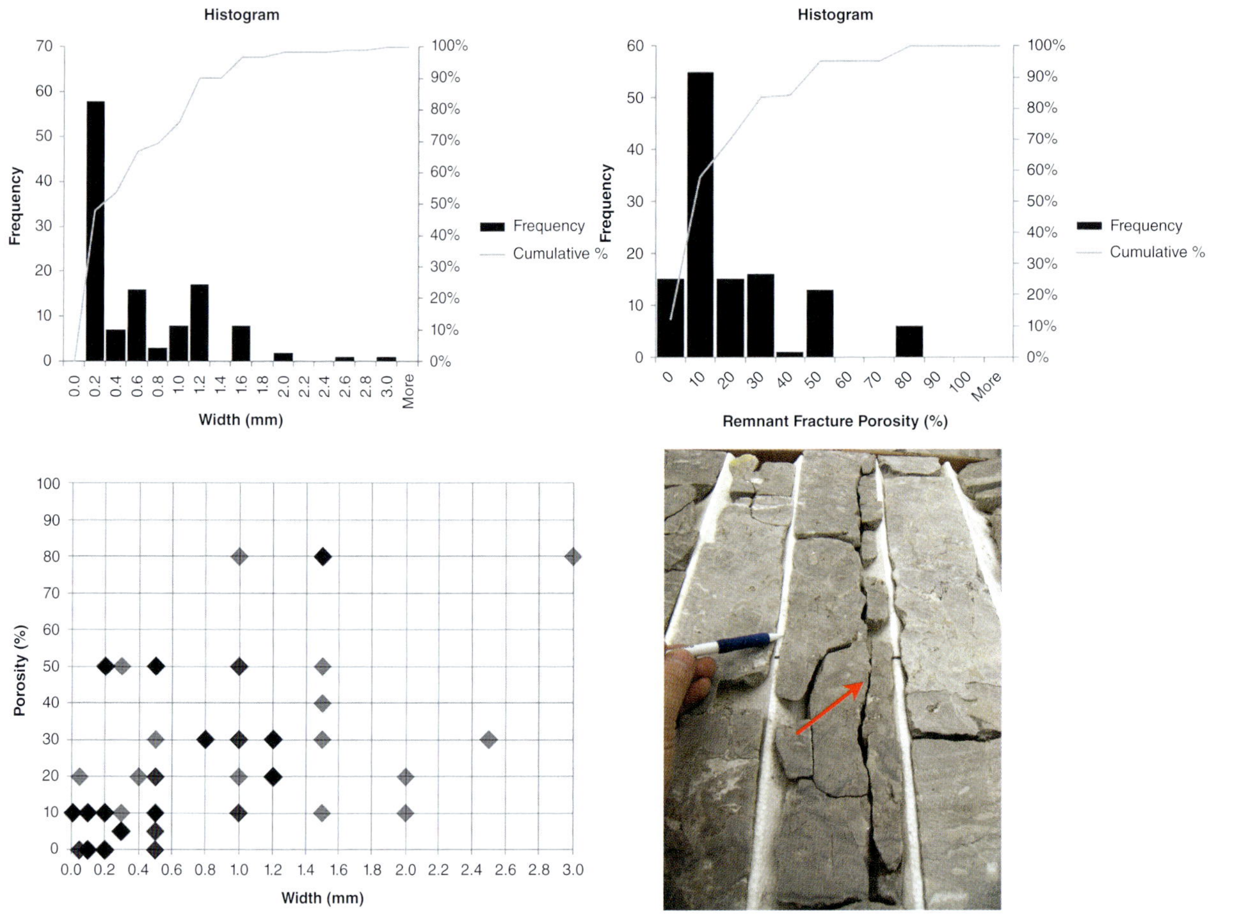

Figure 2.25 Histograms of fracture data from other carbonates can be more regular. *Top left*: histogram showing the distribution of widths for vertical extension fractures in a dolomite (n = 121, average width 0.58 mm). *Top right*: histogram showing the distribution of remnant fracture porosity within these fractures (n = 121, average 20%). *Bottom left*: cross plot of fracture widths and remnant fracture porosities (n = 121) for the same fracture population suggesting that there is no correlation but also that nearly all fractures retain some remnant fracture porosity. Darker points highlight overlapping data. *Bottom right*: photo showing one of the irregular vertical extension fractures of this dataset. (Unpublished data.)

and which can be distilled into an average spacing if the measurements are all from the same set of parallel fractures. Unfortunately, geologic fractures do not all have the same effect on permeability since the fractures of a set are not uniform in size or aperture. Spacings along a linear sample, whether an outcrop scan line or a horizontal core, capture all fractures big and small, so actual, measured fracture spacings may *not* be equivalent to the effective spacings that control fluid flow. Nor do natural-fracture systems typically consist of three, mutually orthogonal fracture sets. It is up to the geologist, engineer, and modeler together to transform measurable fracture spacings into the effective spacings that control the capacity of a reservoir to deliver hydrocarbons to a wellbore. Like other fracture parameters, fracture-spacing populations commonly have log-normal distributions with many close spacings and increasingly fewer wider spacings.

2.4.7.1 Spacings from Horizontal Core

The spacings of vertical fractures that cut normal to the axis of a horizontal core are easily measured with a tape measure. Fracture planes oriented oblique to a core axis still provide excellent spacing data if the fractures can be determined to be parallel to each other; the spacings are calculated from the apparent fracture spacings measured along the core and the intersection angle between the core-axis and the fracture planes (Figure 2.28) (e.g. Terzaghi, 1965; Lacazette 1991), and this simple calculation is commonly called the Terzaghi correction. Priest and Hudson (1976) suggested that a minimum of 200 measurements along a scan line are necessary to accurately define a fracture-spacing distribution, but core commonly captures a much smaller fracture population. The context of that dataset and the shape the histogram of spacing distribution may

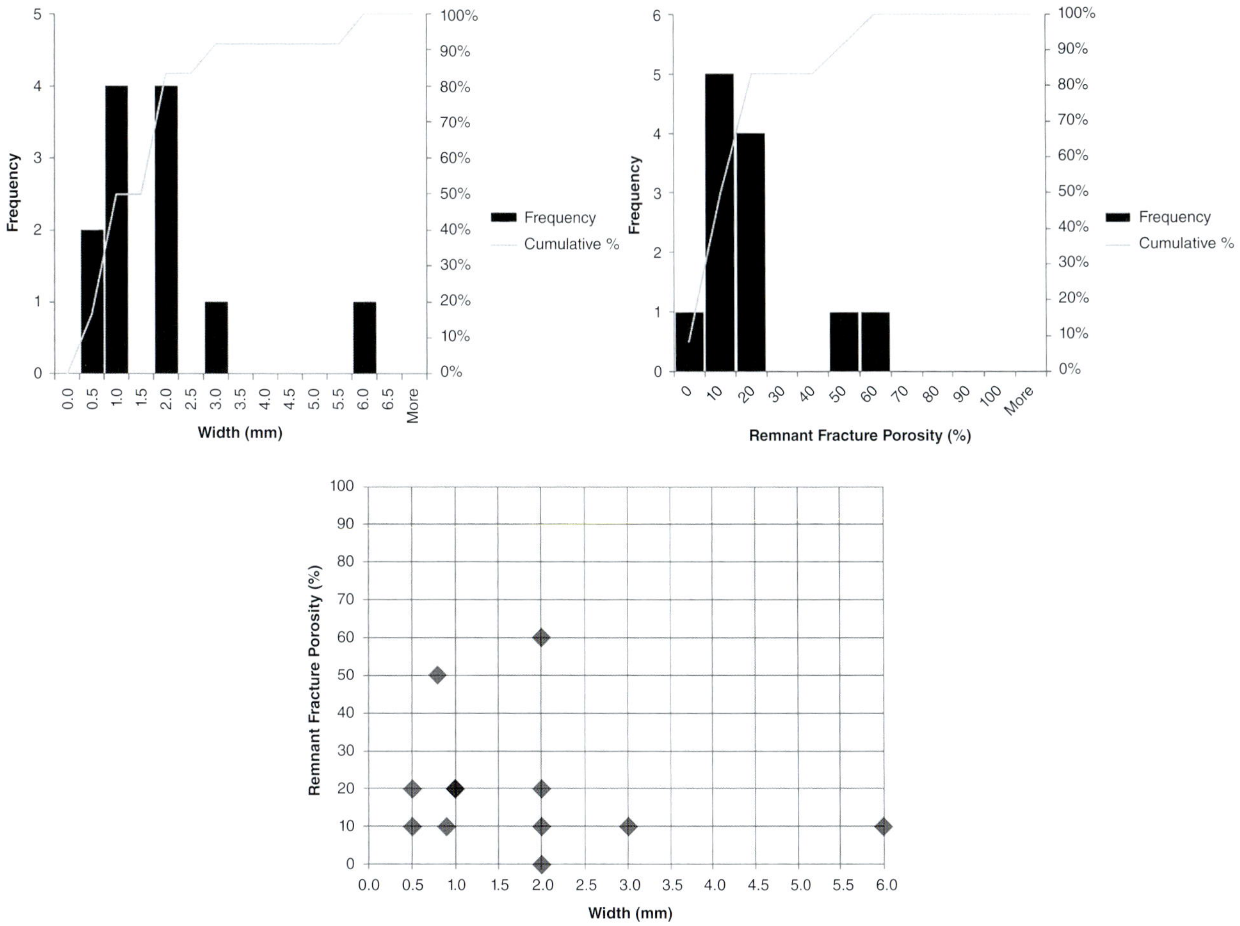

Figure 2.26 A dataset from cored shear fractures. *Top left*: a histogram of the widths of intermediate-dip-angle, dip-slip shear fractures in a limestone in 200 ft/61 m of core (n = 12, average 1.8 mm). *Top right*: a histogram of remnant fracture porosities for the same shear fractures (n = 12; average 20%). *Bottom*: a cross plot of fracture widths to remnant fracture porosities, suggesting that there is no correlation but also that nearly all fractures retain some remnant fracture porosity (n = 12). Darker points indicate overlapping data. (Unpublished data.)

Figure 2.27 Two views of a shear fracture in four-inch (10 cm) diameter core. *Left*: the planar, inclined cross section of the fracture in the core slab. *Right*: the same fracture in the equivalent butt section of the core, showing a slickenlined face with a clayey dissolution residue and patchy calcite mineralization.

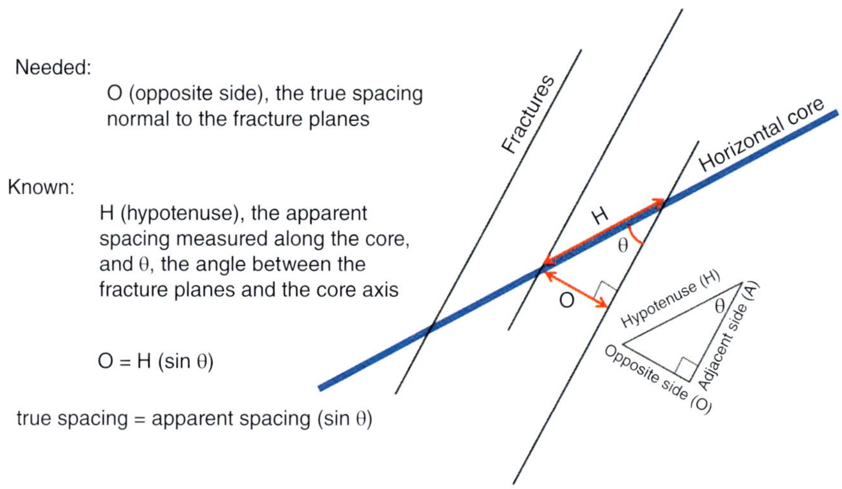

Figure 2.28 The apparent fracture spacings in a core or scan line must be geometrically corrected for the angle of intersection between the core and the fractures to obtain true spacing.

indicate whether the limited dataset is representative of the sampled fracture population.

Image logs from a horizontal well cover more of the wellbore and often provide much larger spacing datasets, but they must be used carefully since they typically cut across multiple lithologies, each with its own characteristic fracture spacing. Image logs also commonly undercount fractures (Nelson, 2001), so calibrating a log with a core run in the same wellbore enhances the value of the log.

One of the better-constrained subsurface fracture-spacing datasets comes from a Paleozoic marine shale where a near-horizontal, 3 inch (7.6 cm) diameter core followed a fractured, 2.2 in (5.6 cm) thick siliceous bed for 26 feet (7.9 m). A population of 46 parallel, bed-normal, high-angle, strata-bound extension fractures in the bed cut across the core at an angle of about 65° to the core axis, providing a unique view and quantitative definition of the fracture spacings, apertures, and heights for this set of strata-bound fractures, but providing almost no information on natural-fracture spacing or intensity in the units above or below this thin bed. The spacings range from just over 1 in (2.5 cm) to 20 in (51 cm), with an average spacing of 6 in (15 cm) and are log-normally distributed (Figure 2.29). The six-inch average fracture spacing is almost three times bed thickness.

Although well constrained, this example is smaller than reservoir scale. These data should not be extrapolated to predict fracture spacings in the thicker, adjacent beds since this thin siliceous bed is brittle and more prone to fracture than the adjacent thicker, more clay-rich reservoir beds. The thicker strata accommodated a similar amount of strain, but did so in part by fracturing and in part by ductile deformation, so fractures in these units have an average spacing that is on the order of 10 ft (3 m).

Datasets from other horizontal cores cut from unfolded strata confirm that subsurface extension-fracture populations commonly have log-normal spacing distributions; if spacing is related to bed thickness it is a loose relationship. The spacings of a different set of 46 mineralized, parallel, vertical extension fractures in deep-marine Paleozoic sandstone of the Spraberry Formation average 2.9 ft (0.88 m), approximating bed thickness (Figure 2.30). In contrast, a dataset of 31 mineralized but partially open vertical extension fractures captured by horizontal core from the Cretaceous Austin Chalk has an average spacing of 6.7 ft (2.1 m) in a 10 ft (3 m) thick bed (Figure 2.31). The average spacing in the latter core is halved when the additional parallel, narrow, completely filled extension fractures are included, doubling the fracture population.

To compare fracture development and spacing in beds of different thicknesses, Narr (1991) used a ratio of spacing to thickness (called the Fracture Spacing Index), and Gross (1993) inverted the ratio, using the ratio of layer thickness to joint spacing (called the Fracture Spacing Ratio). Both ratios normalize fracture spacing to bed thickness, and the authors' measurements show ratios that are similar for beds of different thicknesses in the studied formations, i.e. thicker beds have more widely spaced fractures. Spacing data specific to a given reservoir are not always available, so the fallback oilfield rule of thumb is often that fracture spacing is equivalent to bed thickness, and the Narr and Gross studies suggest that this is sometimes the case. However, this generalization is an expedient to be used only in the absence of data. For the example shown in Figure 2.29, the average spacing of the strata-bound fractures in the thin siliceous bed is nearly three times wider than the thickness of the host bed. More commonly, spacings are much less than thickness: the 40-ft (12 m) thick sandstone illustrated in Figure 2.37 is cut by vertical extension fractures

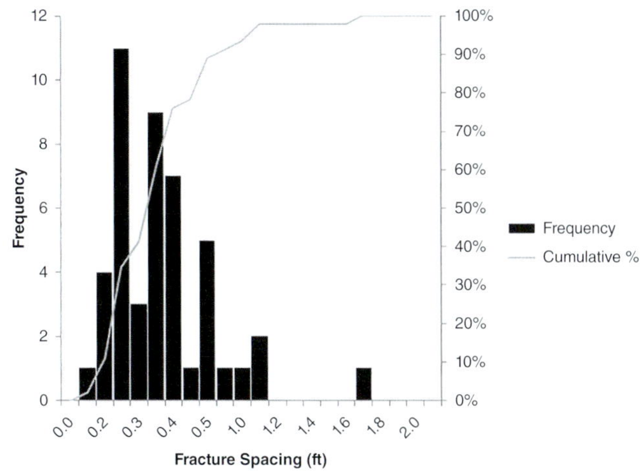

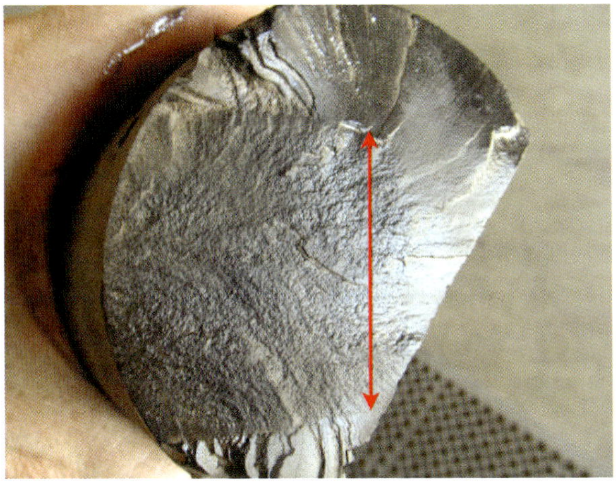

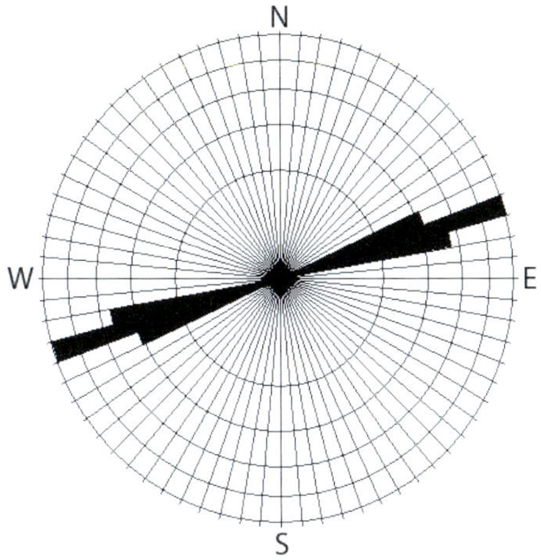

Figure 2.29 *Top left*: a histogram of true fracture spacings (i.e. spacings normal to the fracture planes, corrected for the average 65° intersection angle between the fracture planes and the core axis) for a population of 46 bed-normal extension fractures captured by 26 ft (7.9 m) of near-horizontal core. The average spacing of 6 in (15 cm) is nearly three times greater than bed thickness. *Top right*: photo of one of the cored, strata-bound extension fractures (parallel to the plane of the photo). The fracture height (red bracket) is limited to the siliceous bed but was extended into the twist-hackle zones in the muddier shale above and below the siliceous host bed during coring and handling. *Bottom*: rose plot of the strikes of the vertical extension fractures, with an average of 72°. The core was not oriented, but nearby vertical core showed that the fractures are vertical and bedding horizontal, so fracture strikes were measured relative to horizontal bedding and the known wellbore azimuth. The 180° ambiguity in determining the topside of the core was constrained by the shallow dip of bedding across the core axis. (Unpublished data.)

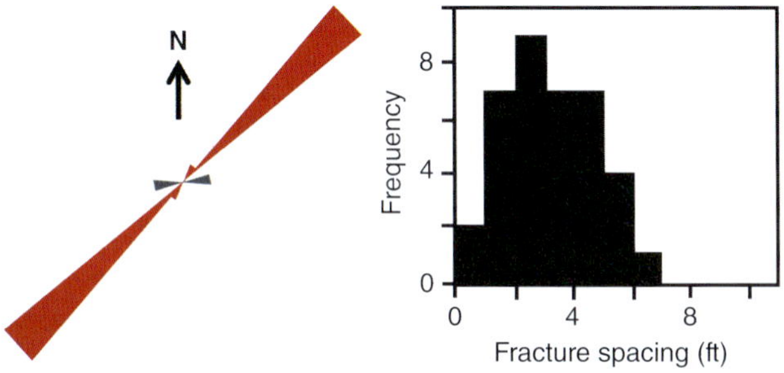

Figure 2.30 *Left*: rose plot showing the orientations of 46 vertical extension fractures in horizontal core from the Spraberry Formation in Texas. *Right*: a histogram of the spacings of these fractures normal to fracture strike; average of 2.9 ft (0.88 m) (adapted from Lorenz et al., 2002).

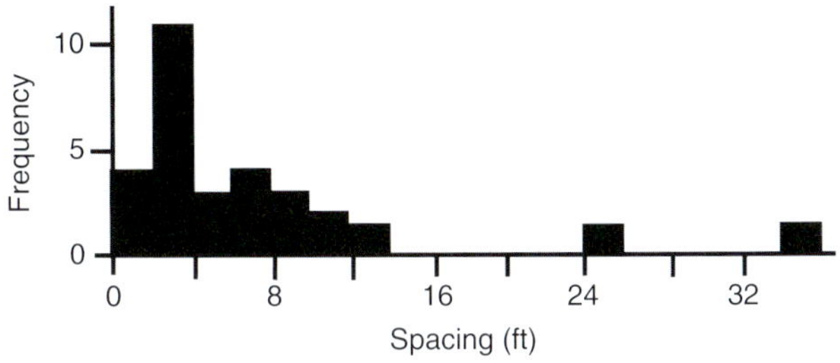

Figure 2.31 The true spacings of 31 vertical extension fractures measured in horizontal core cut from the Austin chalk in Texas. This dataset includes only the wider, incompletely mineralized fractures in the system, omitting an equal number of interspersed parallel, narrow, completely mineralized fractures since they do not contribute to reservoir porosity or permeability (unpublished data).

having an average spacing of only 3 ft (1 m). Thickness is only one of several controls on fracture spacing, and the relationship between them commonly falls apart for reservoir-scale bed thicknesses and for reservoirs composed of heterogeneous lithologies.

2.4.7.2 Spacings from Vertical Core

In contrast to the definitive spacings derived from horizontal core, the measurable spacings of vertical fractures captured by vertical core are limited by the core diameter and therefore rarely useful. Nevertheless, where fracturing is well developed, a vertical core may capture enough closely spaced vertical extension fractures to build a spacing histogram (Figure 2.32). These measurements typically represent only the closely spaced end of the spacing spectrum for that fracture set, but they commonly have a log-normal distribution and can be useful in estimating the larger, complete spectrum of spacings, i.e. if there are enough spacings to form the outline of a log-normal distribution on a histogram, additional data would probably add to the heights of the histogram bars but would probably not extend the tail much further to the right.

True spacings can sometimes be calculated where cored high-angle extension fractures are tall and slightly inclined relative to the axis of the wellbore by using the Terzaghi correction (Figure 2.33), if all measurements can be demonstrated or reasonably assumed to come from one set of parallel fractures.

2.4.7.3 Converting Vertical Observations to Horizontal Fracture Spacings

One would also like to be able to convert the limited data provided by vertical cores to lateral spacing dimensions for populations where spacings that are greater than core diameter. However, capturing a vertical fracture with a vertical core is not an operation with a high-probability of success (Figure 2.34) (Lorenz, 1992). The absence of fractures in a core or image log does *not* imply that a formation is not fractured, whereas the presence of *any* natural fractures suggests that a fracture system is present in the subsurface. Comparisons of fracture populations in paired vertical and horizontal cores illustrate this statement: in one example, 162 feet (49 m) of vertical sandstone core contained one natural fracture in the same interval where deviated core documented the average fracture spacing to be 3 ft (0.9 m) (Lorenz and Hill, 1994).

Narr (1996) offered two formulae by which data from vertical fractures intersected by vertical core could be used to calculate an average lateral fracture spacing, if the core captured an undefined representative fracture population, and if a single set of vertical, bed-normal, parallel-striking fractures with consistent apertures is present in the reservoir.

The two formulae, used alternately depending on whether or not fracture-aperture data are available, are:

1. Average spacing = Average fracture aperture x Core diameter x Feet of fracture-prone lithology, divided

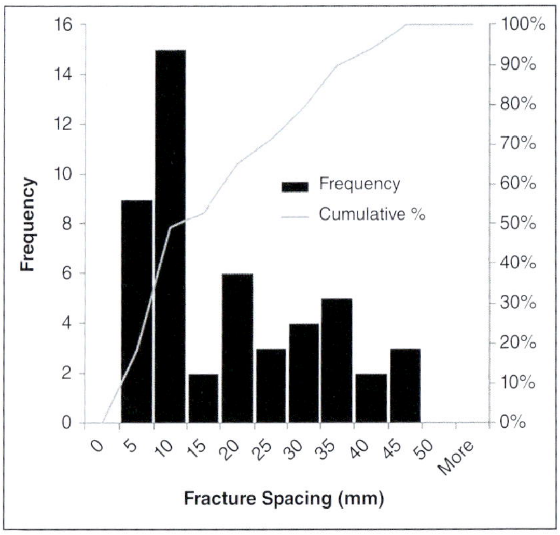

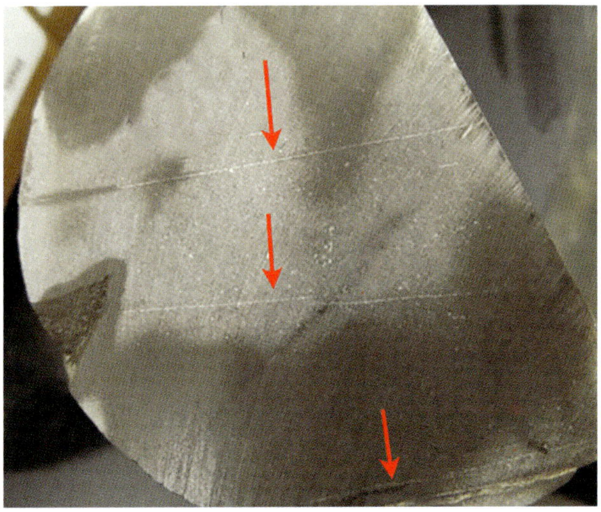

Figure 2.32 *Left*: histogram showing spacings measured between 49 pairs of vertical, parallel, narrow, calcite-mineralized extension fractures captured by a vertical core (unpublished data). *Right*: an example from this core showing two measurable spacings between three fractures (arrows). The maximum captured spacing is only 2 in (49 mm), but the distribution is still log-normal. Since the dataset is relatively large, it may in fact represent much of the range of spacings within a well-developed, closely spaced set of fractures. No examples of intersecting fractures were cored and all fractures probably comprise one set of a parallel-fracture system, yet because the core is not oriented and because the core consists of numerous unlockable intervals, nothing in the dataset precludes the possibility that spacings between fractures of more than one set were measured.

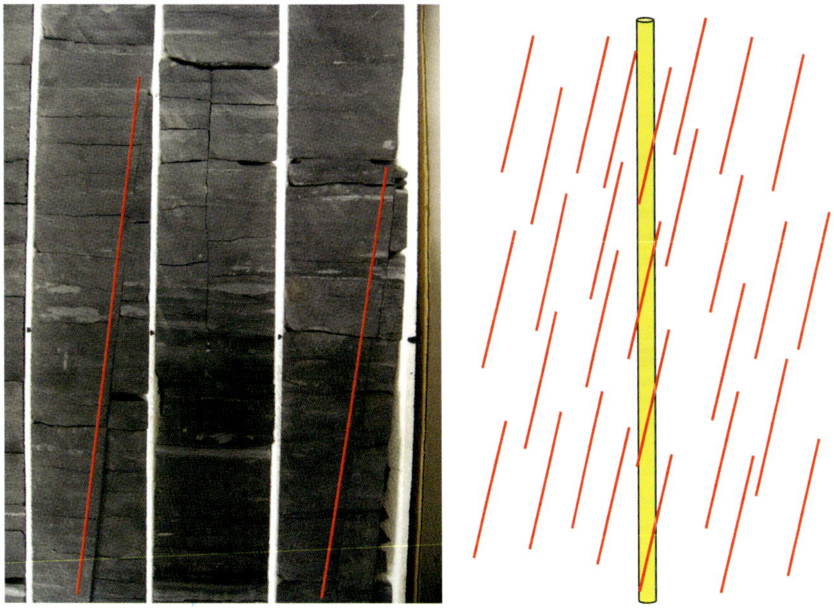

Figure 2.33 An example where 12 near-vertical, calcite-mineralized extension fractures up to 8 ft/2.4 m tall in a deep-marine shale were captured by 180 ft (55 m) of core in a near-vertical well. Several of the fractures could be demonstrated to be parallel to each other in continuous sections of core, and all 12 of the fractures had similar intersection angles with both bedding and the core axis, so the fractures were plausibly assumed to form a single set of parallel fractures. The fractures cross all bedding planes. The average vertical (apparent) spacing along the measured depth of the core is 16.3 ft/5.0 m, which corrects geometrically to an average true spacing, normal to the fracture planes, of 1.4 ft (0.43 m). *Left*: inclined extension fractures (parallel to the red lines) as exposed in core in the slab box (the vertical fracture in the middle slab piece is a drilling-induced fracture). *Right*: a conceptual model of this fracture set, with an exaggerated angle between the fracture planes and the core axis.

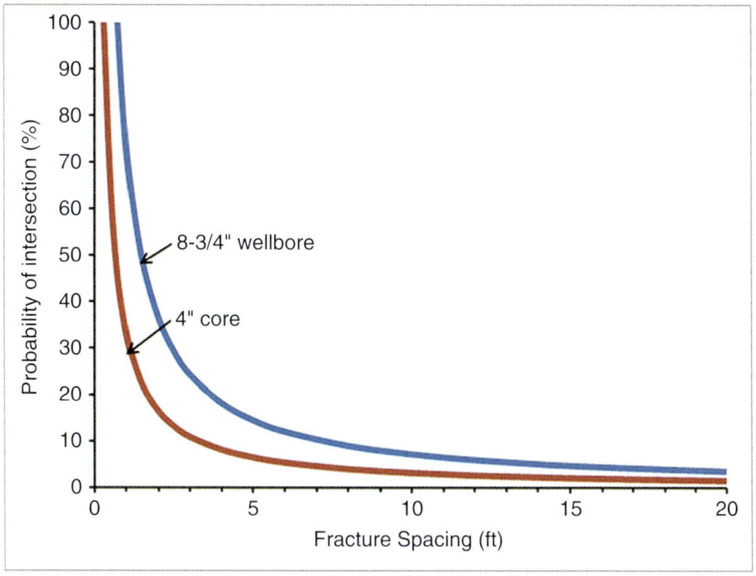

Figure 2.34 Plots of the probability of intersecting vertical fractures with 4-inch (10 cm) diameter vertical core, or with an image log run in a 8.75-inch (22 cm) diameter vertical wellbore, show that the chance of capturing a fracture does not rise above 50% until fractures are spaced less than 8 in (20 cm) or 17.5 in (44 cm) apart, respectively (adapted from Lorenz, 1992).

by the Sum of fracture apertures x Sum of fracture heights.

2. Average spacing = Core diameter x Feet of fracture-prone lithology, divided by the Sum of the fracture heights.

Although seeming to mix unrelated data, the formulae are mathematically correct and can be useful in cores that do not contain multiple intersecting fracture sets or inclined fractures. Given the variety of fracture types, dimensions, and orientations, as well as the range of possible intersections between fractures and wellbores, converting one-dimensional vertical fracture data to lateral fracture spacing is not a trivial problem.

2.4.7.4 Spacings of Inclined and Shear Fractures

Shear fractures, which are likely to be inclined relative to the vertical and to bedding, commonly form as intersecting conjugate pairs, so spacings should be measured normal to the fracture planes of one subset of the pair at a time. Attempts to measure spacings between adjacent fractures of both subsets runs into the problem that the spacing between a fracture of one subset and a fracture of the complementary subset depends on how close to their intersection the measurement is taken. Nevertheless, measurements taken between the parallel shear fractures within each subset can be systematic and can show distributions similar to those of extension fractures (Figure 2.35).

Fracture spacings are also dependent on position relative to a structure; for example, if the radius of curvature of a fold is greater at its crest than on the flanks, closer spacings are to be expected at the crest. Fracture spacing commonly also decreases with proximity to a fault (Figure 2.36), particularly in the hanging wall of normal faults. (e.g. Nelson, 2001),

2.4.7.5 Uses of Spacings

Fracture spacings are easily quantified, but the definitive, measurable data are not always or even usually directly applicable to modeled reservoirs. The more closely spaced and clustered fractures of a system are effectively single conduits for fluid flow in a reservoir, so the spacing value most applicable to a fractured-reservoir fluid-flow model may not be the mathematical average of fracture spacing measured in a core but rather the average spacing of fracture groups or clusters.

In one example (Figure 2.37), in the same sandstone reservoir where horizontal core shows that the average fracture spacing is 3 ft (0.9 m), gas shows in the mud log suggest that the *effective* fracture spacing is on the order of 6–8 ft (1.8–2.4 m). Moreover, the reservoir engineer who modeled the production capacity of the reservoir indicated that best results were obtained using a model where fracture spacing is on the order of 10 ft (3 m). (Multiwell Experiment Project Groups, multiple dates).

2.4.8 Measuring and Using Fracture Strikes

A population of fracture strikes from a core or image log provides a wealth of information including insights into the likely orientation and aspect ratio of drainage ellipses in a fracture-controlled reservoir permeability system. In conjunction with knowledge of the in situ stress system, fracture strike also contributes to predictions of the interactions between a natural fracture system and hydraulic stimulation fractures. Methods for determining stress orientations from core were summarized by Warpinski et al. (1993a), and one of the easiest and most reliable methods turns out to be the use of induced petal and centerline fractures.

Fracture strikes relative to north are most desirable and can be obtained directly from oriented core, indirectly by orienting a core with an image log, or again indirectly by using the orientation references provided by stress-controlled petal/centerline fractures.

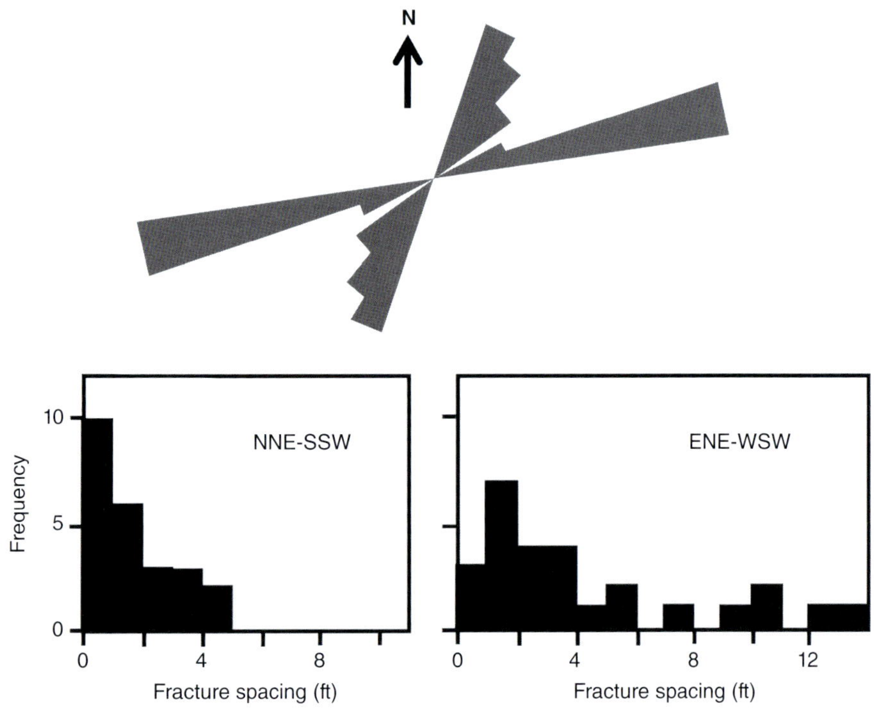

Figure 2.35 *Top*: a rose plot of the strikes of 57 vertical, strike-slip shear fractures measured in horizontal core from sandstones of the Spraberry Formation, Texas. *Lower left*: a histogram of the 19 measurable fracture-normal spacings for NNE-SSW striking shear fractures (average 1.6 ft/0.5 m). *Lower right*: a histogram of the 24 measurable fracture-normal spacings for ENE-WSW striking shear fractures (average spacing 3.8 ft/1.2 m; reprinted from Lorenz et al., 2002, with permission from AAPG, whose permission is required for further use).

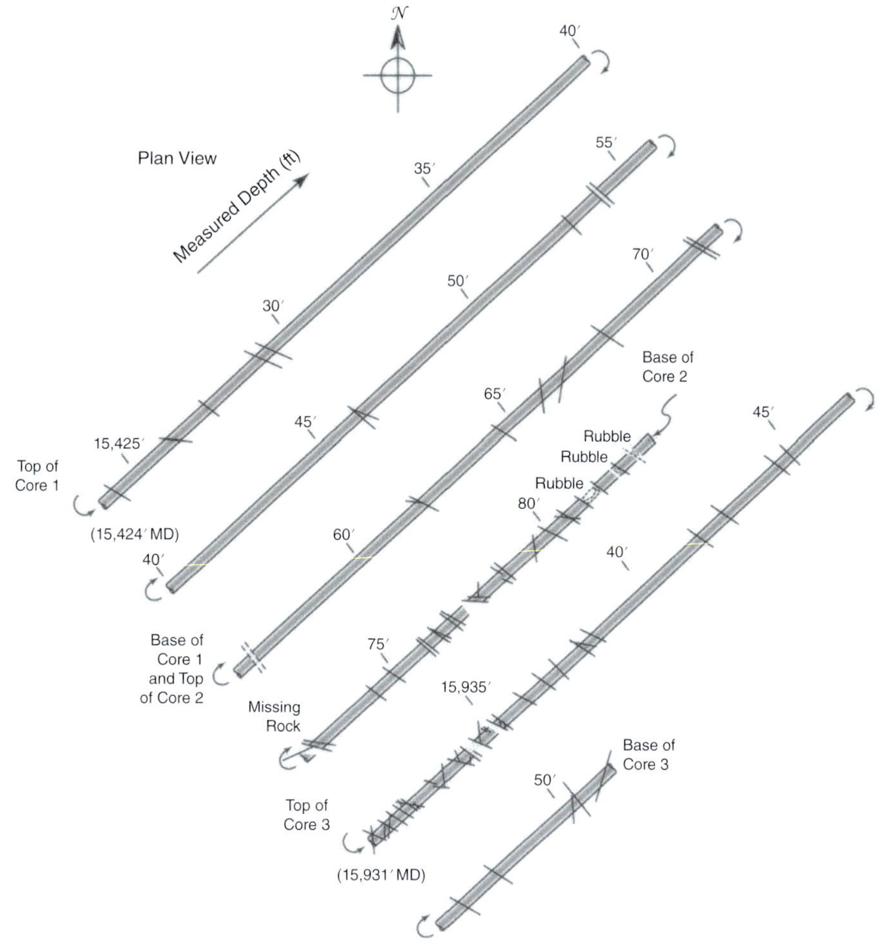

Figure 2.36 Map view of fracture locations and strikes along approximately 80 ft (24 m) of horizontal core cut from the Cretaceous, marine, Frontier Sandstone in a faulted structural setting in Wyoming. This core captured a system of shear fractures and shear-reactivated extension fractures. Some of the shears show associated fault breccia and horizontal to oblique slickenlines, and the core is more heavily fractured near the rubble zones in the core that the associated image logs show to be small faults. Fracture spacing decreases irregularly towards the faults (from Lorenz et al., 2005).

Natural fracture strikes relative to each other and relative to the in situ stresses are important pieces of information that can often be obtained even from an unoriented core, so a fracture analyst should log all fracture intersection angles captured by a core. Multiple natural-fracture pairs with consistent, oblique intersection angles are evidence that more than one fracture set exists in the subsurface, and if the two sets are equally effective in conducting fluids, drainage is likely to be relatively isotropic.

If on the other hand, the measured angles between fracture pairs are all 0°, i.e. all fractures of the pairs are parallel, it may be that only one set of parallel fractures is present in the reservoir, creating highly elliptical drainage. Such a dataset does not preclude the possibility of two sets of intersecting fractures, each developed well enough to exhibit parallel fractures in any given piece of core, so the ambiguity between the two possibilities must be resolved, commonly by using fracture orientations relative to the stress-reference induced fractures.

For example, the two sets of an intersecting fracture pair should have different strikes relative to the induced fractures, which typically have consistent orientations and provide an orientation reference even if the orientation relative to north is unknown: if multiple measurements show natural fractures striking both 30° clockwise and 60° counterclockwise to petal fractures then there are two sets of natural fractures in the reservoir and they have orthogonal strikes. Since petal fractures strike parallel to the maximum horizontal in situ compressive stress, both sets of natural fractures are oblique to that stress and are likely to accommodate some degree of strike-slip shear during production even if they are extension fractures. Other fracture planes, oriented parallel or normal to the maximum horizontal compressive stress may close without shear during production. Because the petal and centerline reference fractures strike parallel to the maximum horizontal compressive stress, natural fracture strikes relative to the petal fractures will also help the analyst assess whether a hydraulic stimulation fracture will cut across, oblique, or parallel to a natural fracture system.

2.4.8.1 Measuring Fracture Strikes in Vertical Core

Circular protractors are the tool of choice for measuring fracture strikes, and can be machined out of plastic or metal at your friendly machine shop, or perhaps

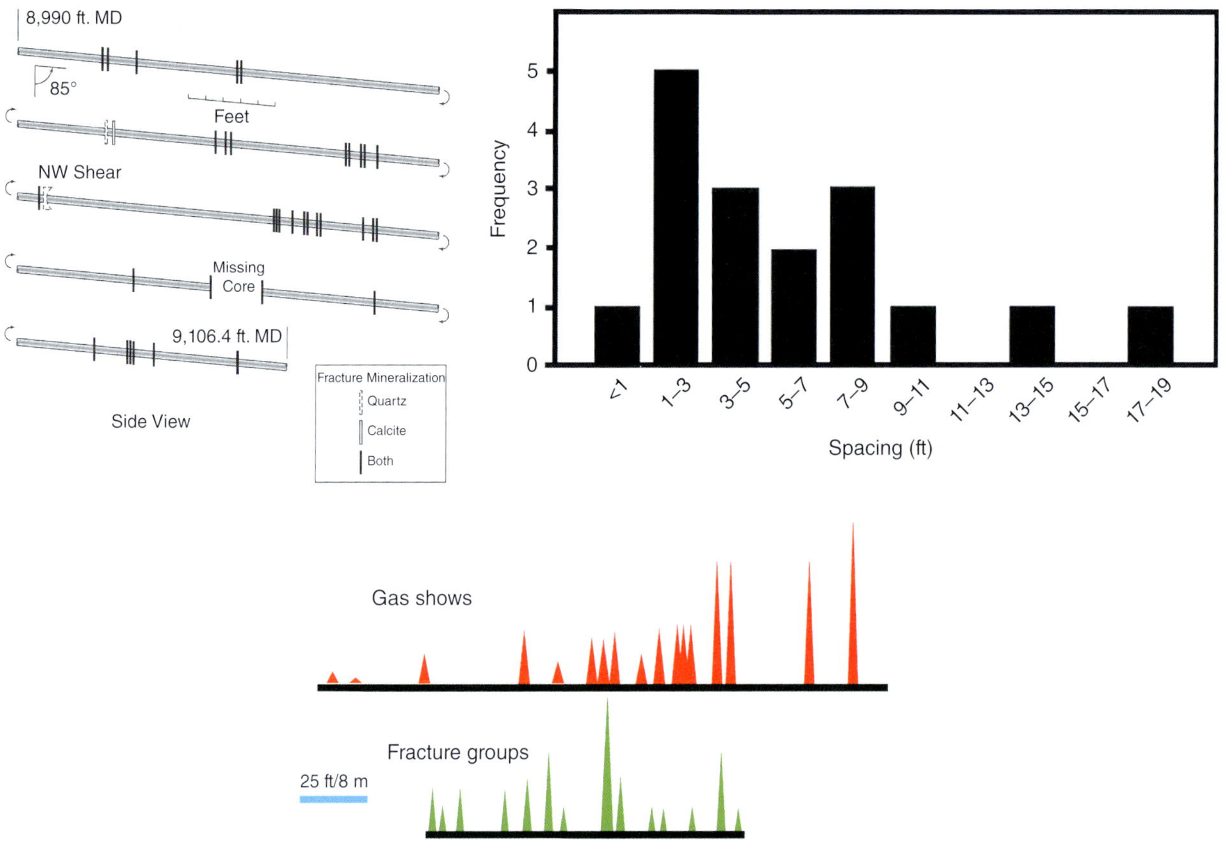

Figure 2.37 *Top left*: side view of 117 ft (36 m) of near-horizontal core from the Cretaceous, marine, Cozzette Sandstone in Colorado, and the locations of 36 cored vertical, partially mineralized extension fractures that strike WNW-ESE, nearly normal to the core axis (from Lorenz and Hill, 1994). *Top right*: a histogram of the measured fracture spacings in this core. Average spacing is 3 ft (0.9 m), much less than the 40 ft (12 m) thickness of the host sandstone. Spacings are not uniformly distributed, and clusters of closely spaced fractures are effectively single conduits at the scale of a reservoir model. *Bottom*: the lower bar shows the locations of the cored fracture groups from the same core (the triangle size is roughly proportional to number of fractures that comprise the group, ranging from one to seven) and the upper bar shows the spacings of gas shows in the mud log of the uncored offset lateral well (triangle height is roughly proportional to the gas units in the show, ranging from 40 to 800), showing that gas shows have spacings that are similar to the fracture *groups* in the core (adapted from Lorenz and Hill, 1992).

constructed with a 3D printer. The circular protractors we use were designed by Jim Clark back in the 1980s, and have an inside diameter that is 1/16 inch (1.5 mm) greater than the diameter of the core being logged. They are marked 0°–360° in 10° increments (Figure 2.38), with degrees increasing clockwise on one side for work when looking downhole, and increasing counterclockwise on the other side for situations requiring work on an inverted core piece (the irregularities of fracture exposures and core shapes often make it more convenient to measure fractures when the core is being held upside-down).

The relative strike of two fractures, however, is often easy enough to measure with a simple carpenter's protractor when both fractures are vertical and exposed on a horizontal section cut across a vertical core (Figure 2.39). The logger needs to be more creative when one or both fractures are exposed only on the sides of a core, or when a fracture is inclined relative to the core axis, and that is when the circular protractor becomes most useful. Descriptions of all possible core and fracture geometries would be tedious, but the idea is to visualize the two fracture planes in three dimensions and measure their intersection angle, or the intersection angle where the fracture planes are projected to a hypothetical intersection outside the core volume.

Commonly, particularly where a fracture does not split a core directly in half, it is necessary to project the fracture plane to the middle of the core to get a fracture strike, or to project a line normal to fracture plane from the midpoint of a fracture out to the core surface, and, using a circular protractor, compare that to the position of the midpoint line of another fracture plane. If a fracture is inclined relative to the core axis, the upper or lower apex of the arc where the fracture plane intersects the cylindrical core surface provides the midpoint of the

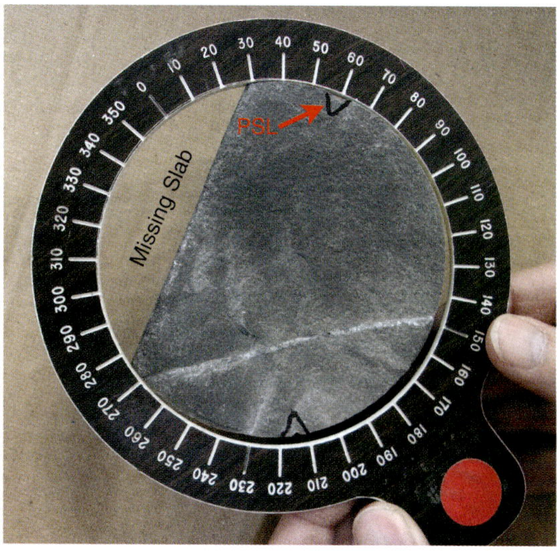

Figure 2.38 Using a circular protractor. *Left*: looking downhole (degrees on the protractor increase clockwise). Placement of the 0° position on the protractor depends on the orientation reference being used; it may be relative to the Principal Scribe Line on an oriented core (the PSL, see Figure 2.52), to the strike of an induced petal fracture, or to the strike of another natural fracture. Here, the orientation survey indicated that the PSL is oriented at 60° relative to north so 60° on the protractor is placed next to the PSL on the core. The fracture strikes approximately 120°–300°. *Right*: holding the core upside-down so that the view is uphole is sometimes necessary because the fractures are not exposed on the top end of the core piece. This view shows the underside of the same protractor; note that the degrees increase in the *counter*clockwise direction when looking uphole. A color differentiation for the two sides of the protractor helps prevent use of the wrong side, easy to do if they are the same color.

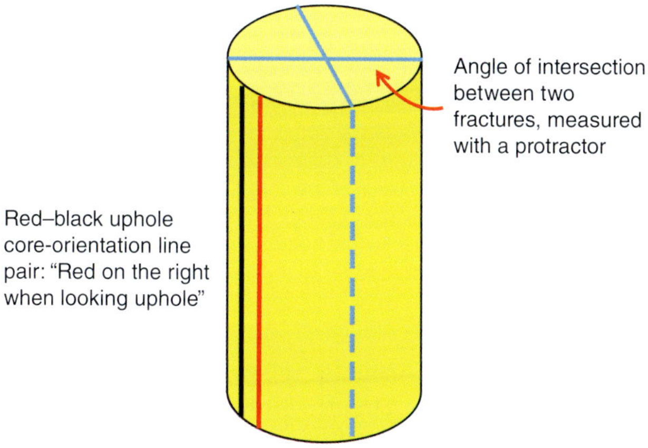

Figure 2.39 Measuring the relative strike/intersection angle between two fractures is straight forward when they are both vertical and exposed on the end of a vertical core.

fracture plane and can be used in the same way: the angle between the apex of one fracture and the next is the same as the intersection angle between the two fracture planes (Figure 2.40).

Fractures have bidirectional strikes relative to north, so half of a rose plot defines their strikes although full-circle plots are more common and show their full bidirectional nature. However, when depicting *relative* fracture strikes, only a quarter of the circular rose plot is needed since typically an analyst measures only the acute intersection angle between fractures (Figure 2.41). If north is not known, 0° on a fracture-intersection rose plot is arbitrarily taken as the strike of one of the two fractures and is *not* north.

If one fracture set is well developed and widely distributed in a core, it may provide an orientation reference against which other fractures can be measured. For example, a pervasive fabric of hairline fractures or one consisting of petal and centerline fractures may dominate a core (Figure 2.42). In this case, however, other fractures need to be described not only in terms of their intersection angle to the reference fractures but also in terms of whether the acute angle is clockwise or counterclockwise relative to the reference. Full-circle plots are best for depicting these relationships, and the 0° position on the rose plot depicts the strike of the reference fractures.

If two pieces of core each contain a fracture, the strikes of the fractures relative to each other can only be determined if the core is continuous between them or if the two fracture orientations can be measured relative to a common reference. In oriented core that reference is the Principal Scribe Line, described in Section 2.6, "Oriented Core," but a Master Orientation Line reference, as described in Section 2.3.6, "Making and Using a Master Orientation Line," can also be marked on an unoriented core, with the two fractures being measured relative to the orientation reference and then related to each other via that reference.

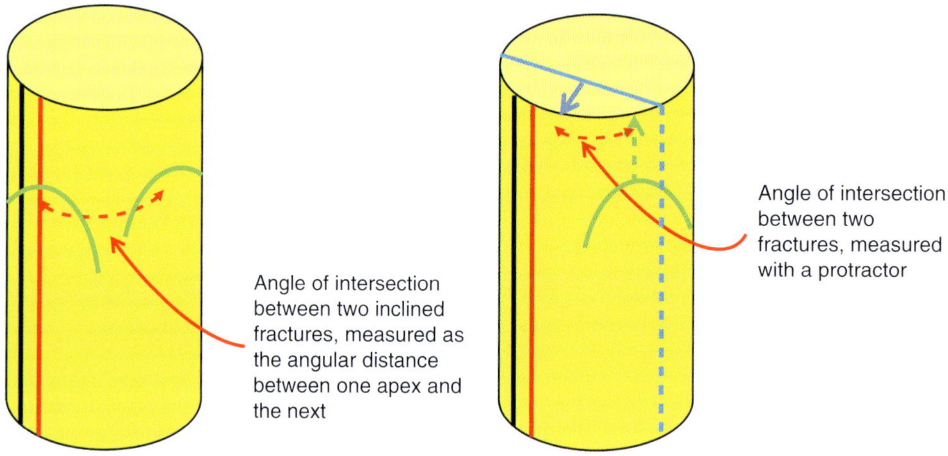

Figure 2.40 Measuring intersection angles between fractures that are not exposed on the end of a core or that do not cut across the middle of the core takes some thought. Often it is easiest to measure strikes by using the apex of a fracture that exits the side of a core, or by using a line projected out to the core surface from the middle of the fracture at 90° to the fracture plane, remembering that these more-easily used strike references are 90° to actual fracture strike. *Top*: the position of the apex of a fracture where it exits the core, as for both the calcite-mineralized natural fracture and the unmineralized petal fracture system below it, provides an orientation that is 90° to the fracture strike. In this case both fractures exit the core at about the same position on the core surface, indicating that the natural and induced fracture planes are nearly parallel to each other. Planar fractures that enter and exit the core appear as parabola on the cylindrical core surface. *Lower left*: the two sketched fractures (green arcs) intersect at nearly 90°. *Lower right*: only one fracture (blue) is exposed on the end of the core, but its intersection with the fracture depicted by the green arc can be determined by measuring the angular distance between the apex of the green arc and a line drawn 90° to the middle of the blue fracture plane.

2.4.8.2 Measuring Fracture Strikes in Deviated or Horizontal Cores

Fracture strikes in horizontal cores are calculated by measuring the angle between the core axis and the fracture plane, provided that one knows both the wellbore azimuth for the cored interval and the in situ rotational orientation of the core around its axis relative to up. The in situ orientation of the core in the hole is not always obvious, but a fracture analyst needs to determine the "top", "high", or "dorsal" side of the core, otherwise the orientations of cored fractures are unconstrained (Figure 2.43). There are also several subtle but important geometric considerations for slabbing a horizontal core to optimize the collection of fracture data as described in Appendix 2.C.

High-side determinations must be made for each interval of core between breaks in the core continuity such as rubble zones, spinoffs, and sequential core runs. Once determined, the high side is commonly marked with a blue line on the top of the core, but experience suggests that previously marked blue lines on archived core should be used cautiously.

If a horizontal core has not been oriented and there is no correlative image log, other techniques can sometimes be used to orient horizontal and deviated cores. Bedding can provide a reference if the core cuts obliquely

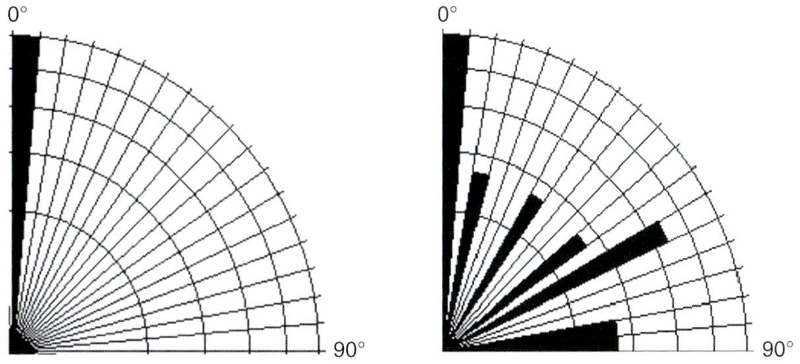

Figure 2.41 *Left*: an example of a plot of 126 pairs of vertical extension fractures in a core, showing that the relative strikes of the two fractures of each pair are parallel. This strongly suggests that the fractures form a single set of parallel fractures in the reservoir; an alternate, possible but less likely interpretation is that two or more fracture sets each display parallelism within the set and that none of the intersections were captured in the core. If they are indeed all parallel, this fracture system should create highly anisotropic reservoir drainage. *Right*: a plot of the intersection angles for ten shear-fracture pairs in a core, ranging from parallel to orthogonal. This fracture system will be well interconnected, and if all fractures are open and effective it should create a relatively isotropic drainage pattern in the reservoir. (Unpublished data.)

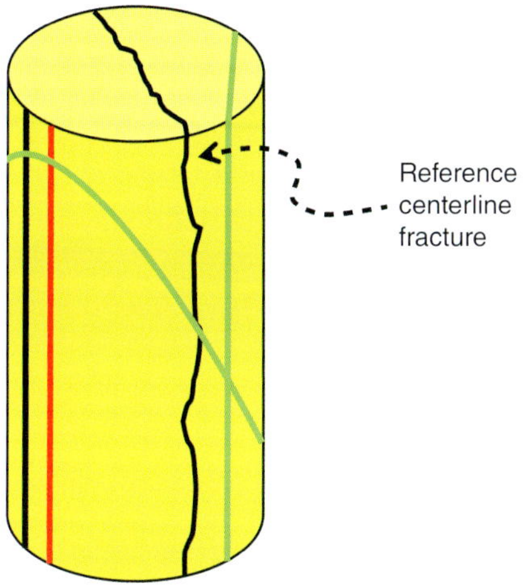

Figure 2.42 Two natural fractures (green) strike about 40° clockwise and 45° counterclockwise to the reference provided by an induced centerline fracture that extends along the axis of the vertical core. If only the angles of intersection between the induced and natural fractures are recorded and not their direction (clockwise or counterclockwise), the two fractures could be misinterpreted as being nearly parallel to each other when plotting relative strike on a rose plot, or when reading the intersection-angle data from a numerical data table.

across a horizontally bedded formation. Beds indicate the horizontal plane if they are indeed flat-lying in the subsurface, but there is still a 180° ambiguity between which side of the bedding is stratigraphic up, and this ambiguity must be resolved since otherwise there are two possible fracture strikes (Figures 2.44, 2.45).

Geopetal sedimentary structures such as ripple marks or graded bedding can provide information on the up direction of a core, but commonly bedding cuts at an angle to the core axis and this is a more consistent indicator of the high and low sides of the core

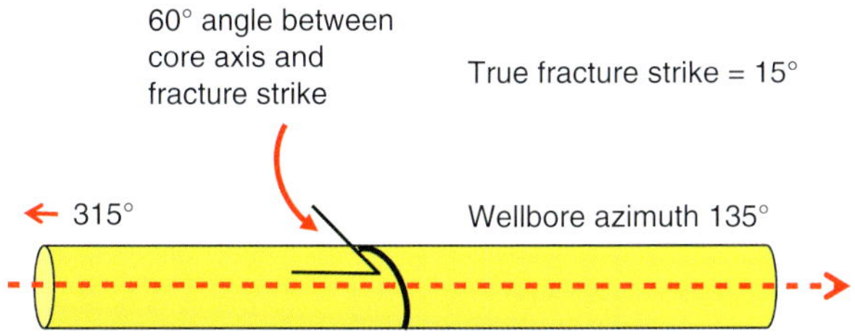

Figure 2.43 Plan view, looking down on the high side of a horizontal core: The strike of the black fracture can be determined to be 15°–195 ° by measuring the 60° angle between the fracture plane and the core axis, and combining that with the wellbore azimuth, known from the wellbore deviation survey to be 135°.

(Figures 2.46, 2.47). One must still decide whether the core cut up- or down-section across bedding, which depends on both the wellbore deviation from vertical and the dip of bedding. Use the relationship with caution since the relative geometry of the wellbore and bedding may change if bedding undulates along a wellbore or if the wellbore porpoises gently through the stratigraphic section.

2.5 New Core vs. Archived Core

Core newly removed from a core barrel provides the most complete fracture database, but usable fracture data can be obtained from core in almost any condition. Robustness and completeness of the fracture data are degraded each time the core is handled, processed, sampled, reboxed, stored, etc., and in some cases fracture logging may be limited to collecting data on fracture type, mineralization, and perhaps aperture and approximate depth since evidence for fracture heights, spacings,

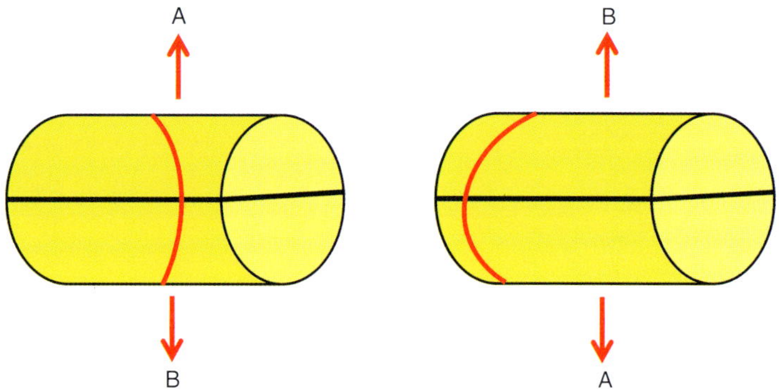

Figure 2.44 If the cored bedding is parallel to the axis of a horizontal core there are two possible fracture strikes depending on which side of the bedding, A or B, is stratigraphic up, unless the fracture strikes normal to the core axis.

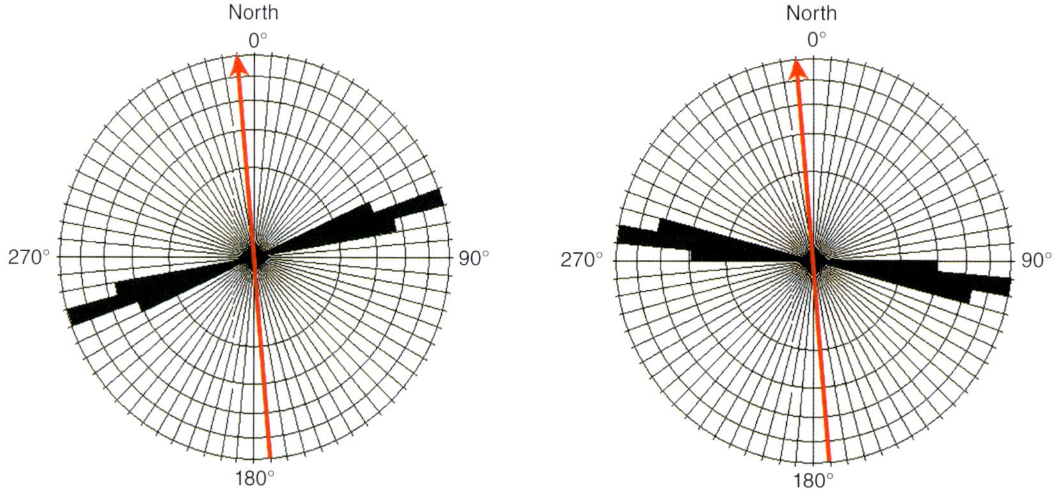

Figure 2.45 Bedding and 47 parallel, vertical extension fractures were captured by a horizontal core drilled along an azimuth of 355° (indicated by the red line). The mineralized, bed-normal fractures cut the core at an angle of 75°–80° to the core axis. Bedding in the core is horizontal and nearly parallel to the core axis so it is horizontal in two positions as the core is rotated about its axis. Depending on which of the two possibilities is the correct in situ position, the average fracture strike is either 70°–250° (*left*) or 100°–280° (*right*). The ambiguity can be resolved if sedimentary geopetal structures can be recognized in the core, or if the core is known to cut up-section or down-section across bedding.

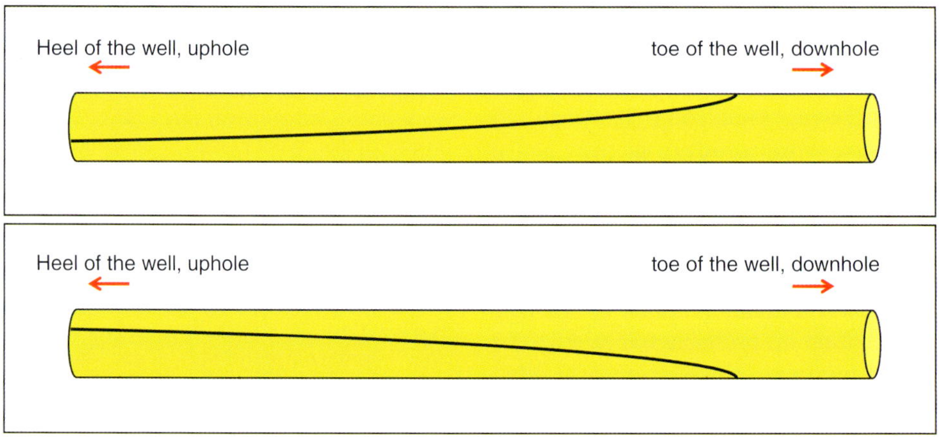

Figure 2.46 The near-horizontal bedding captured by this deviated core, viewed from the side, suggests that the core is cutting either up- or down-section at a shallow angle, depending on which side of the bedding is interpreted to be its top side. Bedding, the wellbore, or both may be inclined relative to horizontal. Wellbore deviation surveys, run in all wells, provide inclination angles for wellbores, measured in degrees from vertical, through the cored interval, but they are not always available to a fracture analyst.

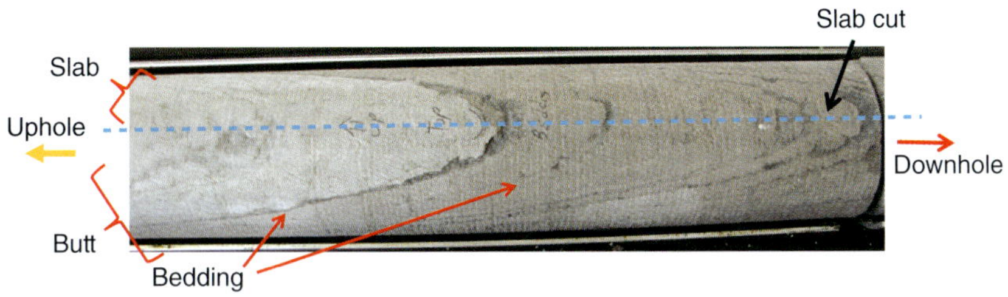

Figure 2.47 Near-horizontal core cutting down-section across near-horizontal bedding. Map view, looking down on the high or dorsal side of the core, with bedding exiting the core downhole on the high side of the core. If the core is slabbed in equal halves, the apparent dip of the bedding in the slab box depends on which of the two halves is put in the box (see Appendix 2.C).

and intersection angles are long gone. Nevertheless, most archived cores are reasonably intact and both butts and slabs are often available together for examination, so archived cores should not be dismissed as a good source of fracture data.

The well and coring records rarely accompany an archived core, so an analyst may not have access to information such as the wellbore deviation survey. Likewise, a core may have scribe-line grooves on the core surface indicating that it had been oriented, but the associated orientation reports seem to disappear early and with disappointing frequency.

The fracture analyst is often asked to log core consisting only of slabs, where the associated butts have been lost, discarded, or are stored in another location. Some archived cores consist primarily of spacers representing samples that were removed for study and never replaced. We examined one core where a single 4-in (10 cm) sample was boxed and preserved from each 3-ft (1 m) interval, and another where the core was slabbed twice so that only a quarter-round section of the core was preserved. Sometimes the fractures in a core have been the target of an extensive fracture-sampling program, leaving little for the next fracture analyst to assess; we found a paper note in one core box to the effect that the "Frightful Fracture Snatcher" had removed a sample; we logged a fracture in the interval to contribute to the overall fracture-density and distribution calculations, but we had no idea of the fracture type, width, mineralization, etc. Another time a one-foot fractured core section had been removed, leaving tantalizing, half-inch-tall remnants at the fracture terminations in the core above and below the missing sample.

Despite this, archived and incomplete cores can support a fracture assessment if the remaining fracture fragments are assessed carefully. For example, core fragments containing small, remnant sections of fracture faces can offer information on probable fracture widths and remnant fracture apertures. The fracture faces often also indicate fracture type, from which an analyst can start to build at least a conceptual model of the fracture distribution and network. Qualitative information and data can be obtained from even the most fragmented cores as long as some of the pieces contain remnants of the cored fractures (Figure 2.48).

Figure 2.48 *Top*: fragmented cores in stacked wooden boxes, even after having lain outside in the weather for a couple decades, can provide fracture information if fracture faces are present in the core rubble. *Bottom*: mineralized, planar, natural-fracture faces (N) in core rubble: in this case a 73° dip angle could be measured relative to a remnant of the edge of the core surface, and the angular relationship of a natural fracture striking 85° counterclockwise from a petal fracture ("85° CCW from P") could be measured in a different piece of rubble. The fracture fragments, however, could not be reassembled. In some cases, if all the fragments are present, the core can be pieced together like a large 3-D puzzle. This requires time but it is usually rewarding since core is commonly most fragmented in the intervals containing multiple fractures.

2.6 Oriented Core

A relatively small percentage of cores are oriented, but orientation adds significant value to any core taken for fracture assessment because fracture strikes are so important. Orienting a core involves extra cost, and fragmentation of the core as the scribe knives cut the reference grooves is a possibility, but the latter is typically only a concern in small-diameter core cut from brittle rock since the scribe grooves are only a millimeter deep. Nelson et al. (1987) described the basic operations for orienting a core, and although the technology has changed and improved in the intervening decades, archived oriented cores are still being pulled from warehouses for study so it is useful to know both the historical mechanics of orienting a core and the present-day techniques.

Core was originally oriented using a "multishot" process, where a 35 mm film camera at the top of the core assembly took photographs of a compass which was locked in its orientation relative to scribe knives in the core bit 30 ft (9 m) below at the bottom of the assembly. The camera and compass were housed in a non-magnetic, nickel-alloy Monel collar at the top of the core barrel. A timer on the camera was set to take photographs at predetermined times, the intervals between photographs depending on the expected rate of penetration and core length. The driller actually stopped the drill-string rotation and shut down the mud pumps for a short time before and after the specified time when the camera was set to take a photo, so that the compass in the core barrel settled down and the camera could operate without the drill-string vibration. Drillers detested this procedure since it provided opportunities for a stuck drill-string as cuttings floated to the bottom of the hole, filling in around the core bit and core barrel, due to the lack of circulating mud. It also resulted in a limited number of orientation "shot" points along the core.

A limited number of shot points works if the scribe lines are reasonably straight. However, friction between the clockwise-rotating drill string and the non-rotating inner core barrel commonly drags the inner core barrel and the attached scribe knives clockwise with depth. The resulting rotation of the orientation scribe lines on the core ranges from nil to excessive; rotation can be up to about 10° per foot (Figure 2.49) and still be used with caution if the rotation is constant so that the fracture analyst can interpolate reliably between the photographic shot points, and if a reliable depth correlation can be established between the core and the orientation report. Greater amounts of rotation can occur locally (Figure 2.50) or over the entire length of a core, and the scribe can jump or be dragged sideways

Figure 2.49 This interval of vertical core, viewed with uphole to the left, shows the Principal Scribe Line (red arrow) rotating slowly clockwise with depth, increasing in deviation from the green Master Orientation Line (MOL). The amount of rotation of the scribe relative to the MOL will be measured and compared to the toolface-orientation report. The Principal Scribe Line is sometimes referred to as the Main Reference Line or Groove.

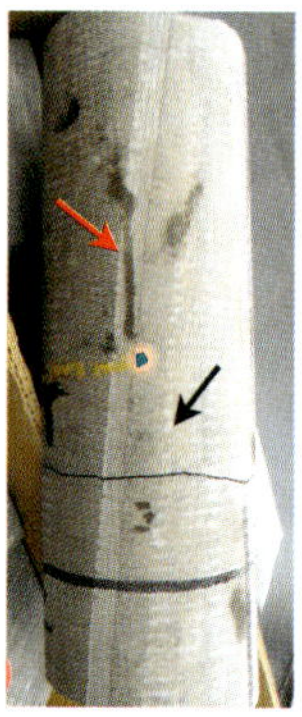

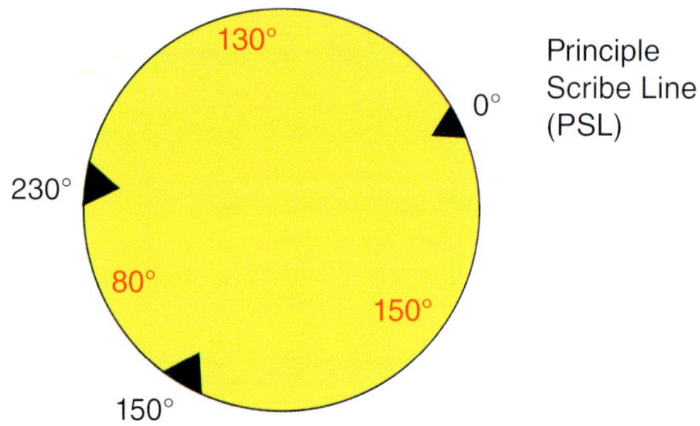

Figure 2.50 *Left:* the scribe groove (red arrow) on this core rotates excessively, about 35° per foot, and vibrational chattering has widened it to about 10° on the core surface. Fracture orientations from this core would only be valid if core depth can be exactly correlated to the depths listed in the orientation report. Note also the scar left by a dragging core catcher (black arrow). *Right*: sketch of a core, looking down hole, scribed with the three scribe knives oriented asymmetrically at 150 ° and 230° clockwise from the Principal Scribe Line (PSL). The orientation tool orients the PSL relative to north at each foot of drilled depth. Asymmetry of the knives provides an up-down reference for the core as well as a means of potentially keeping the core oriented after it is slabbed.

for tens of degrees over very short depth intervals before resuming a slower rotation, so the common assumption of constant rate of rotation between camera shot points is not always valid.

Accelerometers, which are more reliable and which provide a constant stream of orientation data, have replaced the camera-and-compass multishot arrangement. Accelerometers eliminate the need to stop drilling in the middle of coring operations, and they provide many orientation points per foot of core. The data points are culled during processing to provide one core orientation at each foot of penetration in the service-company's final core-orientation report.

There are three scribe knives in the core-orientation shoe mounted within the core bit at the base of a core barrel, and they cut three grooves or scribe lines on the core. One of the lines, called the Principal Scribe Line or PSL, is oriented; it is typically referred to as the "toolface" in an orientation report. A core-orientation survey must be quality-checked by measuring the amount of scribe rotation on the core with depth, noting the locations of scribe jumps, and comparing them to the foot-by-foot toolface orientations provided in the core-orientation report (Figure 2.51). Blind use of the reported orientations can miss important core-to-survey depth shifts, scribe-line jumps, and unreported failures of the orientation tool. We worked with one orientation report where the toolface was reported to have rotated slowly clockwise with depth as expected, but where the scribed grooves on the core were straight, invalidating the core-orientation report.

A fracture analyst must determine the exact depth shift to apply in correlating core depths to an orientation report. If the depth shift is off by plus or minus two feet and the scribe lines rotate by 10° per foot, any measured fracture strike could be off by plus or minus 20°. Most oriented cores will contain several places where the scribe jumps or skips, often coinciding with a drill-string connection or a spinoff in the core, or where the core cut a fault and consists of rubble. These minor glitches in the orientation and coring processes are valuable since they provide depth-correlation points: there should be a shift of equivalent magnitude in the reported toolface orientation near the core depth of the observed shift in the orientation groove on the core, and the two depths can be shifted accordingly to make them match.

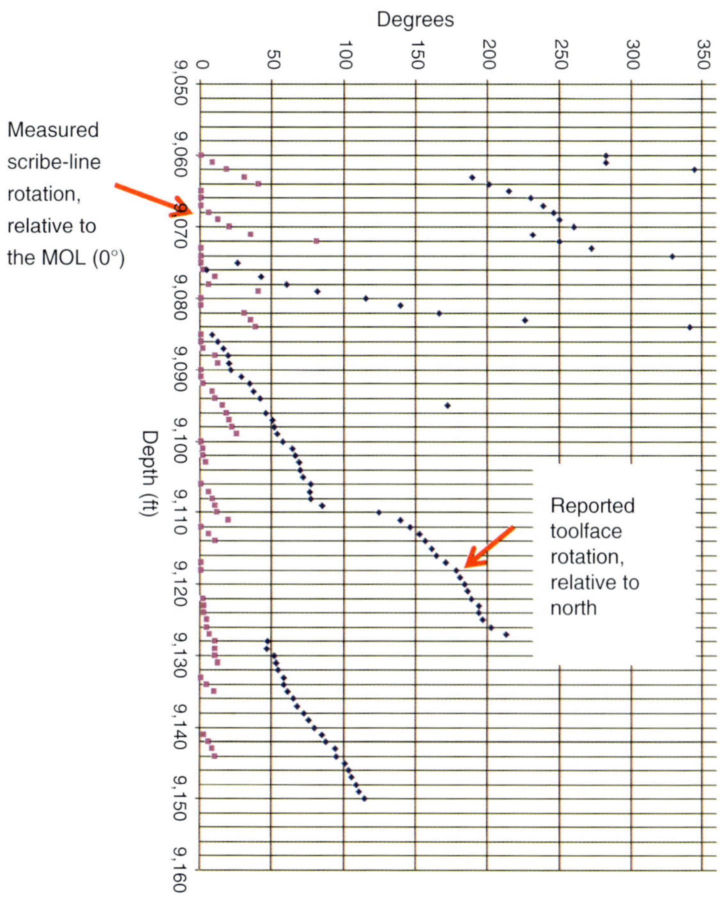

Figure 2.51 Data comparing 1) the measured rotation with depth of the Principal Scribe Line on the core (the magenta data points) relative to a Master Orientation Line drawn on the core surface (an arbitrary line drawn on the core and taken as a 0° reference); and 2) the reported rotation of the toolface (blue data points) in the orientation survey report as plotted relative to north (0°). This figure shows a reasonably good correlation between the two datasets, especially for the lower two-thirds of the core where the scribe drifts clockwise at the same rate as the reported toolface and at an acceptable rate of about 4° per foot. The major offsets in the two datasets have the same vertical spacings but occur at different reported depths, indicating that a depth shift *must* be applied before orienting fractures in the core with this orientation report. There are more discontinuities in the core than in the orientation report since additional core breaks were created during processing.

The scribe grooves on some cores rotate *counter-*clockwise for unknown reasons, possibly due to downhole vibrational harmonics much like a hui (sometimes "hooey") stick, where vibrations made by rubbing one stick along the notches on another make a propeller attached to the notched stick rotate, but equally well in either direction. Nelson et al. (1987) noted that the orientation surveys associated with these cores may be invalid, but the newer accelerometer data are usable for cores with counterclockwise groove rotations if the reported toolface data also rotate counterclockwise with depth and match the measured rotation of the scribe grooves on the core.

The Principal Scribe Line, for which the orientation relative to north is listed in the core-orientation report, is no wider or deeper than the other two scribe lines. It is identified as the line that is relatively isolated from the other two, more closely spaced lines cut into the core surface (Figure 2.50). Different service companies use different scribe configurations, but many presently use an asymmetric scribe geometry where the secondary scribes are located at the 150° and 230° positions clockwise around the circumference of the core from the PSL when looking downhole, taking the PSL as the 0° position. In this configuration, the two secondary scribes are only 80° apart from each other, but the PSL is distant by 150° from one secondary scribe and by 130° from the other. Other companies use symmetric configurations (a 0°–130°–230° arrangement used to be common), but the PSL is still the isolated scribe line on the core.

For some reason, scribe configuration is rarely listed on an orientation report so the core analyst usually has to measure it from the core, which can be difficult or impossible after the core has been slabbed since only one or none of the scribes may be present on the back of the slabs. The advantage of an asymmetric scribe configuration, however, is that if both the scribe configuration and the uphole orientation of the core are known (the latter from the red-black line pair drawn on most cores), one can still determine the position of the PSL if only two of the three scribe grooves are present on the available section of the core, usually the butt.

Some service companies specifically slab cores so that the PSL is on the back of the slabs figuring that it will provide an orientation of the slab. In this case, that scribe line should be conspicuously marked as the PSL in numerous places on the slabs, but it rarely is. It also means that the slab plane usually rotates downhole to maintain a consistent orientation relative to the spiraling PSL.

Once the PSL is identified and the orientation survey has been validated, fracture strikes can finally be measured. The core-orientation report lists the orientation of the PSL relative to north for each foot, i.e. if the orientation of the PSL was listed as 62°, a vertical core can be restored to its in situ position by holding it in a vertical position and rotating it so that an arrow drawn from the center of the core outward and through the PSL on the outer core surface points east-northeast (62°). Using a circular protractor, some analysts first measure fracture strike relative to the PSL and then calculate actual strike using the reported orientation for the PSL from the orientation survey. Others take a shortcut and measure actual fracture strike directly, placing the circular protractor so that the PSL is matched to the reported orientation for that depth, which makes 0° on the protractor equivalent to north.

Unless a fracture cuts across the middle of the core, the intersection of the fracture with the protractor is not its strike, but rather provides the information for calculating strike, as illustrated in Figures 2.52 and 2.53.

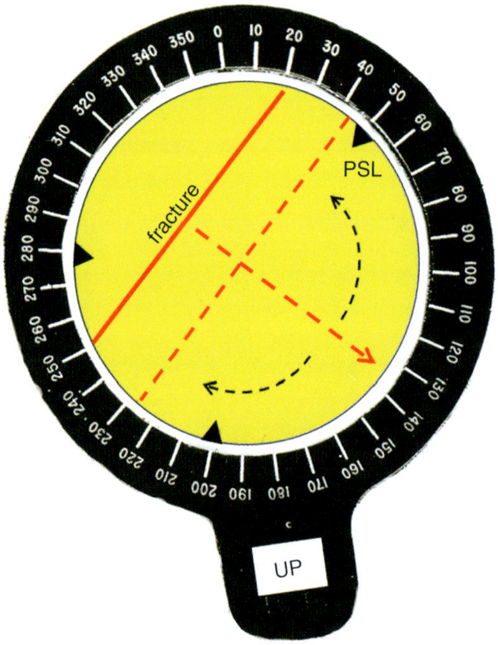

Figure 2.52 A circular protractor placed around the core, looking downhole. If the orientation survey reports that the toolface is oriented at 50° at this depth, the Principal Scribe Line (PSL) is placed next to 50° on the protractor so that fracture strike relative to north can be measured. The fracture does not cut through the middle of the core so its strike is not the 20°–250° intersection points with the protractor. Rather, the fracture can be visualized as a parallel line that does cut through the middle of the core, giving the actual fracture strike (45°–225°). Alternately, a line drawn at 90° to the fracture across the middle of the core intersects the protractor at 135°, normal to the fracture strike, which can then be converted to true strike by adding/subtracting 90°. "UP" in this figure indicates that this side of the protractor must be used when looking downhole at a vertical core. The degrees are in the same positions on the underside of the protractor (which is labelled "DOWN"), but they increase in the counterclockwise direction when that side is facing the viewer; that side must be used if the core is measured while viewing it from the bottom up.

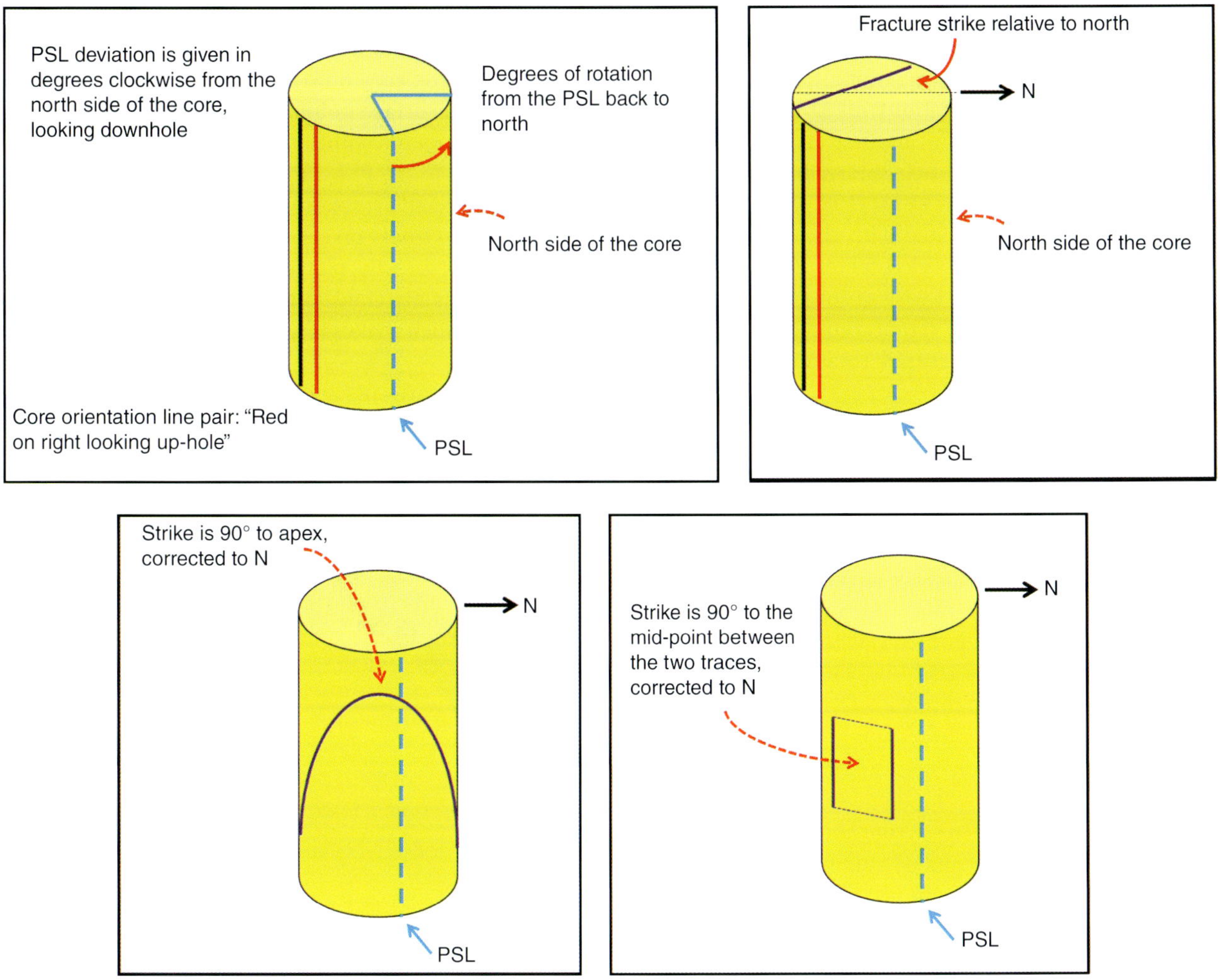

Figure 2.53 There are various ways of determining fracture strikes relative to the Principal Scribe Line, depending on the configuration and exposure of the fracture. *Upper left*: the core must first be oriented relative to north using the position of the PSL reported in the orientation survey. *Upper right*: once the core is oriented, fracture strikes are easy to measure relative to north if they are exposed on the end of a core. *Lower left*: if a fracture exits the side of the core upward or downward, the position of the apex of the fracture on the core surface is normal to the fracture plane, and strike can be calculated by adding/subtracting 90° from the intersection of the apex with the protractor. *Lower right*: if the fracture is only exposed on the sides of the core it may be necessary to take the intersection points of the fracture with the protractor and average them, then add/subtract 90° to obtain the strike. For example, if the fracture in Figure 2.52 were only exposed on the sides of the core, the calculation to obtain the true fracture strike of 45°–225° would be:

$$(20^\circ + 250^\circ)/2 = 135^\circ, \text{minus } 90^\circ = 45^\circ \text{and plus } 90^\circ = 225^\circ$$

Each of these steps for orienting fractures are simple, but it is easy to omit a step and come up with an incorrect measurement.

Horizontal cores can also be oriented, but for these cores the orientation survey measures the position of the PSL relative to up, i.e. relative to the in situ high side of a core, rather than relative to north. The toolface is given in degrees clockwise from the high side of the core while looking downhole. Fracture dips, correcting for wellbore deviation if the hole is not perfectly horizontal, can be measured easily with a protractor once the core is oriented. Calculating fracture strikes relative to north requires additional knowledge of the wellbore azimuth, as shown in Figure 2.54. Orienting a horizontal core with an orientation assembly may be the only way to obtain fracture dips and strikes if the core cuts through an un-bedded, massive lithology or if stratigraphic up cannot be determined from bedding, although comparisons to fractures in image logs from the same part of the wellbore will also often provide the needed information.

2.6.1 Other Ways of Orienting a Core

If the subsurface bedding dips consistently in a known direction and is cored with a vertical well, the core can

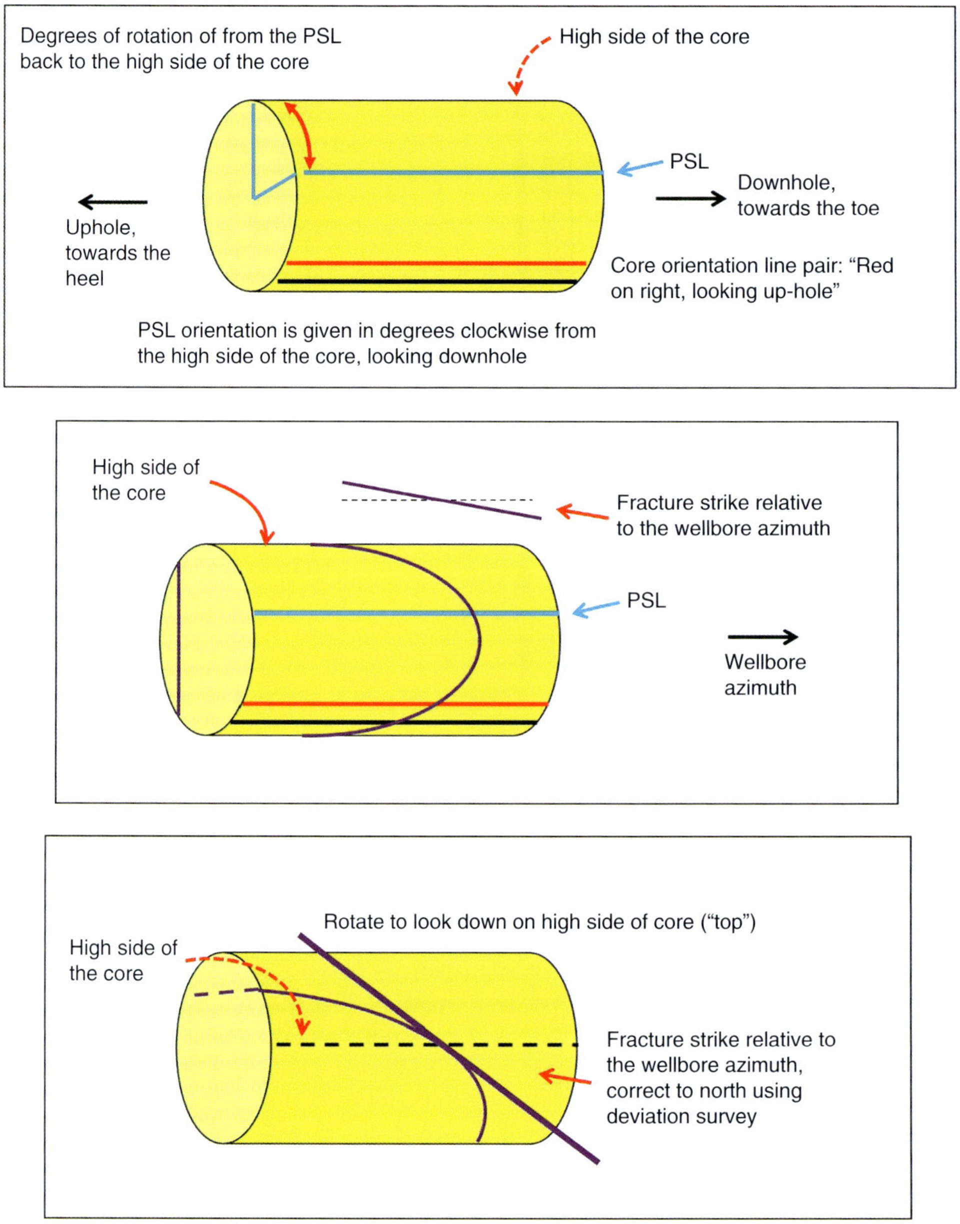

Figure 2.54 *Top*: orienting a horizontal core and determining the high side. In this example the position of the PSL is given as 50° from vertical, so the PSL is placed against the 50° mark on a circular protractor, with degrees increasing clockwise looking downhole, whereupon the 0° position on the protractor defines the high side of the core. The core is rotated around its axis so that the high side of the core faces upward on the layout table. *Middle and bottom*: fracture dip and strike then can then be measured using the core azimuth (obtained from the wellbore survey) and a carpenter's or grade-school protractor.

be oriented using the observed bedding dips. Likewise, if a formation has a horizontal dip but is cored by a wellbore that is deviated as little as 10° from vertical, the core can be oriented using the apparent dip of bedding in the core and the wellbore survey which shows the amount and direction of wellbore deviation through the cored interval. Cores taken from strata with other structural configurations can be oriented if the analyst has definitive knowledge of the subsurface structure and the wellbore deviation. Both vertical and horizontal cores can often be oriented by comparing the major features in the core to the fracture and bedding signatures shown on an image log run through the same interval of the well (Figure 2.55; e.g. Ameen and Hailwood, 2008; Fernández-Ibáñez et al., 2018).

Un-oriented vertical core can also be reliably oriented using the same paleomagnetic techniques developed to reconstruct the drift paths of continents across the

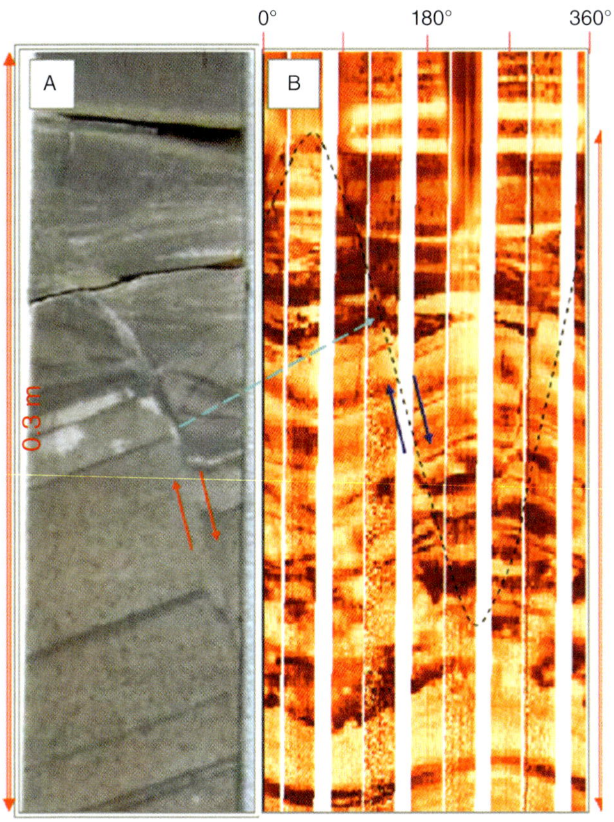

Figure 2.55 A syn-depositional fault in core (*A, left*) creates a recognizable signature in the image log at nearly the same depth (*B, right*). Two or more structures, recognizable in both the core and the image log and spaced vertically at similar intervals in each dataset, would be definitive confirmation of the correlation. The SSW dip azimuth measured for this structure in the image log can be used to orient the section of core containing the fault, as well as any contiguous sections of core above and below the fault (Reprinted from Ameen and Hailwood, 2008, with permission from AAPG, whose permission is required for further use). Core would have to be re-oriented across each unlockable break in the core for other core intervals.

surface of the earth. For core orientation, the technique uses the stronger magnetic signature imposed on the rock by the present-day magnetic field rather than the weak paleomagnetic signature that was locked into the rock at the time of deposition, but the term "paleo" is entrenched and widely understood.

Like the Master Orientation Line, each paleomagnetic determination is good for only one continuous interval of core, so numerous measurements would be needed if a core is badly broken up. Eight plugs are typically taken from the core in each interval to be oriented, and they alternate on opposite sides of the core to compensate for any magnetic signature imparted to the rock by the rotating plugging bit. An arbitrary orientation line is drawn down the length of the interval to be oriented, and the orientation report provides the orientation of that line relative to north so that it can be used like the Principal Scribe Line on a conventionally oriented core.

2.7 Using CT Scans

Computed X-ray Tomography (CT) scans can be invaluable to a fracture analyst since the images provide a view of fractures in three dimensions that is not available from fracture exposures on the core surfaces or from image logs. Moreover, since the CT scans are run before the core is extruded from the inner core barrel and before the core has been slabbed, plugged, or sampled, the core is commonly more continuous and intact in a CT scan than it is lying in the core boxes. CT scans can also provide information on subtle lithologic changes that are not always apparent on rough core surfaces.

However, like image logs, CT scans provide an indirect look at the core and the cored fractures. The scans cut across fractures and do not provide information on the fracture faces, so it can be difficult to distinguish between shear and extension fractures, or even between natural and induced fractures. CT scans can make insignificant fractures look important if the fracture is filled with a mineral that contrasts with the host rock (Figure 2.56), they can minimize important, open fractures (Figure 2.57), and it may be difficult to distinguish between drilling mud and mineralization filling a fracture aperture. CT scans are important contributions to a fracture analysis, but they supplement a detailed hands-on core examination, they do not replace it.

One subtle but important aspect of CT scans, inherited from its medical origins, is that cross-sections across the core are presented with the viewer looking uphole rather than the conventional geologic downhole view. This is apparent when trying to match CT fracture orientations with those seen in core, and becomes important when

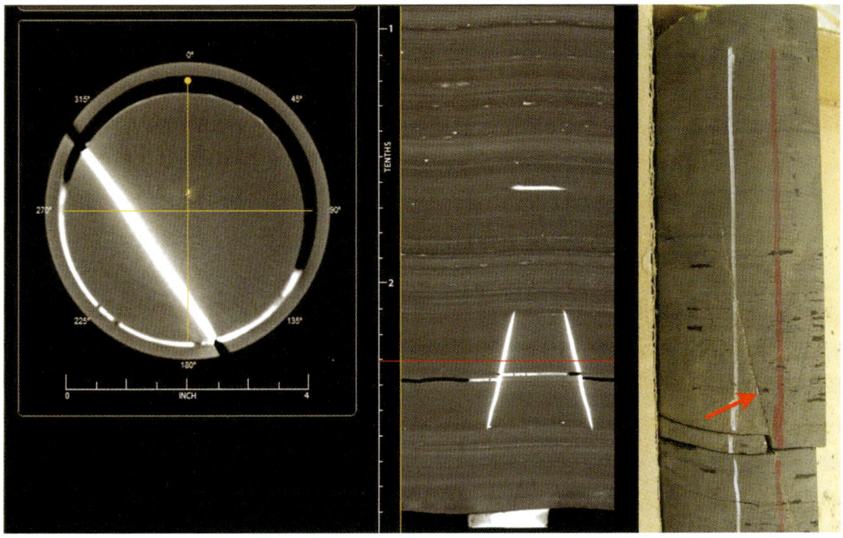

Figure 2.56 A wide, gypsum-filled fracture in a calcareous mudstone has a bright signature on the CT scans but is obscured by drilling mud in the core. Although eye-catching, this fracture is occluded and contributes little to reservoir permeability. *Left and middle*: horizontal and vertical sections through the intact core show cross sections of the fracture at the location indicated by the red cursor line across the middle image. Cross sections are typically presented looking uphole. Note the equally bright signature produced by drilling mud, both along the horizontal core break (middle image) and locally between the core surface and the inner barrel (left image). *Right*: this same fracture as seen in the core butt (red arrow). It is not present in the equivalent slabs due to the orientation of the slab cut.

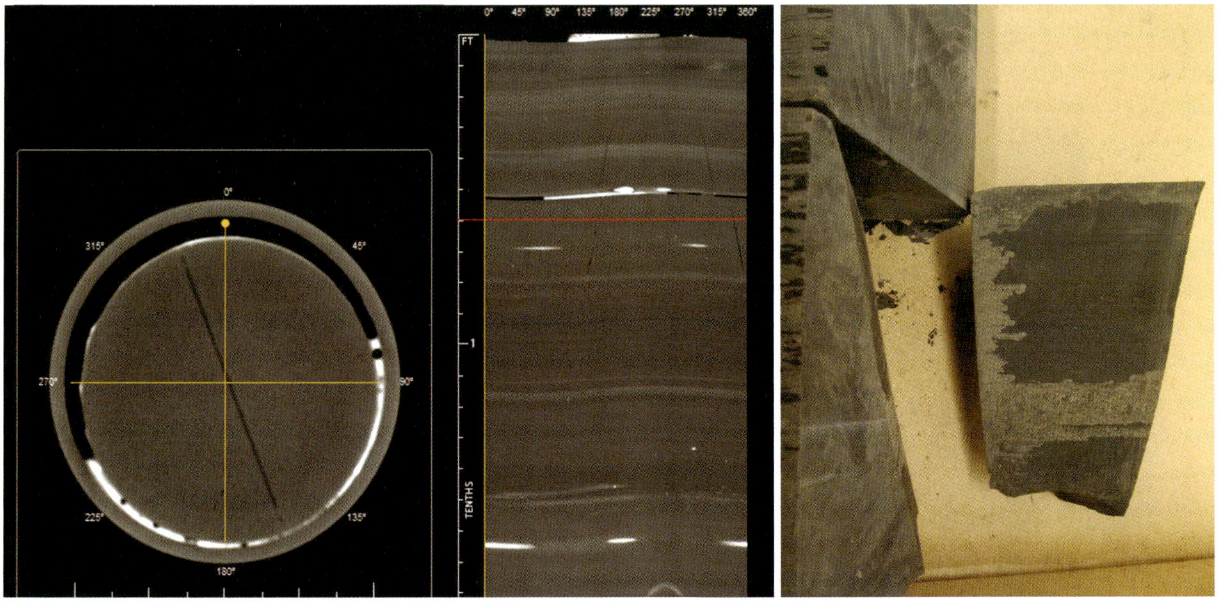

Figure 2.57 A much fainter fracture signature in the CT scan (*left and middle*) a foot lower in the same core than the gypsum-filled fracture shown in the previous figure, records a more important, high-permeability, unmineralized fracture. *Right*: the core shows the fracture face is partially invaded by tan drilling mud.

measuring natural fracture strikes clockwise vs. counter-clockwise relative to induced fractures using CT scans. The analyst also needs to recognize that the degree scale on CT cross sections is applicable only to relative fracture strikes and only for comparisons of strikes within each continuous core interval (Figure 2.58).

2.8 Fracture Data from Image Logs

The hydrocarbon industry and geology in general have benefited significantly from image logs since they have provided a wealth of subsurface data, including information on the previously little-known fracture systems

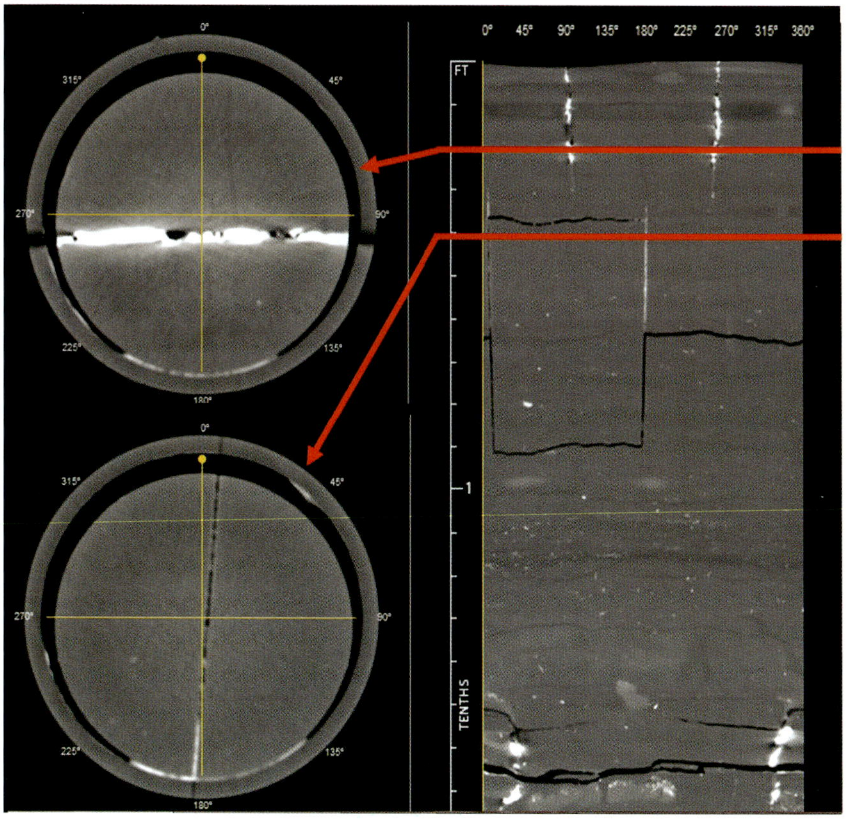

Figure 2.58 *Right*: the two fractures shown in the CT scan of a continuous piece of core would intersect at a right angle if extended vertically towards each other. *Left*: the two cross sections show the correct intersection angle since they occur in the same section of core. The upper fracture is a calcite-mineralized, ptygmatically folded fracture; the lower one is an unmineralized, vertical extension fracture.

in reservoirs. Image logs provide significant volumes of fracture data over large intervals of a wellbore. The fracture data are oriented and can easily be manipulated and analyzed with the associated software programs (Figure 2.59). Running and processing an image log are not inexpensive, but logging a well requires significantly less rig time than coring one. Fractures imaged by logs can also usually be easily assessed for their strike relative to the in situ stresses. This volume does not offer an exhaustive treatment of image logs (the reader is referred

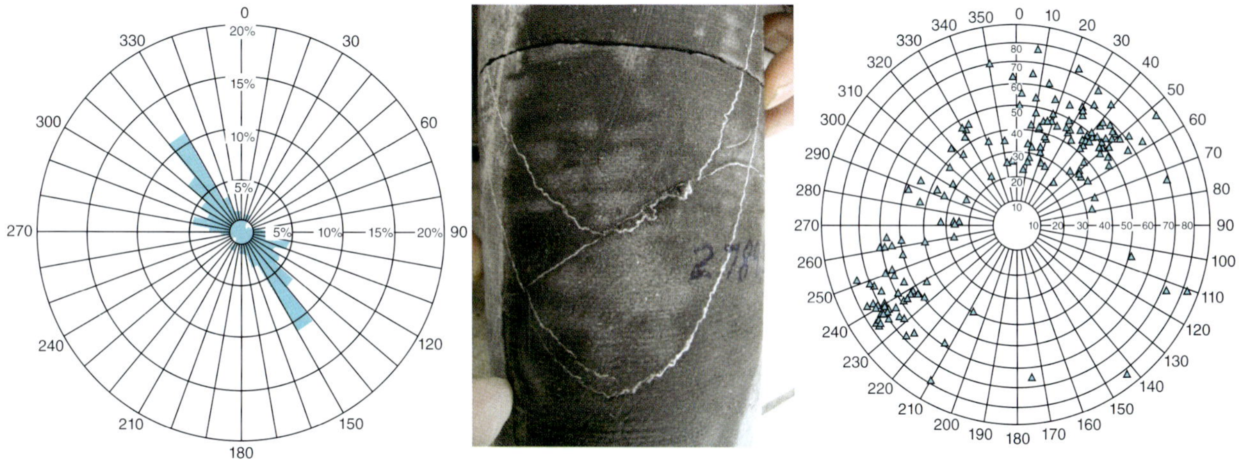

Figure 2.59 Fracture data from an image log are easy to manipulate using service-company software. *Left*: a rose plot of resistive fracture strikes in an image log shows the dominant NW-SE pattern, but the subordinate strikes are lost in the noise since the lobes on the rose represent percentages of the fracture population and the oblique fractures are not numerous. *Middle*: core from the same interval shows definitively that the calcite-mineralized fractures have intersecting strikes and dips, forming an interconnected network, but that they are occluded. *Right*: a stereoplot of the image-log fracture orientations offers a better representation of the intersecting dips and strikes of the fractures in this reservoir.

to descriptions such as those in Ruehlick, 2015; Nelson, 2020), but it does examine correlations between core and image-log fracture data (see Figure 2.55).

The quality and precision of the fracture signatures on an image log are influenced by fracture geometry, mud properties, formation fluids, and whether or not drilling mud has invaded the fracture apertures. Borehole rugosity is another influence, and fracture studies using image logs should start with an assessment of the caliper logs plotted on the same charts to check the condition of the borehole walls. Fracture populations measured from image logs are typically much larger than those gathered from short core intervals, so plots of the image-log fracture distributions are commonly smoother and more systematic, more representative of the larger fracture population, than are the smaller fracture datasets derived from core.

An image log in an 8¾ in (22 cm) diameter wellbore samples a surface area that is over twice that provided by the surface of a 4-inch (10-cm) diameter core, yet when core and logs are both available, the log commonly images fewer fractures than are captured by core. Neither an image log nor a CT scan provides a view of the fracture face and its fractographic markings. Bed-parallel fractures are difficult if not impossible to distinguish from bedding in an image log, and vertical natural fractures that parallel a vertical wellbore can be misidentified as drilling-induced fractures. Shear fractures and fractures that have been subjected to dissolution have rough, irregular planes that can be difficult to interpret in a log where they may not create regular sinusoidal signatures.

Letting an analyst become familiar with a core cut from the formation being logged allows for a more informed image-log fracture interpretation. Fracture picks from an image log may be "automatic," but the analyst still subjectively sets the thresholds and contrasts for picking signatures, and a geologically oriented analyst is more likely to use appropriate settings and make geologically plausible interpretations.

Fracture classifications used by image analysts are based on electric or acoustic properties of the fracture signatures rather than on the mechanical origins of fractures, so comparing datasets between core and image logs can be an apples-to-oranges assessment. Rather than shear or extension, the common image-log fracture classes such as "conductive" and "resistive" imply fracture aperture and effectiveness as conduits.

An image log adds significant value to a core taken from the same hole, often allowing the core to be oriented and allowing an extrapolation of a cored fracture system beyond the limits of the core. In turn, core adds significant value to an image log by providing a calibration of the logged fracture signatures and a means of quantifying the imaged fracture populations. Nevertheless, comparing image log fractures to core fractures from the same well never seems to be as easy as it should be. Surprisingly, it is common for only a few fractures, and not always the larger ones, to be definitively identifiable as equivalent between a core and an image log. Even then it may require an inconsistent core-log depth shift since some image logs use software to automatically adjust depth for wireline stretch due to stick-slip of the imaging tool in a hole. This correction is intended to provide accurate depths, but it can obscure depth correlations to fixed-length cores. An analyst cannot be dogmatic in correlating cored and imaged fractures by depth alone.

Fracture dip histograms offer a relatively easy way to assess the similarity of an image-log fracture population to a correlative core fracture population, since unlike strike, fracture dip angles can be measured whether or not a core is oriented. The two datasets shown in Figures 2.60 and 2.61 illustrate both good and poor agreement between core and image-log fracture populations. The comparisons suggest that the two sources of fracture data are complementary and should be used together to support each other where possible.

Deformation-band shear fractures are sometimes considered to have poorly defined signatures on image logs since the comminuted material that defines the shear plane consists of the same basic material as the host rock. Nevertheless, a recognizable signature is commonly provided by difference in fluid saturation between the matrix

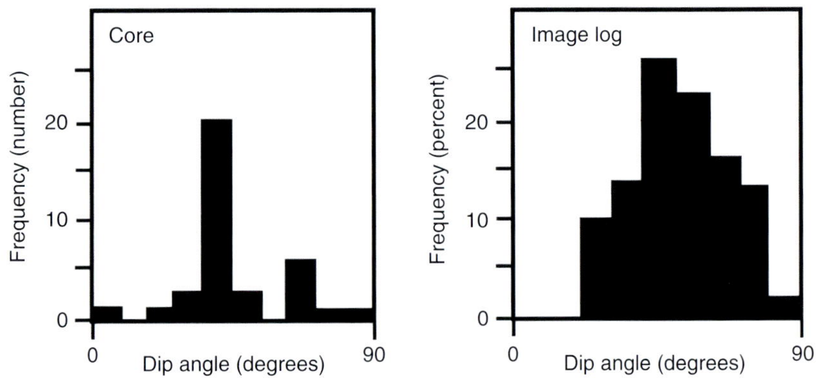

Figure 2.60 An example of correlative fracture dip-angle histograms from a vertical well. *Left*: dip data for calcite-mineralized fractures in 16 ft (5 m) of core (n = 36). *Right*: Image-log data from the same well, for 98 fractures from 246 ft (75 m) of borehole that includes the cored section. The larger population from the image log has a smoother data distribution, but the histograms show compatible dip-angle distributions. On a per-foot basis, however, the core contains five times more fractures than the image log. (Unpublished data.)

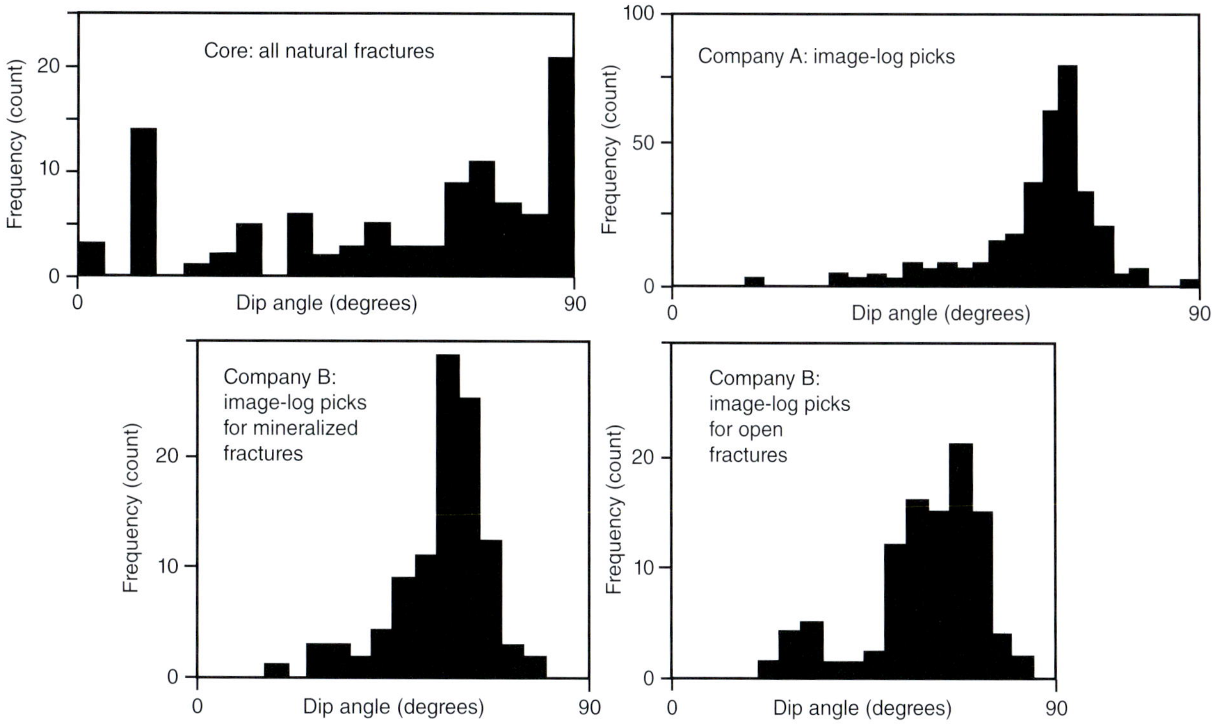

Figure 2.61 Another example comparing fracture dip-angle histograms from a vertical well. *Top left*: a histogram of 101 dip angles measured from core. *Top right and bottom*: dip angle histograms compiled by two service companies, each interpreting the same image log which included the cored interval. The image log, run over a much larger interval of the wellbore than the cored interval, measured a fracture population that is larger by an order of magnitude than that provided by the core. Company A (*top right*) lumped all fractures, whereas company B (*bottom left and bottom right*) distinguished between "mineralized" and "open" fractures. The image log did not distinguish between bedding and a population of horizontal and near-horizontal shear fractures that could be identified in the core. Likewise, most of the distinct, mineralized, vertical natural fractures in the core were interpreted by both image-analysis companies to be induced fractures. (Unpublished data.)

and the fracture-plane gouge, and deformation bands have image-log signatures in many formations. The corollary, however, is that deformation-band signatures on image logs look very much like mineralized fractures. The distinction is an important one since deformation bands are detrimental to reservoir permeability.

2.9 Comparing Fracture Data from Outcrops, Core, and Logs

2.9.1 Introduction

Outcrop fracture-study techniques are well documented and described in the literature (e.g. Watkins et al., 2015), so rather than present a review, this section discusses and compares the types of fracture data that can be obtained from outcrops, core, and image logs, which provide independent but mutually supporting pieces of the fracture-characterization puzzle. An integrated study using all three data sources, supported by seismic and engineering data, can provide a more complete picture of a fracture system than can the use of any one of the sources by itself.

Outcrops offer the opportunity to collect most of the parameters listed earlier for measuring fractures in core, with the added benefit that the fractures can be viewed in two and often three dimensions so that an analyst can obtain better information on fracture lengths and distributions, as well as on the interrelation of fracture sets. Three-dimensional conceptual models of fracture systems can be more easily formulated from outcrop, but with the caveat that outcrops have been subjected to a different stress history than that imposed on the same formation at depth and may therefore contain fractures that cannot be extrapolated into the subsurface. Moreover, weathering alters fracture characteristics. In northern Iraq, for example, the generally arid but seasonally wet climate widens fracture apertures in the chalky limestone outcrops of the Eocene Pila Spi Formation due to dissolution during the wet season, and fills them in again during the dry season with windblown dust that gets cemented into the fracture apertures by mobilized carbonate from the host rock. Some fracture sets in outcrops are related to stress release during uplift and exposure, or to gravitational stress along valley walls, and have no equivalent in the subsurface.

Numerous outcrop fracture studies have been carried out in support of efforts to understand and model reservoirs. Three published large-scale studies that illustrate some of the benefits of such studies of are summarized below. More specific aspects of three smaller-scale studies are also presented, the more limited studies illustrating techniques and offering caveats for extrapolating outcrop fracture data into the subsurface.

2.9.2 Large-Scale Outcrop Studies

De Keijzer et al. (2007) and Wennberg et al. (2007) used outcrop data from fractured carbonates, in Oman and Iran respectively, to complement the limited fracture data obtained from associated reservoirs. Both studies were used to develop conceptual fracture models that could in turn be used to calibrate and constrain fluid-flow fracture models and thereby improve the ability of the engineers to history-match production patterns.

De Keijzer et al.'s outcrop data, derived from tilted but not folded limestones on the flank of a broad fold, document two pervasively developed sets of bed-normal fractures with systematic and roughly orthogonal strikes and multiple scales of development (strata-bound to vertically throughgoing), forming an interconnected fracture-permeability system. In addition, the outcrops are cut by numerous, more localized fracture corridors with a range of sizes. A combination of the pervasive fracturing and the corridors provided the best fit when modeling reservoir production patterns. The corridors also explained enhanced local production along specific trends, and were used as analogs to expand similar structures observed in image logs into three dimensions. Moreover, the presence of corridors explained otherwise ambiguous seismic anomalies.

Wennberg et al. (2007) studied fractures in limestones on a more distinct fold. In addition to developing a conceptual model, Wennberg et al. (2007) documented quantitative fracture parameters including length distributions, crosscutting or abutting fracture relationships, and fracture distributions relative to lithology and bed thickness, data that were used in generating Discrete Fracture Network models. As did De Keijzer et al., Wennberg et al. found that the outcrop fractures could be segregated into a system of pervasive fractures and a system of localized fracture swarms or corridors, but in this case, all of the structures could be related to the stresses and strains created by folding and fold-related faulting. Wennberg et al.'s pervasive fracture system consists of four bed-normal, strata-bound fracture sets, one striking normal and one parallel to the axis of folding, and two striking oblique to that axis, creating a well-interconnected fracture-permeability network. Fracture density and height are controlled by both structural position and by the lithology-dependent mechanical stratigraphy. Vertically extensive fracture corridors are present in the outcrops, but they are not as common as those found in the De Keijzer et al. study, and they therefore play a lesser role than in the Wennberg et al. model.

A third example outcrop study of fractures developed on folds, also in Iran, was likewise undertaken to improve understandings of reservoir fracture systems. Superseding the important McQuillan (1973) study of folded Asmari Formation limestones, Stephenson et al. (2007) used outcrops to document the development of vertically extensive fracture corridors along and parallel to the crest of an anticline, and extensive bedding-parallel shear on the flanks. Both systems were created by the stress anisotropies produced by folding. Additional fracture corridors are oriented oblique to the fold axes, but these formed over reactivated basement faults and are unrelated to the fold.

These structures, combined with an understanding of the reservoir quality controlled by sedimentology, were useful in understanding production trends from nearby reservoirs, particularly the local zones of high production which are inferred to be located along similar, subsurface fracture corridors. Stephenson et al. also documented a segregation of fracture patterns that depends on mechanical stratigraphy, where shorter, strata-bound fractures developed in the more thinly bedded units whereas tall corridors are more common in the thicker, more massive units. Bed-parallel shear, common on the fold flanks, developed in both lithologies during folding.

2.9.3 Local Outcrop Studies

2.9.3.1 Raton Basin

One surface/subsurface study that allowed a specific comparison of three fracture datasets from different sources comes from the Raton Basin of Colorado and New Mexico (Lorenz et al., 2004). Outcrops in this basin provided extensive data on fracture strikes and their three-dimensional distributions, but most of the outcrop fractures are vertical extension fractures, whereas inclined conjugate fractures are prominent in the core and the image logs in equivalent strata that were also assessed as part of the study. In part this is because the vertical core and image data are biased towards intersecting inclined fracture planes, and in part it is because the inclined planes are best developed in muddy lithologies that are well represented in the core and logs but which are obscured in outcrop by weathering.

The dominant fracture trend in the basin is nominally east-west, normal to the Laramide thrust belt at the western margin of the basin, so it is inferred that thrusting created an east-west maximum horizontal compressive stress orientation as well as the stress anisotropy that fractured the rock in the adjacent basin. Outcrops offer no information on the present-day in situ stress orientation or anisotropy, so this piece of information had to be obtained from wellbore breakouts in the image logs, and the north-south orientation of the present-day maximum horizontal compressive stress, as measured by consistent wellbore breakouts, was unexpected.

Since the maximum horizontal stress is normal to the generally E-W strike of the fractures, the fractures are susceptible to closure during drawdown of the reservoirs, and hydraulic stimulation fractures will cut across the dominant natural-fracture trend. The cores and image logs also provided a much better idea of the fracture distributions with respect to lithology, since rock in a wellbore is not weathered and since image logs are typically run across long sections of a wellbore, cutting across and sampling fractures in multiple lithologies.

In the Raton Basin study, the image logs were used to determine differences in fracture density and strike in the three basic lithologies (Figure 2.62). Image-log data often provide better information on fracture distributions than cores since cores may target specific, reservoir lithologies whereas image logs are often run continuously over large intervals of heterogeneous formations.

In the Raton study, the image logs offered more precise, and more easily obtained, strike data than cores oriented indirectly by petal fractures. Interestingly, coals, which comprise only 18% of the logged section in the example described here, contain 63% of the 111 logged fractures, whereas 9% of the logged fractures were in sandstones which comprises 38% of the logged section. To compare the fracture intensities of the different lithologies, the percent of the total logged fracture population that occurs in each lithology can be divided by the percent of the logged section consisting of that lithology, providing a dimensionless but normalized number for comparing fracture development by lithology. Fracture intensities normalized in this way show that the relative fracture intensities in coals, shales, and sandstones are 3.5, 0.6, and 0.2, respectively. Stated another way, coals have a simple fracture density that is 17.5 times greater than that of the sandstones and 6 times that of the shales, helping to explain why the Raton basin has been primarily a coalbed-methane play.

The image-log data also show that fracture strikes in the coals and sandstones are primarily NNW-SSE (this well was drilled on an anticline overlying a N-S wrench fault, a local structure where fracture strikes do not conform to the regional E-W trend). In contrast, the

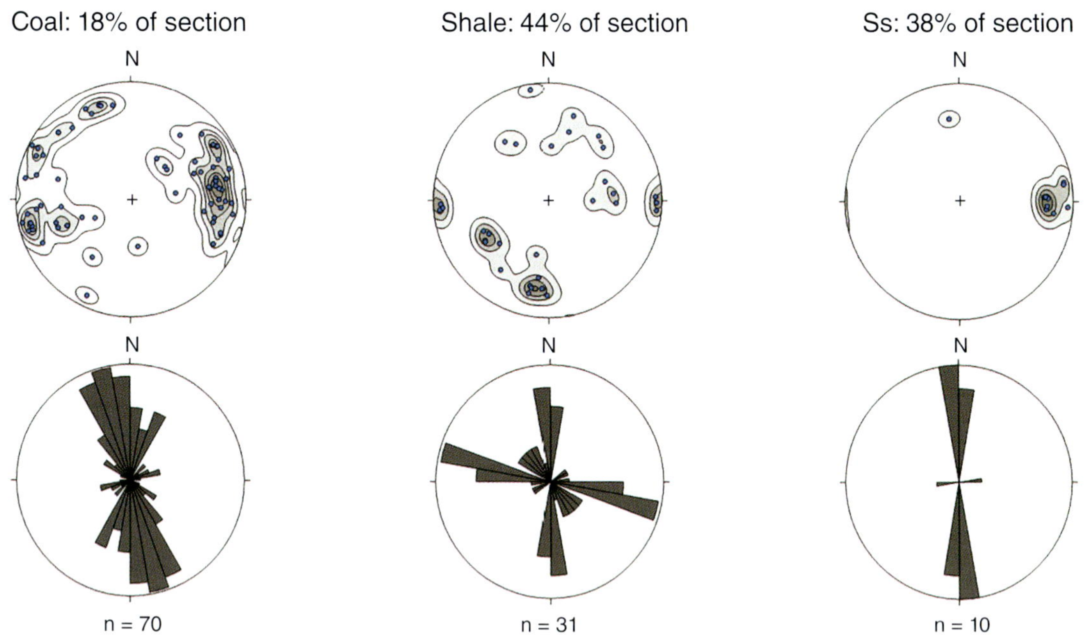

Figure 2.62 Determining fracture distribution by lithology from image logs. Stereonets and rose plots depicting all 111 fractures shown in an image log run in one well in the Raton Basin show that 70 fractures (63%) occur in the small percentage of the section that consists of coal (*left*), whereas only 10 of the fractures (9%) occur in the sandstones that comprise just over a third of the section (*right*). This suggests that fractures are better developed in coals than in the sandstone by a factor of 17.5, and better in the shales (*middle*) than in the sandstones by a factor of three. This kind of fracture-distribution analysis is harder to do in outcrop where the fine-grained lithologies are not well exposed. It can also be problematic in cores that target specific stratigraphic intervals and that therefore do not provide a representative sampling of other strata (from Lorenz et al., 2004).

logged fracture population in the shales shows composite population consisting of a WNW-ESE striking conjugate fracture pair (probably dip-slip conjugate shear fractures) and a set of N-S striking high-angle fractures (probably extension fractures). The first set is not apparent in the sandstone or coal fracture datasets.

When considered separately, the three fracture datasets derived from outcrop studies, downhole imaging, and core examination, suggest slightly different conceptual fracture models that, to be compatible, must be considered in terms of the exposures of the different lithologies from which the datasets came, and within a framework of the structural evolution of the basin.

Fracture Model Based on Outcrop Observations

1. Vertical extension fractures and vertical, strike-slip conjugate shear fractures dominate the sequence.
2. Strikes of extension fractures are dominantly E-W, implying an east-west orientation for the maximum horizontal compressive stress during fracturing.
3. Strikes of the acute-angle bisector for the sets of conjugate, strike-slip shear fracture pairs are also E-W, implying the same east-west orientation for the maximum horizontal compressive stress.
4. Fractures are concentrated in brittle rock types, primarily sandstones.
5. Thrust-oriented shear fracturing also occurs in outcrops, the thrusts again imply an east-west compressive stress.
6. Outcrop study allowed recognition of the influence of structure (anticlines and faults) on fracture patterns.

Fracture Model Based on Borehole Imagery

1. Fracturing is dominated by E-W striking, inclined conjugate fracture sets but with a significant N-S set developed locally.
2. Fractures are concentrated in finer-grained strata and coals.
3. The present-day maximum horizontal compressive stress in the basin is oriented N-S.
4. There is some evidence for thrust-oriented shear fracturing.

Fracture Model Based on Core Examination

1. Both near-vertical extension fractures and inclined conjugate shear fracture sets are present, but the strikes of fractures generally cannot be determined.
2. Fractures are more abundant in mudstones and siltstones than in coals.
3. Compaction fractures are ubiquitous in the finest-grained lithologies.
4. There is minor evidence for thrust-oriented shear planes.

Biases and Differences in the Three Fracture Datasets The most obvious difference between the three fracture datasets, and one that can create significant differences in conceptual fracture models, is the difference in fracture abundances suggested by the outcrop and subsurface observations. Fractures in outcrop are predominantly in brittle rock types, specifically sandstones. The abundant fracturing in mudstones and coals, including numerous ineffective compaction fractures, that is apparent in cores and image logs is not obvious from outcrop data where the muddy strata are poorly exposed. A second difference is that fracture-orientation data are only available from outcrops and from image logs.

Both the core and the image-log datasets are biased against intersecting the near-vertical fractures since all of the wells are vertical. This bias may account for the paucity of vertical extension fractures in the cores and in the downhole geophysical imagery compared to their abundance in outcrops. One-dimensional, vertical wellbore data are also biased against recording vertical fractures in thicker beds where fracture spacing is typically wider than it is in thinner units. Of the three datasets, only cores offer reliable information on the in situ fracture mineralization, measurements of in situ fracture widths, and estimates of the remnant fracture porosities.

Another major difference is in the inferences regarding the maximum horizontal compressive stress. Most of the fracture information suggests that this stress should trend generally E-W on the assumption that the stress system has not changed significantly since fracturing. However, consistent image-log breakout data show that the present-day maximum horizontal compressive stress in fact has a N-S trend. This makes sense in the context of the basin location on the margin of the presently active E-W extension associated with the Rio Grande Rift system. The stresses have not "rotated," rather the N-S compressive stress, unchanged in magnitude, became the maximum horizontal compressive stress when the E-W compressive stress associated with basin-margin thrusting was reduced to become the minimum compressive stress during E-W extension.

2.9.3.2 Rifle Gap

An example of the need to filter outcrop data for application to the subsurface comes from Cretaceous sandstones of the Mesaverde Group in Colorado, where fractures in outcrop can be compared directly to those in an extensive population of cored fractures obtained from wells drilled seven miles (11 km) to the southwest. This study was completed before the advent of image logs, but the outcrop and core data were compared to engineering tests conducted on the fractured reservoirs, providing three mutually supporting datasets.

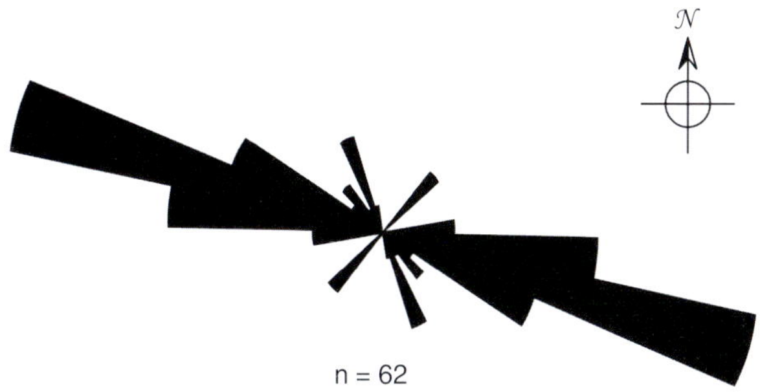

Figure 2.64 The strikes of 62 vertical extension fractures in oriented core from the MWX project, cut from sandstones that are equivalent to the outcrop sandstones shown in the previous figure. These fractures also strike WNW-ESE. The 65 additional fractures captured in horizontal core at the same site corroborate this trend (from Lorenz et al., 1989).

The outcrop fracture data come from exposures at Rifle Gap, where the 4,000 ft (1,220 m) thick section of Mesaverde strata has been uplifted to the surface and tilted 90° so that the sandstones in the formation form tabular ridges. Uplift and tilting impose significant stress on the rock, so caution must be used in extrapolating outcrop fracture systems into the subsurface, and in fact, many of the outcrop fractures at this location are secondary and have no subsurface equivalent. After the strata have been rotated to their pre-uplift/pre-rotation orientation, however, the dominant fracture pattern is resolved into a set of long, throughgoing, systematic extension fractures striking WNW-ESE, and a set of short, irregular, non-systematic fractures striking approximately NNE-SSW and terminating against the systematic fractures (Figure 2.63).

The throughgoing, regional, WNW-ESE fractures are best expressed on the west side of the Rifle Gap canyon where they form bed-normal planes. On the opposite side of the canyon the opposing hill slope is oblique to these fractures, which have been obscured by superimposed irregular fractures created by weathering and gravitational processes.

The WNW-ESE pre-tilt outcrop fracture strikes are consistent with the strikes of 62 bed-normal extension fractures captured by oriented core cut from the Mesaverde strata between the depths of 4,000–8,000 ft (1,220–2,440 m) in nearby wells (Figure 2.64). Strikes of an additional 65 fractures in core from a deviated wellbore cut through the same interval confirm the WNW-ESE strike. The cores were taken as part of the MWX Experiment, a research project funded by the U.S. Department of Energy in the 1980s and 1990s (see Multiwell Experiment Project Groups, final reports, various dates).

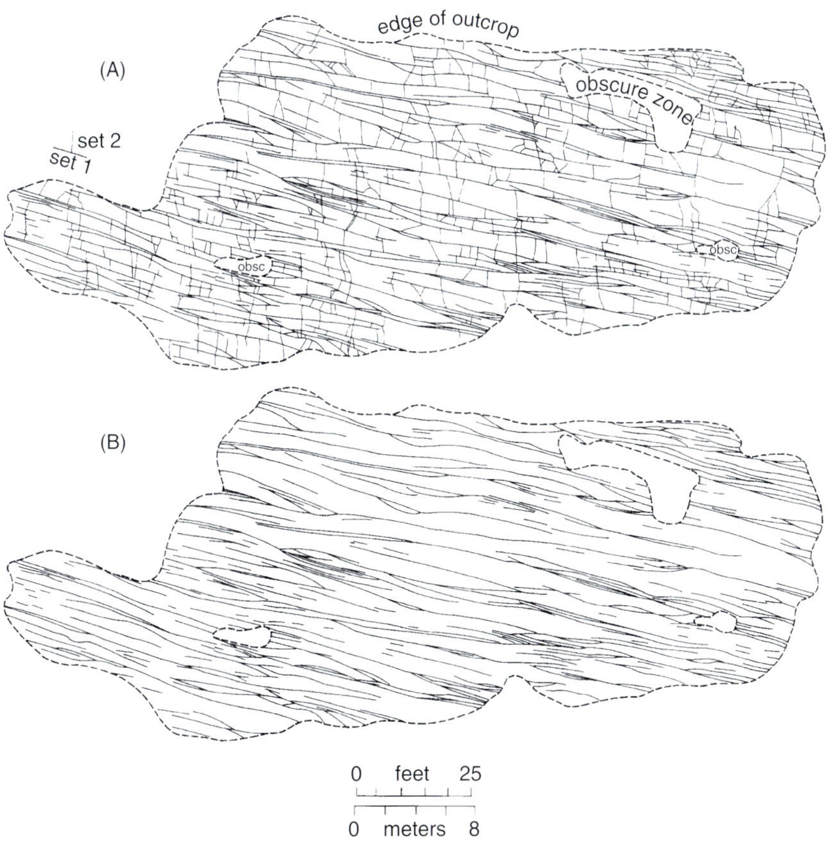

Figure 2.63 Outcrop data and filtered outcrop data. Bed-normal extension fractures, exposed in map-view on the upper bedding surface of a Cretaceous sandstone, in a vertical sandstone flatiron at Rifle Gap, Colorado. The dashed outlines indicate the limit of the outcrop. Bedding strike, horizontal across the figure, is east-west. The throughgoing fracture set has a WNW-ESE strike when the strata are restored to a horizontal position. *Top (A)*: All fractures in this outcrop, traced from a photomosaic. *Bottom (B)*: Only the throughgoing fractures, with the younger, abutting, secondary fractures removed. Fracture strikes in equivalent strata in the subsurface all strike WNW-ESE, and engineering data from reservoirs in this formation indicate a strong permeability anisotropy where k_H also trends WNW-ESE. (Reprinted from Lorenz and Finley, 1991, with permission from AAPG, whose permission is required for further use.)

Engineering data from well tests in the natural-gas, low-permeability sandstone reservoirs at this site, and laboratory measurements, showed that the natural fractures enhance reservoir permeability over matrix permeabilities by several orders of magnitude. Moreover, well-interference tests showed that the horizontal permeability is highly anisotropic, being from 10 to 100 times greater in the WNW-ESE, fracture-parallel direction than in the direction normal to fracture strike. Significantly, there is little subsurface evidence in either the cores or the well tests for the set of short, irregular, cross fractures that are well-developed in outcrop.

These mutually supporting datasets point to the presence of a single set of parallel extension fractures with WNW-ESE strikes in the subsurface (Lorenz et al., 1989; Lorenz and Finley, 1991; Lorenz and Hill, 1994). The cross fractures that are present in outcrop are secondary, surficial features and are not present in the subsurface.

Outcrop fracture data are vital to understandings of reservoir fracturing but must be used with caution. In some cases, the outcrop fracture fabrics may bear no relationship to subsurface fracture patterns in equivalent strata at depth. Even in that case, however, outcrop data can at least provide indications of the fracture susceptibility of various lithologies in a formation.

2.9.3.3 San Ysidro

Another example of the need to use outcrop fracture data cautiously when reconstructing fracture systems at depth comes from outcrops near San Ysidro, New Mexico, in the southeastern corner of the San Juan Basin. Techniques such as linear and circular scan lines on bedding-plane "pavement" surfaces (e.g. Mauldon et al., 2001), and traverses along the irregular edge of an outcrop, can gather slightly different fracture datasets. This case shows that fracture patterns reconstructed from outcrop measurements, even in well-exposed strata, can vary with the method used to sample fractures, and offers a caution against overinterpreting the fracture populations provided by one-dimensional samples from horizontal core or from image logs run in a horizontal wellbore.

In this example, three fracture-strike datasets were acquired from one outcrop of nearly flat-lying Cretaceous Dakota sandstone using three techniques: 1) a circular scan line crossing fracture traces on the bedding-plane pavement; 2) two linear scan lines of equal length on the same pavement surface where each line was intentionally oriented normal to one of the two dominant fracture sets; and 3) measurements taken from the irregular edge of the same pavement where the sandstone is exposed in vertical outcrop (Figures 2.65, 2.66).

Prior to taking these measurements we had determined that the oldest, NNE-SSW-striking fractures originated as a set of widespread, regional extension fractures (Lorenz and Cooper, 2003). A younger, oblique horizontal compressive stress was later superimposed on this fracture system, creating a second, NE-SW-striking set of left-lateral shear fractures. Some of the original extension fractures were reactivated in right-lateral shear under the new stress system, but offset was minimal. This resulted in what appears to be a pair of intersecting conjugate shear fractures, but the two fracture sets have strikingly dissimilar planarities, and one set consistently either abuts or refracts where it approaches or crosses the other set. The two sets formed sequentially, not penecontemporaneously; they did not form as a conjugate shear pair even though one of the strike roses might be interpreted that way based on the conjugate geometry.

All three techniques picked up two dominant fracture sets, but the technique using two linear scan lines

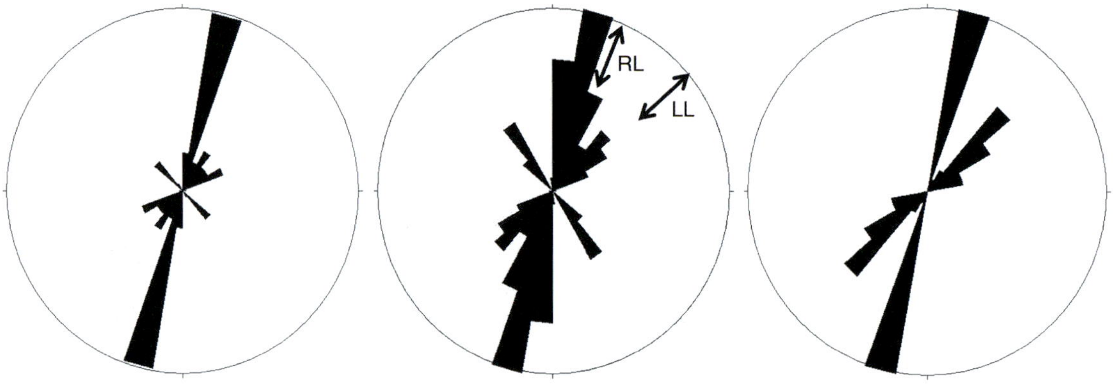

Figure 2.65 *Left*: a rose plot of fracture strikes measured in an 8 m (26-ft) diameter circular scan (n = 54). *Middle*: a rose plot of fracture strikes collected from the irregular outcrop edge immediately adjacent to the site of the circular scan line, including right-lateral (RL) and left-lateral (LL) kinematic indicators; n = 63). *Right*: a rose plot of fracture strikes collected from two 10-m (32 ft) tape scans, each normal to one of the two dominant fracture sets in the same outcrop (n = 39; unpublished data).

Figure 2.66 The sandstone bedding surface from which the rose plots of the previous figure were derived. View is towards the northeast, showing the older planar, thoroughgoing fractures (upper left to lower right) and younger, less regular cross fractures (top to bottom across the photo). A broad-brimmed hat to the left of the plant in the middle right of the photo gives the approximate scale.

entirely missed a third, NW-SE striking fracture set since this set is not well developed and since the two scan lines were deliberately oriented normal to the dominant fracture sets and were thus inadvertently oriented at shallow angles to the third, subordinate set. This deficiency in the scan-lines could have been remedied by measuring fractures along three or even four scan lines, with orientations that intersect at 60° or 45° respectively (LaPointe and Hudson, 1985; Hudson and Priest, 1983). Alternately, the third fracture set would probably have been captured by measuring each fracture within a specified area on the pavement (a "box count"), provided that area was sufficiently large.

Measurements from the edge of the outcrop were the only ones to offer a view of the fracture faces and the kinematic indicators they display, showing subtle but consistent right-lateral and left-lateral shear steps and local horizontal slickensides. The two dominant fracture sets are most easily distinguished by the technique using two separate linear scan lines, but this is almost a predetermined outcome since the orientations of the scans were specifically chosen to be normal to the two dominant fracture sets. In contrast, the two sets are not distinct on the circular scan line, and the NW-SE striking set that is well developed and distinct in the outcrop-edge dataset might easily be dismissed as scatter in the rose plot derived from the circular scan line. Only the two straight scan lines suggest that the two dominant fracture sets are equally well developed numerically. The strike rose plotted from the linear scanlines also highlights the distinct difference in planarity between the two fracture sets, capturing the narrower strike distribution of the older, throughgoing NNE-SSW fracture set, consistent with its origin as a set of parallel extension fractures formed in rock with few mechanical heterogeneities, whereas propagation of the younger, more irregular fracture set was influenced by the presence of the older fracture set.

The three datasets were taken within 100 m of each other, so the variability is unlikely to be due to structural variations. Larger datasets might bring the three datasets into better agreement, but these results suggest that there can be substantial differences in fracture data acquired from outcrops depending on the techniques used. Any sample and any sampling technique can be limited and/or biased. More important to hydrocarbon reservoirs, this example shows that horizontal cores and image logs can provide excellent but specific and localized fracture data. They are essentially one-dimensional scanlines and provide the same type of data that would be provided by a single linear scanline on an outcrop pavement surface, subject to the same limitations.

Fracture data acquired by remote techniques (satellite, aircraft, etc.) can also provide important insights into gross-scale fracture patterns, but they must be supported by observations and data collected at a more detailed scale. An early broad-brush study of fracture patterns on the Colorado Plateau conducted by Kelley and Clinton (1960) and based on high-altitude aerial photography is important in showing that fracture domains can vary significantly over scales equivalent to the footprint of typical reservoirs. The reasons for variation between fracture domains were not apparent in the remote data, however, and Kelley and Clinton specifically noted that they might have had a better understanding of the origin of the different fracture patterns if they had been able to visit the outcrops and distinguish shear fractures from "stretch" (i.e. extension) fractures. Any remote-sensing fracture study should be ground-truthed.

2.10 Fracture Data from 3D Seismic Surveys*

Three-dimensional (3D) seismic data volumes are routinely used in the petroleum industry to define such things as subsurface structure, stratigraphy, lithology, porosity, and hydrocarbon content. Various 3D seismic-based techniques have also been developed to help define the presence, intensity and orientation of fracture systems. Hart (2011) provided a summary of these techniques (their physical basis, data acquisition, processing, and interpretation), noting that

* Dr. Bruce Hart provided significant and much-appreciated input into this section.

seismic-based fracture-detection methods fall into two broad categories: direct and indirect approaches.

Tsvankin and Grechka (2011) and Far et al. (2013) discussed the theoretical and applied aspects of the direct-detection approach to seismic fracture characterization, which relies on the relationships between natural fractures and seismic wave propagation. Some of the methods aim to characterize fractures by examining changes in wave-propagation velocity as a function of propagation direction (e.g.velocity variation with azimuth – VVAZ). Compressional waves (p waves) travel faster parallel to natural fractures than perpendicular to them, and so 3D seismic datasets can be analyzed to see whether seismic energy preferentially travels faster in one direction than in another during data acquisition (Figure 2.67). In addition, shear waves (s waves) become polarized in the presence of oriented natural fractures, with a fast set traveling parallel to the fractures and a slow set traveling perpendicular to the fractures.

These and other velocity-based analyses are performed on the seismic data prior to stacking, and they detect the body properties of the rocks through which the seismic signal is traveling. Other types of direct detection methods examine changes in amplitude along seismic horizons that might be due to the presence of oriented natural fractures at bed boundaries (Figure 2.67). These methods (e.g. amplitude variation with azimuth – AVAZ – or amplitude variation with

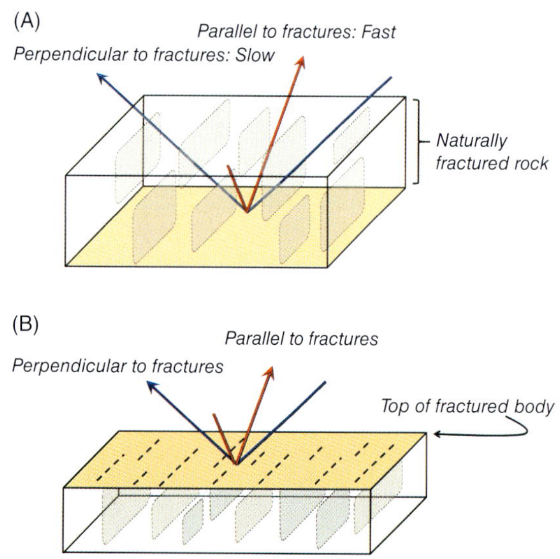

Figure 2.67 Conceptual diagrams illustrating the physical basis of direct detection methods for natural fractures (figure courtesy of B. Hart). *Top*: seismic waves travel through a naturally fractured rock body faster in the direction parallel to the fractures than perpendicular to the fractures. *Bottom:* the strength of reflections from the top of a naturally fractured bed varies depending on whether the seismic waves are traveling parallel or perpendicular to the fractures at that interface.

offset and azimuth – AVOAZ) analyze interface properties in pre-stack seismic datasets. For any of these direct-detection methods to succeed, the seismic data need to be acquired (e.g. wide-azimuth acquisition) and processed with the goal of natural fracture detection in mind.

Indirect methods for seismic natural-fracture detection are usually based on some "geometric" attribute such as curvature or coherency that detects a structural feature (a fold or a fault), and the presence, intensity, and orientation of fractures are then inferred based on assumed relationships between the larger structures and associated natural fractures. For example, the width of a fractured fault damage zone might be estimated from the fault throw as defined by reflection offsets in the seismic data, and different orientations and/or types of fractures might be predicted for different parts of a fold. The seismic attributes are generated from the stacked and migrated 3D seismic data volumes, and wide-azimuth seismic acquisition is not a prerequisite. Chopra and Marfurt (2005) and Hart (2011) discuss various aspects of this approach.

Both the direct and the indirect fracture-detection approaches can provide non-unique or even misleading interpretations. For example, although oriented natural fractures can produce velocity anisotropy in rocks, a stress differential imposed on rock can create a similar velocity differential in otherwise homogeneous, unfractured rock (Tutuncu, 2010). Seismic anisotropy may not be detectable if multiple, intersecting, open fracture sets are present in the subsurface, and it is not possible to determine whether the fractures inferred to be associated with seismically defined structures are open or mineralized. Geometric attributes (e.g. curvature) are unlikely to detect regional joint sets in otherwise undeformed strata. Finally, data quality can have a significant impact on fracture detection using either direct or indirect methods.

For these and other reasons, seismic-based fracture interpretations should be integrated with, and/or calibrated by, other attributes and datasets such as natural-fracture characteristics obtained from core, image logs, and outcrop analogs. Production data can also be helpful for defining the presence, orientation and hydraulic characteristics of seismically detected natural fractures.

One example of successful seismic definition of a fracture domain and its influence on production comes from the San Juan Basin in New Mexico, where seismic-horizon curvature attributes highlight linear, NNE-SSW-striking anomalies (Figure 2.68). Hart (2006) documented a correlation between these seismically delineated features, production anomalies, and observations of inter-well interference (see Figure 3.19). The seismic curvature attribute defines a narrow,

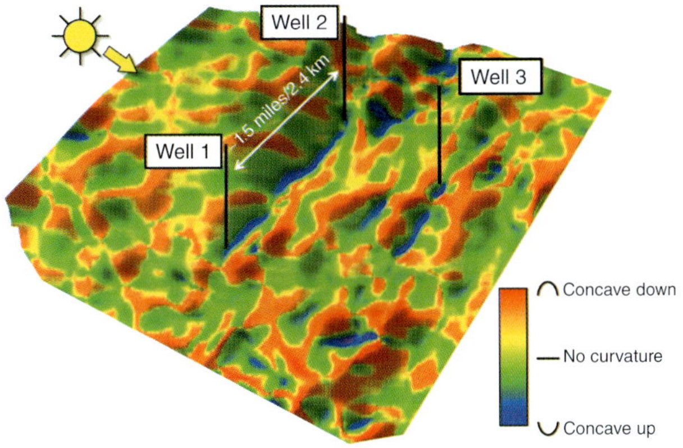

Figure 2.68 NNE-SSW trending concave-up structural features (blue) defined on the top of a Cretaceous tight-gas sandstone in a 3D seismic dataset from the San Juan Basin. The map illustrates strike curvature (color) draped over a shaded relief map of the time-structure horizon. All three wells have anomalously high production rates relative to several tens of off-trend wells (not shown), and interference between wells 1 and 2 indicates a shared high-permeability pathway. Outcrop study shows that zones of intense fracturing are associated with NNE-SSW-striking faults and grabens in nearby strata, providing an interpretation for the seismic anomalies as well as explaining the high production rates and interference along the trends. (Modified from Hart, 2006, with permission from AAPG, whose permission is required fore further use).

elongate, concave-up structural feature that probably represents a fault-bounded graben rather than a fold/flexure, even though the displacement on the faults is small enough that it cannot be seismically resolved by discrete reflection offsets. The correct seismic interpretation of these structures only became apparent when the seismic data were combined with production data and nearby outcrop observations of analogous faults. A small-scale outcrop analog of one of these faults is shown in Figure 2.69.

Two published seismic interpretations of natural fracture systems at depth in the reservoirs at Teapot Dome, Wyoming, illustrate some of the techniques and interpretations of more general seismic studies for natural-fracture characterization. Ouenes et al. (2010) and Schneider et al. (2016) used the same Teapot Dome 3D seismic dataset and followed similar approaches to the problem, using attributes such as coherence, curvature, and texture to create maps of fracture strikes, dips, and intensities across the study area.

Ouenes et al. (2010) published fracture maps for several of the Teapot Dome reservoirs, deriving the maps from combinations of post-stack fracture-related seismic attributes, and validating the maps with both reservoir simulations and data from multiple "blind" wells. The authors incorporated geologic "fracture drivers" such as structural curvature, lithology, and proximity to faults into their model, and calibrated the model with "fracture

Figure 2.69 An outcrop of Cretaceous Dakota sandstone showing a 10 ft (3 m) scarp formed by a normal fault (to the right of the geologist), and an intensely fractured zone in the hanging wall below the geologist's feet. This example is below seismic resolution, but it provides an analog for larger, parallel, and similarly fractured faults in the area.

indicators" such as image logs, production logs, and core descriptions, iteratively integrating the drivers and indicators with neural-network techniques to develop a Continuous Facture Model (CFM). Figure 2.70 shows one of their composite fracture maps, derived from compiling fracture attributes and calibrating them with core and image-log data.

In contrast, Schneider et al. (2016) focused on four post-stack seismic attributes (coherence, curvature, coherence on spectral decomposed seismic cubes, and textural attributes), and published separate fracture maps for these attributes. Each map defines the major faults that cut across Teapot Dome; there is less overlap with fracture patterns at smaller scales. The authors concluded that attribute analysis can provide volumetric estimations of fracture intensities that can be used to construct Discrete Fracture Network (DFN) models.

2.11 Fracture Data Acquired by LiDAR

Terrestrial-based LiDAR (Light Detection and Ranging), has been used successfully to acquire fracture data from large, well-exposed outcrops. LiDAR equipment utilizes a pulsed laser (Light) and measures the reflected pulse (Detection and Ranging), collecting millions of data points with centimeter-scale resolution from an outcrop. LiDAR can be used to acquire data from relatively remote outcrops such as tall and otherwise inaccessible cliff faces. The fracture data that are acquired by LiDAR are primarily the geometric characteristics of fractures such as strike, height, length, spacing, and

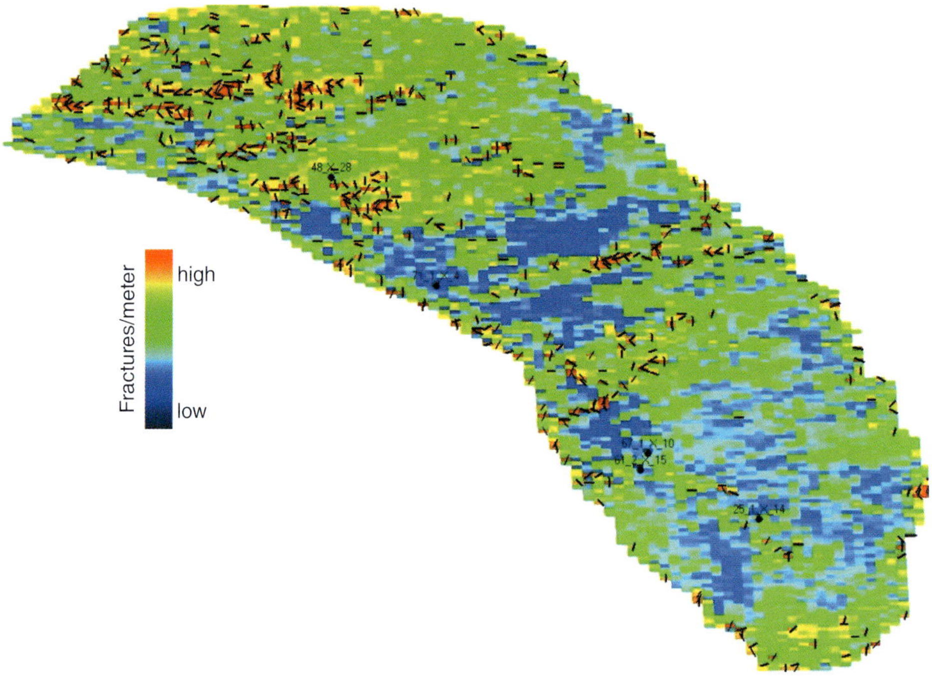

Figure 2.70 A fracture intensity and orientation map for one stratigraphic layer of a 3D continuous fracture model derived from fracture attributes. The map shows areas of high fracture intensity (warm colors), and the dominant fracture trends inside the high fracture-intensity pixels (black tick marks). Geologic and geomechanical fracture drivers were correlated to local fracture densities, derived from image logs, using artificial intelligence, to construct the 3D continuous fracture model from which this map was extracted. (From Ouenes et al., 2010.)

distribution. Such datasets are usually enhanced with smaller-scale nose-to-the-outcrop studies that define other characteristics such as fractography, aperture, and mineralization. LiDAR data can be combined with high-resolution photography, stratigraphic, seismic, and ground-penetrating radar datasets, and examined via various attribute analyses (Lee et al., 2007; Heritage and Large, 2009).

LiDAR-based fracture studies require an investment in equipment, the expertise to run it, and analysis of voluminous datasets. LiDAR scans take time to prepare reference targets on the outcrop and set up the equipment, but scanning an outcrop takes only a few hours, and the scans can be done in almost any condition of ambient lighting since the system provides its own light. Most LiDAR studies use overlapping scans to enhance data quality. Processing the data typically takes longer than acquiring it.

Two complementary examples of land-based LiDAR studies, utilizing the same dataset, examined a reservoir-scale fold of the Pennsylvanian Tensleep Formation at the Alcova Anticline, Wyoming (Figure 2.71). Two vertical outcrops, each on the order of 100 m (300 ft) tall and 500 m (1,600 ft) wide, expose parallel, opposing cross sections of the formation's interbedded eolian sandstones and interdune dolomites on the sides of a 140 m (460 ft) wide canyon. The extensive LiDAR dataset was used by Gilbertson (2006), and by Zahm and Hennings (2009), to define fracture distributions and characteristics and ultimately to create fracture models of the anticline. The data define both a pre-fold fracture system and fracture sets related to folding, and show that fracture distributions and intensities are controlled by lithologic variation, structural domain, and diagenetic differences in the strata across the large fault-propagation fold.

Shear fractures can only be distinguished from extension fractures in these studies by physical calibration to the gross differences in geometry of the fracture types in the outcrop. However, a LiDAR study can in part circumvent the need for that distinction, important in determining whether fractures cross bed boundaries to enhance vertical permeability, since the LiDAR data provide direct information on height relative to bed thickness. The distinction is still be important, however, in assessing individual fracture permeabilities since the typical apertures of the two fracture types can be very different. The authors of the Alcova studies also noted that it was difficult to detect the deformation-band shear fractures and compaction bands that are common in some parts of the structure since the rock rarely breaks

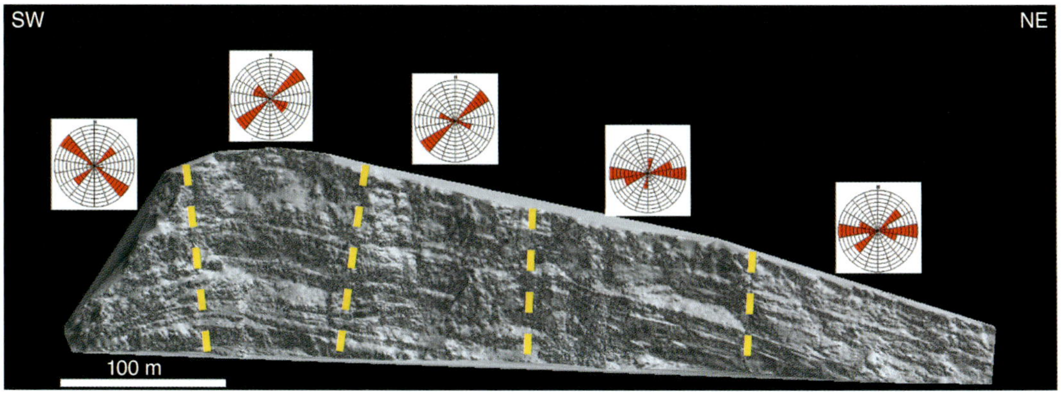

Figure 2.71 The Alcova Anticline consists of folded eolian sandstones and dolomites of the Tensleep Formation draped over a basement fault-propagation fold, which verges towards the southwest (to the left in this image). LiDAR imaging of the large outcrop collected millions of data points with centimeter-scale resolution, which were used to pick hundreds of fractures, defining several fracture sets as well as variations in fracture intensity, spacing, strike, and height across the fold. (From Gilbertson, 2006, with permission of the author.)

along these shear-fracture planes, thus they create few of the planar faces that are detected by LiDAR.

2.12 Fracture Data from Engineering Tests

In the context of this volume, the main reason for characterizing and analyzing fractures is to help predict and understand fracture-controlled flow in a reservoir. Conversely, production and well test data such as flow rates and decline curves can provide insights into fracture characterization. Like seismic data, the interpretations of engineering data are commonly non-unique since in addition to the fracture system, the flow rates from a well are governed by numerous other factors including pressure, reservoir heterogeneity, matrix permeability, and fluid viscosities and saturations. Interpretations of the degree to which fractures are important depend on how well these other parameters are constrained, but well tests and production records can provide significant pieces of the puzzle and, where available, should be included when assessing the natural-fracture plumbing of a reservoir. Fracture analyses constrain the interpretations of well tests, and vice versa.

At the first level of interpretation, the difference between the production capacity of a well and the commonly more limited capacity of the reservoir matrix is the effective, reservoir-scale *fracture permeability* of the system (Figure 2.72) (e.g. Belfield, 1988). However, flow from a single well does not constrain the degree or orientation of anisotropy imposed on a reservoir by a fracture system, so single-well testing cannot, for example, suggest placements for development wells, and hands-on fracture analyses still have their place early in a field development.

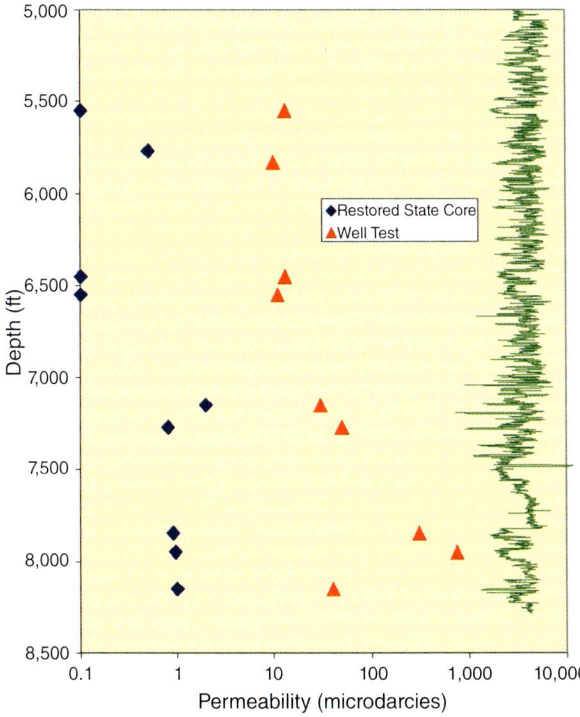

Figure 2.72 Natural-gas sandstone reservoirs of the Mesaverde Group in Colorado deliver at rates indicating that the system permeability is on the order of 10–100 millidarcys, whereas the measured reservoir matrix permeabilities measured under restored-state conditions in the laboratory are less than a microdarcy. Natural fractures make the difference. (Reprinted from Warpinski and Lorenz, 2008, with permission from AAPG, whose permission is required for further use.)

If the system permeability of a reservoir measured from production or well tests is equal to the restored-state matrix permeabilities measured in the laboratory from core plugs, then either the reservoir is not fractured or the fracture system is plugged and ineffective.

Multiple wells, and well-interference tests or tracer tests, are needed to determine whether in situ horizontal permeability is significantly greater in one direction than another. If so, then the reservoir probably contains a system of open, interconnected fractures of a type and orientation that provide enhanced permeability in only one horizontal direction. Well tests are also one of the ways of determining whether the permeability enhancement provided by a fracture system is dynamic, changing with changes in the in situ stress system during production, and if so whether that would damage the fracture-permeability system. Nevertheless, pre-production predictions of damage are more useful than observations of damage during production, so laboratory work should precede production if damage to the fracture system is a possibility.

Further discussions and descriptions of the effects of fractures on fluid flow in reservoirs are provided in Part 3, so this is only a brief introduction to the importance of incorporating engineering tests into fracture analyses. A well-documented example of the effects of natural fractures on production was provided by the Multiwell Experiment in Colorado, where detailed tests and an extensive public database showed that natural fractures in low-permeability sandstones create millidarcy-scale reservoirs out of otherwise un-producible microdarcy-scale matrix rock (Figure 2.72). Fractures are the primary production mechanism for these reservoirs, and well tests have shown that their permeability is sensitive to changes in the in situ stress during production, as well as susceptible to damage by injected stimulation fractures (Multiwell Experiment Project Groups).

Early attempts to stimulate these reservoirs with large volumes of viscous fluids and gels (e.g. van Poollen et al., 1977), and even with nuclear devices (Lorenz, 2001), merely plugged the natural-fracture permeability system, as documented by pre- and post-stimulation well tests showing that production rate could be halved by the intended stimulations. However, the reason for that reduction, i.e. damage to the pervasive natural fractures, was not recognized until the reservoirs had been characterized by an extensive core and logging program and related well tests (see the Multiwell Experiment Project Groups reports), and only then could non-damaging stimulations be properly designed.

2.13 Case Studies in Estimating Fracture Effectiveness from Core

2.13.1 Introduction

The following case studies illustrate how different kinds of fracture data from cores can be analyzed to assess the effectiveness of the cored fracture systems and their likely effects on reservoir plumbing. All core studies and fracture analyses are not the same; they depend on how much and what kinds of data are available for analysis. The cores described below provided datasets of different sizes and types, constraining the possible analyses yet offering different opportunities. Collected fracture data from cores may not all fit directly into a fracture-controlled fluid-flow model, but they are all useful in a qualitative understanding of the permeability system and thereby focus and constrain models.

Relatively long cores intercepting numerous fractures provided robust datasets for the following case studies. One case study describes an archived core using only the core butts, one describes a core where both the slabs and butts were available, and one describes an unslabbed horizontal core fresh out of the core barrel. The cores were not oriented by core-orientation tools but the top side of the horizontal core could be determined from the wellbore survey, bedding, and the shallow angle between bedding and the core axis.

The cores and core data are proprietary so they are not identified by well name, company, location, or formation. Complete disclosure is not essential since the fracture data and interpretations are largely independent of context.

2.13.2 Case Study 1: Archived Vertical, Unoriented Core

In Case Study 1, a population of 177 natural fractures was logged in approximately 240 ft (73 m) of vertical core. The core had been warehoused for almost four decades and had been slabbed and sampled prior to our study. The slabs were not available but the butts were relatively complete, and only about 5% of the core was missing. The natural fractures in this core had a high overall density of 0.75 fractures per foot of core, and a high ratio of 0.49 feet of fracture height per foot of core. However, most of the fractures are occluded and ineffective, so the fracture porosity-permeability system in the reservoir is controlled by a set of open fractures that comprise a relatively small percentage of the total fracture population.

Most (74%) of the logged fracture population in this core are extension fractures, consisting of a set of 131 calcite-mineralized, vertical fractures (Figure 2.73). Although numerically dominant, this fracture set contributes minimally to the permeability system because the fractures are narrow, averaging only 0.6 mm wide, and they are largely occluded with calcite. Only 9 of the 131 logged fractures of this set retain visible remnant fracture porosity, and they provide only 10% of the total open fracture volume in the system. Moreover, the extension fractures are strata-bound, limiting

Figure 2.73 Calcite-mineralized extension fractures. *Top*: a 0.05 mm wide, 1.2 ft (36 cm) tall extension fracture (between the arrows) with no visible remnant fracture porosity. *Bottom*: a 0.4 mm wide, 1.4 ft (43 cm) tall extension fracture with an estimated 30% remnant fracture porosity within the partially occluded fracture width.

their contribution to vertical permeability, and they are parallel, non-intersecting fractures so that what little permeability enhancement they do offer is only in the direction parallel to fracture strike.

The extension fractures strike parallel to coring-induced petal fractures, therefore the extension fractures strike parallel not only to each other but also to the trend of the maximum horizontal in situ compressive stress. Hydraulic stimulation fractures, controlled by the in situ stresses, will propagate parallel to the extension fractures in this reservoir.

In fact, the fracture-permeability system in this reservoir is dominated by a much smaller population of 12 wide, incompletely mineralized, intermediate-angle shear fractures with variable strikes (Figure 2.74). Numerically, these fractures comprise only 7% of the fracture population, yet they contribute 80% of the total open volume of the fracture system since they average nearly 2 mm wide and the open void space within the mineralization (the remnant fracture porosity) averages 20% of the fracture widths. Moreover, these fractures cut across bedding, connecting the reservoir vertically across the interbedded mudstones. The shear fractures have multiple, intersecting strikes, documented where the core could be pieced together for intervals up to 10 ft (3 m) long. This population of relatively open shear fractures creates a network of wide, mutually intersecting, vertically extensive fractures that intersects and interconnects with the other cored fracture types.

Several small sets of compaction and syn-sedimentary shear fractures complete the natural-fracture population in the Case 1 core (Figure 2.75), but these fractures are narrow, short, and contribute little to the reservoir permeability system.

2.13.3 Case Study 2: New, Un-Slabbed Horizontal Core

2.13.3.1 Introduction

In Case Study 2, approximately 450 ft (140 m) of horizontal core captured 162 natural vertical extension fractures that can be divided into three distinct sets striking NE-SW, E-W, and NW-SE (Figure 2.76). The three fracture sets occur in all three cored lithologies (fine-grained limestones, coarse-grained limestones, and mudrock), creating an effective and well-connected permeability network.

Fracture data obtained from a horizontal core are similar to data from an outcrop scan line and are subject to the same limitations, but this core appears to have sampled representative populations of all three fracture sets because the core is long and the fractures are closely spaced, and because the core cut oblique to fracture strikes. The core also cut oblique to bedding and so provides information on fracture distributions relative to lithology. The robust dataset allowed for a detailed assessment of overall fracture effectiveness as well as a comparison of the relative effectiveness of the three cored fracture sets. Because the wellbore cuts downward oblique to bedding (Figure 2.77), the high side of the core could be determined and fracture strikes could be measured relative to north (Figure 2.78).

Fractures are most closely spaced in the fine-grained limestones (Figure 2.79), where 37 fractures were captured by 20 ft (6 m) of core, and where the average spacing between pairs of parallel fractures for all three fracture sets combined is 0.46 ft (0.14 m). The average fracture spacing for the combined fracture sets in the coarse-grained limestones is somewhat wider, 1.4 ft (0.43 m), where 30 ft (9 m) of core captured 32 fractures.

Figure 2.74 Three views of one of the intermediate-angle shear fractures in the Case Study 1 core. *Upper left*: the fracture was intersected at right angles by the slab plane, showing its true 48° dip angle. *Upper right*: the fracture face displays dip-slip slickenlined calcite mineralization and a significant percentage of remnant void space within the mineralization. *Bottom*: photomicrograph of the fracture face showing the local pockets of euhedral calcite crystals that grew into open void spaces along the fracture plane, as well as dark patches of solid bitumen within the fracture aperture. The gray-tan material is drilling mud that was not completely removed from the fracture face despite scrubbing with a toothbrush.

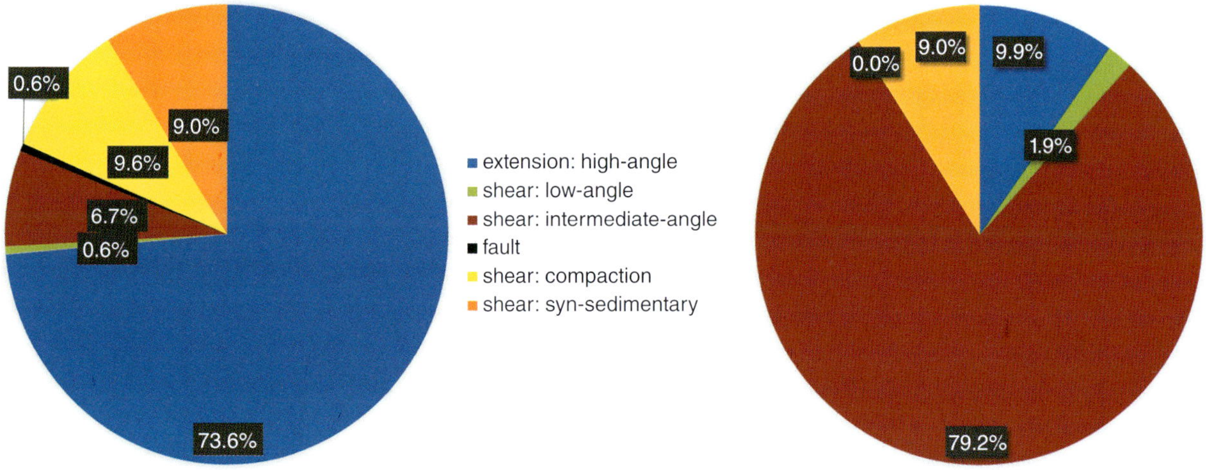

Figure 2.75 Pie charts of the fracture population in Case Study 1, showing that the set of 131 high-angle extension fractures dominates the total fracture population of 177 measured fractures (*left*), but that the population of 12 wider and more open inclined shear fractures dominates fracture effectiveness (*right*).

Figure 2.76 An interval of closely spaced NE-SW and NW-SE striking vertical fractures (adjacent to the red lines) that are two of the three fracture sets captured in this horizontal core. These fractures are incompletely mineralized with calcite. View is looking down on the high side of the core (i.e. "map view"). Fracture strikes ("NE" and "NW", as written on the core surface) were obtained by using the known wellbore azimuth through the cored interval, the measured angles between the fracture planes and the core axis, and a determination of the high-side of the core based on the low-angle intersection between the core axis and bedding planes.

Figure 2.77 View of the side of the Case Study 2 core, showing a bedding plane (just above the red line) that cuts obliquely across the core as the core cuts down-section. Downhole, towards the toe of the wellbore, is to the left, and the core is rotated so that the high side of the core is at the top as laid out on the table. The core rests in a trough made of angle iron bolted to the table, which keeps the core in place yet affords the opportunity to both push the core up and down along the trough and rotate the pieces for a full view of the core surface.

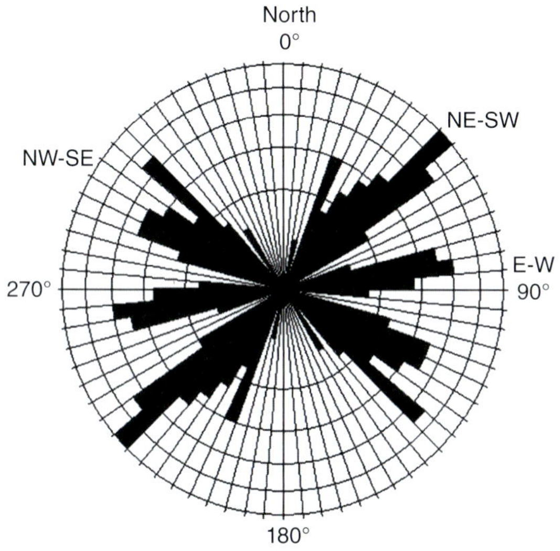

Figure 2.78 High-angle extension fracture strikes relative to north, as measured in the Case Study 2 core, define three fracture sets (n = 162).

Figure 2.79 Two NE-SW striking vertical fractures (arrows) in horizontal core. Crystalline calcite covers the prominent fracture face on the end of the core, indicating that the fracture is incompletely filled, as does the lack of connection between this fracture face and the opposing one. The dark areas are covered with translucent calcite crystals, the white areas consist of calcite mineralization that was crushed during coring and processing. Mineralization in the second, parallel fracture is more complete and has largely healed the rock. The high side of the core is at the top of the photograph.

Table 2.1 Summary of the average parameters for each of three fracture sets in three lithologies*. "Porosity" is the remnant fracture porosity within the fracture widths.

		Width (mm)	Porosity (%)	Spacing (ft)
Fine-Grained Limestone	NE-SW	0.1	38.33	1.87
	E-W	0.07	39	0.49
	NW-SE	0.14	0.48	0.22
Coarse-Grained Limestone	NE-SW	0.38	12.81	5.19
	E-W	0.14	79.67	7.13
	NW-SE	0.44	9.77	8.66
Mudrock	NE-SW	0.09	54.17	0.79
	E-W	0.09	70	2.54
	NW-SE	0.09	21.67	1.21

*Fine-grained limestones: n = 28. Coarse-grained limestones: n = 80. Mudrock: n = 26. The total of 134 is less than the total number of fractures present in the core since the dimensions listed in this table could not all be measured for each fracture.

Average spacing is widest in the mudrocks at 6.4 ft (1.95 m), where over 400 ft (120 m) of core cut across 93 extension fractures. The fractures retain significant open aperture within the partial mineralization, and they have intersecting strikes so they form an interconnected and effective fracture-permeability system, but the three sets have different properties and are not equally effective in enhancing permeability.

Combining the measured spacings of multiple fracture sets provides a quick-look summary of relative fracture development by lithology, but for the full analysis of this core the fracture spacings were separated and listed by set as well as by lithology. The three lithologies and three fracture sets form a nine-box matrix (Table 2.1). The same spacing trends occur for each of the three fracture sets within each of the lithologies, suggesting that during all three fracture events the three lithologies had the same fracture susceptibilities relative to each other.

Fourteen fractures were picked on the image log within the cored interval, all of them within a restricted 200-ft long (60 m) interval of mudrock. The image-log fracture signatures have strikes that are compatible with strikes of fractures observed in the core, but the log did not record the same fracture distributions or intensities seen in core.

2.13.3.2 Fracture Effectiveness

Fracture width measurements and the estimates of remnant fracture porosities (Table 2.1) can be combined to calculate a fracture "effectiveness" parameter for each fracture, for each fracture set, and for the average fracture in each lithology. Effectiveness was calculated as the measured width (in millimeters) multiplied by the estimated "remnant fracture porosity" (the amount of void space within the mineralization, estimated as a percentage of the fracture width) for that fracture, and then divided by 1 mm to remove the length unit. Fracture effectiveness is thus scaled relative to width but these are not parallel-plate fractures. Effectiveness is calculated as a dimensionless number because the numbers are not widths, i.e. they summarize but do not capture the high degree of variation in width along each fracture or account for the effect of rough fracture faces on fluid flow. Effectiveness is best visualized as a scale relative to width adjusted for mineralization and tortuosity.

Fractures with lower effectiveness numbers will have less permeability due to the more tortuous flow pathways. Fractures that are completely occluded by mineralization have width but no effectiveness in most oil reservoirs, although in high-pressure, natural-gas reservoirs even macroscopically occluded fractures may contribute to permeability. The results are plotted on histograms (Figure 2.80) and show log-normal distributions, from which we have taken the average from each fracture set for the subsequent calculations. The histograms summarize all measurable fracture widths in the left column, and the calculated fracture effectiveness in the right column. The histograms show that the average NE-SW fracture and the average E-W fracture both have an effectiveness of 0.07, over twice that of the average NW-SE fracture. However, as shown next, the NE-SW fractures dominate the system due to additional factors including the greater number of fractures that comprise this set.

2.13.3.3 System Effectiveness and Permeability Anisotropy

Without considering the in situ stress system (poorly constrained at this site), the maximum permeability enhancement should be along the most volumetrically effective NE-SW fracture set. However, individual fractures must be interconnected to be effective across a reservoir, thus one must also consider the relative development of the fracture sets in order to assess their effects on permeability. For the Case 2 fractures, the average individual fracture effectiveness is the same for the NE-SW (0.07) and the E-W fractures (0.07), but over twice as many NE-SW fractures (71) as E-W fractures (31) contribute to the system, thus their contribution to the total permeability is double that of the E-W fractures. In contrast, although there are also nearly twice as many NW-SE fractures (55) as E-W fractures (31), their effectiveness (0.03) is only half that of the E-W set (0.07) because they have less remnant fracture porosity.

The influence of each fracture set on permeability directionality can also be considered as a percentage of the cumulative total effective porosity of all fractures. The ternary Excel "radar" plot (Figure 2.81) shows an

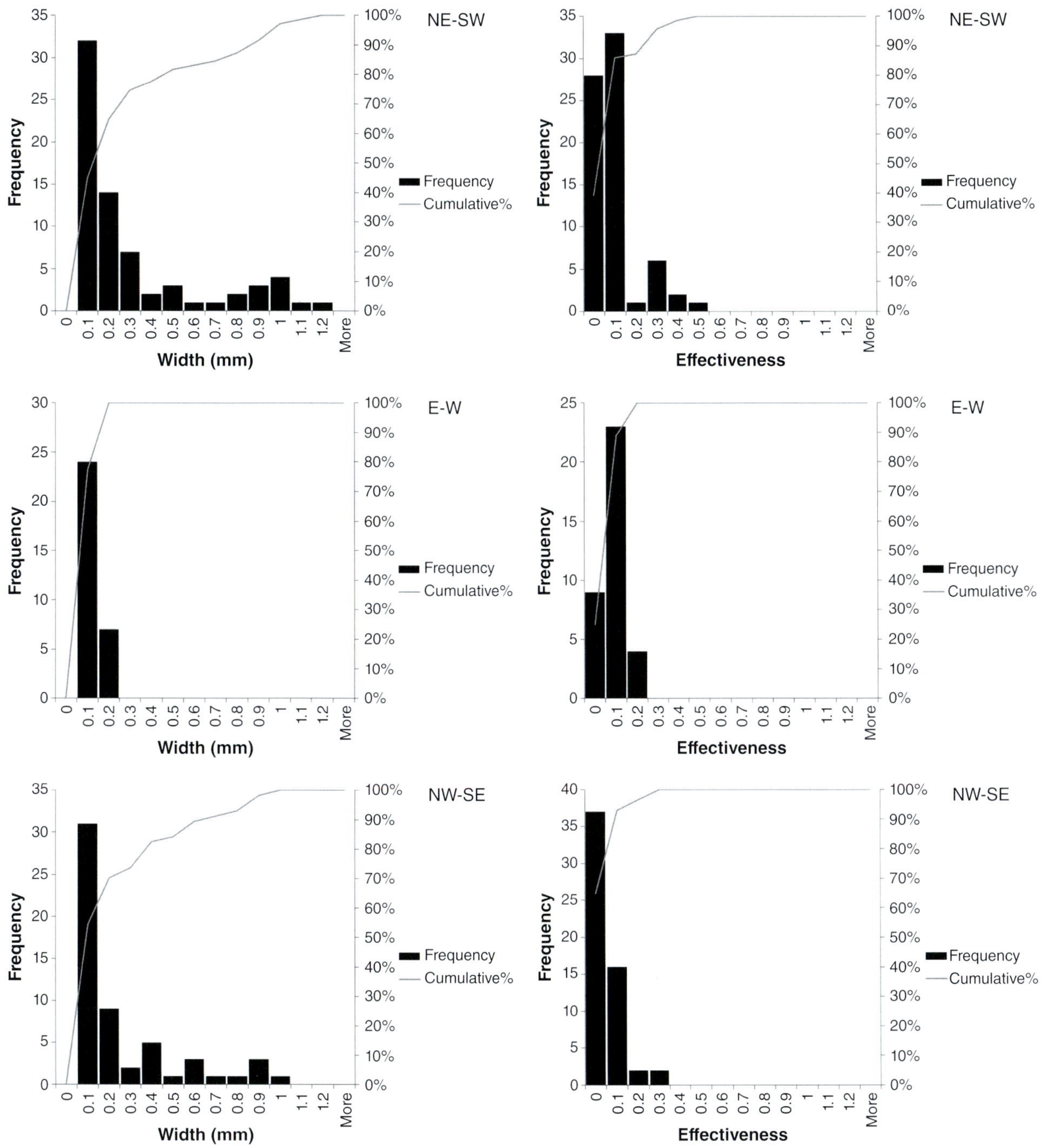

Figure 2.80 *Left column*: high-angle extension fracture width histograms for all three lithologies combined. *Top to bottom*: NE-SW fractures (n = 71; min 0.01, max 1.20, ave 0.31 mm), E-W fractures (n = 31; min 0.01, max 0.20, ave 0.10 mm), and NW-SE fractures (n = 57; min 0.03, max 1.00, ave 0.25 mm). *Right column*: high-angle extension fracture effectiveness histograms for all three lithologies combined. *Top to bottom*: NE-SW fractures (n = 71; min 0, max 0.45, ave 0.07), E-W fractures (n = 31; min 0, max 0.20, ave 0.07), and NW-SE fractures (n = 55; min 0, max 0.25, ave 0.03). Each of the three fracture sets can be broken out by lithology, providing nine such graphs for width and nine more for remnant fracture porosity, but those are not provided here.

approximation of the fracture-controlled permeability directionality in the Case 2 reservoir, calculated from the percentage of the total system effectiveness attributable to each fracture set (effectiveness sums for NE-SW set plus E-W set plus NW-SE set equal a total system effectiveness).

This reservoir has an interconnecting drainage system of three natural fracture sets, but the three sets do not

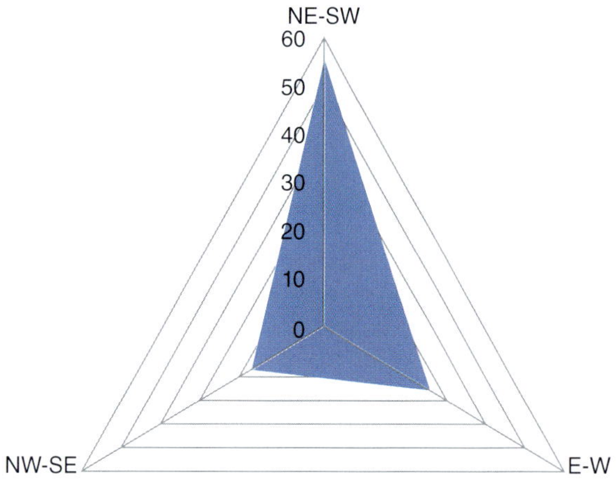

Figure 2.81 Permeability directionality in the Case Study 2 reservoirs based on effective fracture data and the relative degree of development of all three fracture sets in all three cored lithologies combined. The three axes on this plot are the individual fracture sets and not true compass directions, thus this diagram does not show permeability anisotropy directions relative to north.

contribute equally to permeability. For the combined lithologies, NE-SW fractures contribute 56% of the total fracture effectiveness, E-W fractures contribute 26%, and NW-SE fractures contribute only 18%. These plotted numbers show that the NE-SW fracture set dominates the system but that there is an important contribution from the E-W set. In contrast, there is little contribution from the NW-SE fracture set.

Fracture-permeability directionality shows similar NE-SW-dominated trends when assessed by lithology (Figure 2.82). Directionality is strongest in the mudrocks, and weakest in the coarse-grained limestones. The directionality in the mudrocks should dominate the system since 90% of the core consists of this rock type (the core was cut at a uniform angle across the stratigraphy so the core percentages are a reasonable representation of the percentages of the stratigraphic lithologies). Moreover, 57% of the cored fractures occur in this lithology. However, the other two lithologies occur as relatively thin beds and the more closely spaced, interconnected fractures in those lithologies should create high-permeability streaks in the formation.

2.13.4 Case Study 3: New, Slabbed, Vertical Core

2.13.4.1 Introduction

The slabs and butts of 300 ft (91 m) of vertical, unoriented core were both available for this study. The core contains a fracture population consisting of 422 fractures which can be separated into four sub-populations by fracture type. Only one type, consisting of a set of 69 high-angle (i.e. vertical to steeply dipping) extension fractures, is effective and contributes significantly to the permeability of the four cored, lithologically distinct reservoirs.

The effective extension fractures are relatively tall, open, and interconnected. They are present in all four cored formations, but the degree of fracture development is dependent on lithology. Extension fractures are poorly developed in the muddy lithology of formation A, whereas they are well developed in the calcareous strata of formation C (Table 2.2, Figure 2.83). They are moderately well developed in the mixed lithologies of formations B and D.

The other cored fracture types include ptygmatically folded extension fractures, horizontal fractures, and compaction shear fractures. These fractures are lithology-specific in their development, having formed primarily in the muddier lithologies during the early-burial and compaction history of the formation. Although they are well developed locally, and although numerically they outnumber the extension fractures by a factor of five, they are ineffective contributors to reservoir permeability because they are short, localized, poorly interconnected, and typically occluded by mineralization.

2.13.4.2 Calculating Effectiveness

We calculated a fracture effectiveness for each high-angle fracture and combined the results within each formation, since the density of fracturing alone does not provide enough information to fully assess the effect a fracture set on reservoir permeability. The permeability of an individual fracture depends on its width and degree of mineralization (Table 2.3), and the effectiveness of a set of fractures depends on interconnectivity which varies with density and strike variability. To calculate a volumetric effectiveness (distinct from the "effectiveness" calculated in Case Study 2), we used the following parameters, measured for each fracture in the core:

1. Fracture width
2. Fracture height
3. Fracture length (estimated as the diameter of the core)
4. Core height
5. Core volume (calculated)

Converting the measurements to the same units (i.e. converting feet to millimeters) we can multiply the fracture-dimension data to obtain a matrix of parameters for each fracture, i.e.:

1. Width (1-D, length)
2. Width x Height (2-D, area)
3. Width x Height x Length (3-D, volume)

The results can be normalized by dividing them by the height, area, or volume of the host core, creating three possible dimension-less ratios:

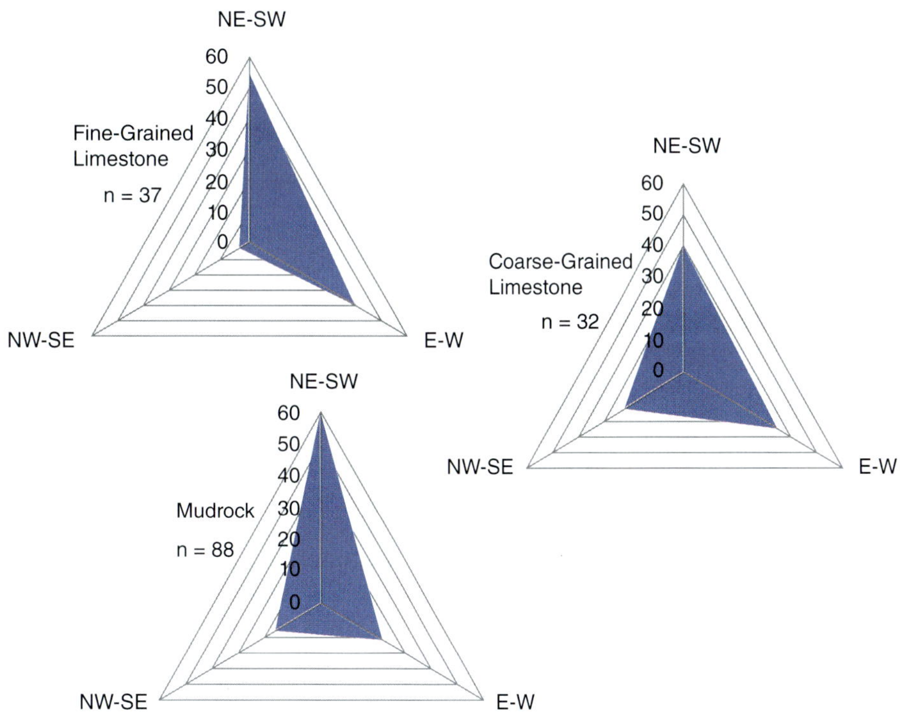

Figure 2.82 A comparison of the predicted fracture-controlled permeability directionality in three lithologies, based on 157 fractures captured by horizontal core and calculations of the effectiveness of the three cored fracture sets. Ternary diagrams for each lithology show that regardless of the relative intensity of the three sets, the NE-SW fractures should dominate the permeability system, supported by the E-W fractures. In contrast, the NW-SE fracture set has minimal influence even though it is well developed. For example, *Upper left,* NW-SE striking fractures are numerically dominant within the thin limestone beds, comprising 57% of the population (21 out of 37 fractures), but they are occluded and ineffective. *Bottom left*: the thick beds of mudrock, comprising most of the section, contain 56% of the total cored fracture population (88 out of 157). However, fracture development in the mudrock is relatively low; when normalized, there are nearly eight times more fractures per foot of core in the limestones, and 5 times more in the calcarenites, than in the mudrocks. *Right*: effectivness of the three fracture sets is more equally developed in the coarse-grained limestone. The thin, relatively brittle, and more intensely fractured carbonate beds should create permeability streaks in the reservoir.

Table 2.2 Summary of extension-fracture density* and effectiveness†.

Formation	Feet of Core	no. Extension Fractures	Fracture Density	Fracture Effectiveness
A	94	5	0.05	0.001
B	107	19	0.18	0.006
C	46	35	0.88	0.031
D	62	10	0.43	0.009

*density = number of vertical fractures per vertical foot of core
†effectiveness is explained in Section 2.13.4.2, "Calculating Effectiveness."

1. width/core height
2. (width x height x 2)/core area
3. (width x height x length)/core volume

Finally, the "effectiveness" of each high-angle extension fracture is calculated by multiplying any one of these three calculations by the remnant fracture porosity remaining in the fracture mineralization. For example, to calculate volumetric effectiveness, the calculation would be: (width x height x length) / core volume) x remnant porosity = effective fracture volume.

Summing the effective volumes of each fracture for each formation provides a quantitative fracture volume for that interval of core (the last column in Table 2.2). This volume can be used to assess the effectiveness of the natural fractures in increasing the porosity in that formation. It also provides a basis for estimating of the effects of fractures in enhancing system permeability, assuming that the greater the effective fracture volume the more likely the fractures are to be both permeable and interconnected. These effectiveness values can also be used to compare the relative impacts of fracturing in adjacent cored formations. The ratios can be converted to percentages by multiplying by 100. The fracture *network* effectiveness is then qualitatively estimated by factoring in an estimate of the probability of fracture intersections, derived from strike and spacing data.

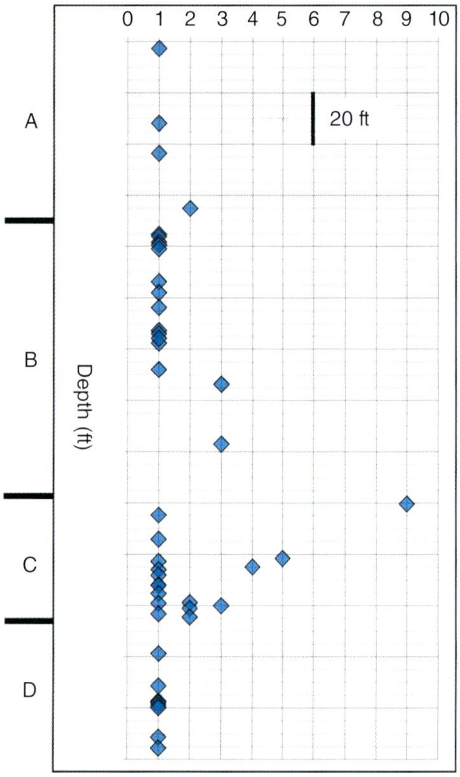

Figure 2.83 The frequency and distribution high-angle extension fractures by depth and formation (n = 69). Overlying data points are darker.

Figure 2.84 Two views of a high-angle extension fracture marked by plume structure. *Top*: face-on in the core butt. *Bottom*: edge-on in the core slab. "X_1" on the butt mates to "X_2" on the slab. The fracture surface is un-mineralized but has a patina, an aged texture indicating minor dissolution.

2.13.4.3 Description of the High-Angle Extension Fractures

This final segment of the case study provides a description of the effective extension fractures. It is easy to reduce fractures to numbers when modeling, but their less quantitative characteristics provide insights and should be kept in mind as they are being modeled.

The faces of the vertical extension fractures are commonly marked by plume structures indicating an origin in extension (Figure 2.84). The plumes on some of the fractures are obscured by mineralization, whereas the plumes on other fractures have been removed by dissolution.

Measurable heights of the high-angle extension fractures range from 0.1 to 2.7 ft, with an average of 0.8 ft (Figure 2.85). This average height is a minimum since the full heights of some fractures were not captured by the core. However, many extension fractures in this core are strata-bound and do terminate at the contacts with adjacent, more ductile beds.

Table 2.3 High-angle extension-fracture summary chart*, all four cored formations combined.

Fracture type		# Fractures	Height (ft)	Width (mm)	Remnant Fracture Porosity (%)	Spacing (mm)	Dip angle (°)
extension: high-angle	**Sum**		56.56	25.85			
	Count	69	69	69	69	18	69
	Min		0.10	0.07	0	5	75
	Max		2.70	3.00	100	82	90
	Average		0.82	0.37	20.4	23.3	87.2

*This summary combines the data from all 69 extension fractures in all four cored formations.

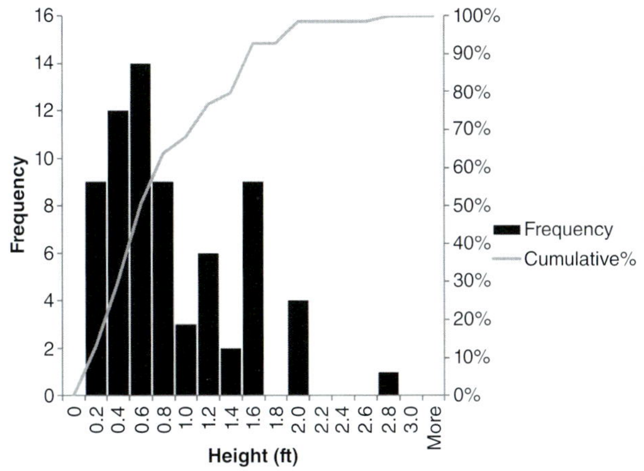

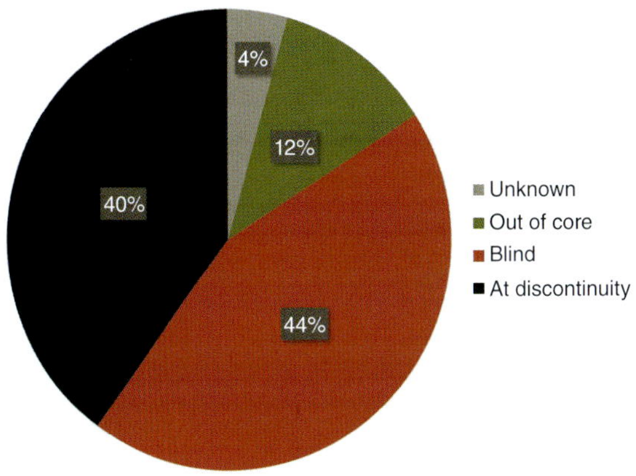

Figure 2.85 *Left*: high-angle extension fracture height histogram (n = 69; min 0.10, max 2.70, ave 0.82 ft). *Right*: fracture termination-type chart (n = 138).

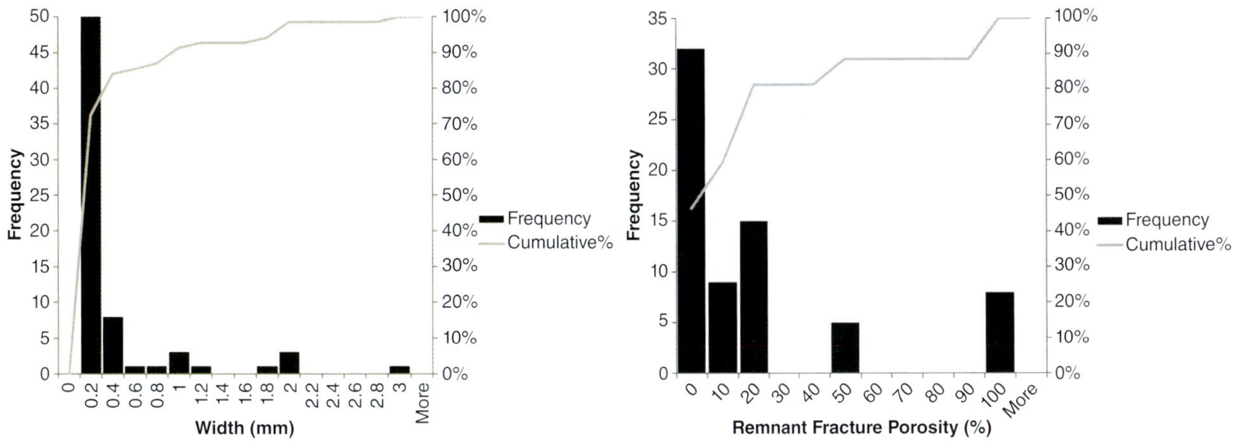

Figure 2.86 Histograms showing the distribution of widths (*left*, n = 69; min 0.07, max 3.00, ave 0.37 mm) and remnant fracture porosities (*right*, n = 69; min 0%, max 100%, ave 20.4%) for the high-angle extension fractures. (Unpublished data.)

The extension fractures have an average width of about a third of a millimeter, and over half have visible remnant porosity despite mineralization (Figure 2.86). Some are completely unmineralized, displaying an aged patina suggestive of minor dissolution along the fracture plane. Calcite is the dominant mineralization type, with gypsum filling a few of the fractures.

None of the wider fractures are completely unmineralized, and even a small percentage of remnant porosity within a wide mineralized fracture creates a significant permeability pathway within a rock. A cross plot comparing fracture width to remnant fracture porosity shows that many of the narrower fractures have significant remnant fracture porosity and will contribute to the reservoir permeability system despite being narrow (Figure 2.87). This cross plot also suggests that there may be two populations of high-angle extension

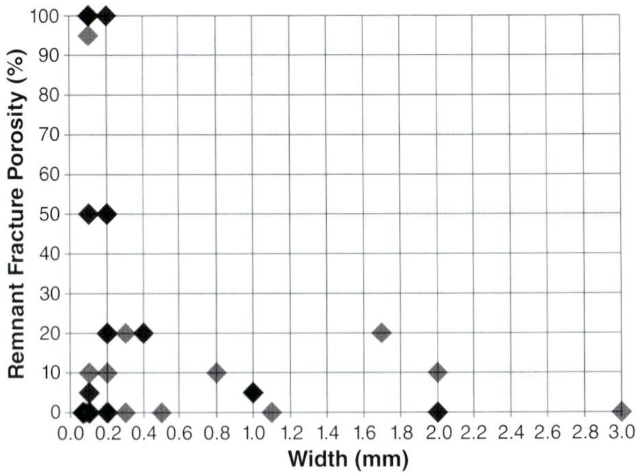

Figure 2.87 A cross plot of fracture widths and remnant fracture porosities (n = 69). Darker points highlight overlapping data. (Unpublished data.)

fractures, i.e. narrow and variably occluded fractures with a range of heights, and wider, shorter, and more completely occluded fractures. Fracture strike data support this observation, showing the presence of two sets of high-angle extension fractures with nearly normal strikes. Mineralization, calcite vs. gypsum, may also support the presence of two fracture sets.

Spacings between 18 pairs of parallel high-angle extension fractures could be measured where two or more occur in the same core piece or in adjacent, lockable pieces of core. The overall average spacing is not quite an inch (23.3 mm) (Figure 2.88), which is a minimum since spacings of vertical fractures wider that the four-inch core diameter cannot be captured by vertical core. Most extension-fracture spacing populations consist of numerous close spacings and increasingly fewer wider spacings, and the close spacings captured by this core probably represent the lower end of such a spectrum. Regardless, the measurable spacings suggest that the extension fractures are closely spaced and well developed, especially in formation C. The presence of 69 high-angle extension fractures in 300 ft of core, or one fracture for about four feet of core, is consistent with this inference.

Most of the high-angle extension fractures that can be related to each other have parallel strikes, suggesting the presence of one parallel, dominant extension-fracture set in the subsurface (Figure 2.89). However, the strike data also show a sub-set of extension fracture intersection angles oriented at right angles to each other, indicating that there is a small set of orthogonal, connecting fractures and suggesting that drainage should be only moderately anisotropic. Most of the intersecting strike relationships, as well as most of the fractures that are mineralized with the subordinate gypsum mineralization, come from formation C, suggesting the

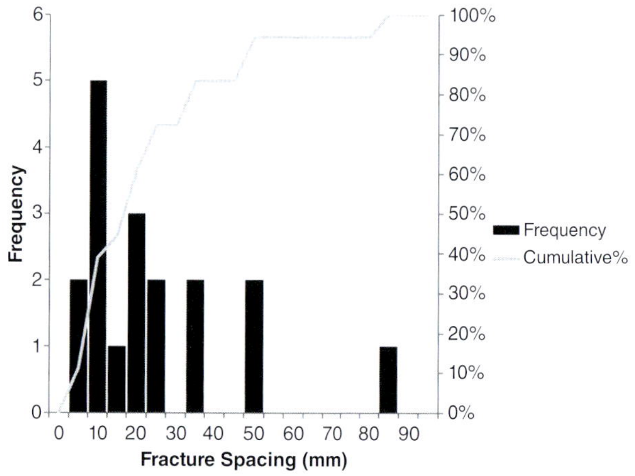

Figure 2.88 Histogram showing the distribution of measurable spacings between pairs of high-angle extension fractures (n = 18; min 5, max 82, ave 23.3 mm).

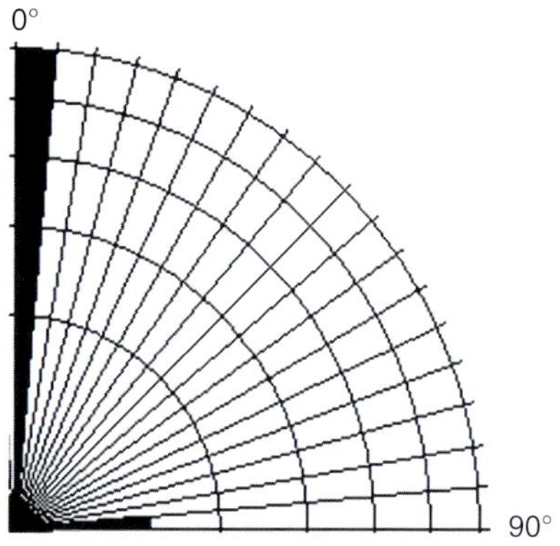

Figure 2.89 Rose plot showing the relative strikes of pairs of high-angle extension fractures (n = 52). 0° is not north but indicates fractures with parallel strikes, and 90° indicates measurements of two fractures that are orthogonal to each other. The fractures of four pairs have orthogonal strikes (the minor lobe) whereas the two fractures of the other 48 pairs have parallel strikes.

presence of a better-developed fracture network and better fracture-system effectiveness in this formation.

The inference of an interconnected fracture network is supported by strikes of the extension fractures relative to the orientation references provided by petal and centerline fractures, which also document intersecting fracture sets (Figure 2.90). The pattern shows a 75–80° rather than an 80–90° intersection, but the discrepancy

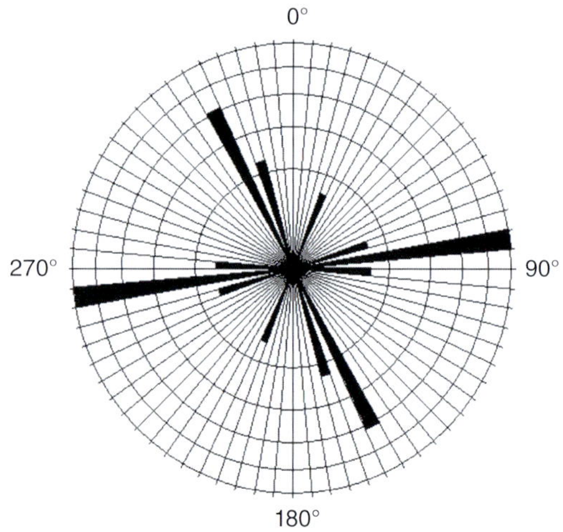

Figure 2.90 Rose plot showing the intersection angles between high-angle extension fractures and nearby drilling-induced fractures (n = 18). 0° is not north but denotes the strike of the reference induced fractures. None of the fractures are parallel to the drilling-induced fractures (0°) and therefore none are parallel to the maximum horizontal in situ stress compressive stress. Fractures oblique to that stress are "critically stressed" and subject to shear during reservoir production. Fractures that are normal to the maximum horizontal compressive stress may close during production. Hydraulic stimulation fractures will intersect the natural fractures.

Figure 2.91 Core boxes are full of surprises.

is minor given the accuracy limits of the two measuring techniques. If the local orientation of the maximum horizontal compressive stress is known, the full-circle rose diagram can be rotated so that the 0–180° line is parallel to that stress, providing the true orientation of the extension fractures relative to north.

The relative-strike measurements suggest that the extension fractures trend oblique and nearly normal to the maximum horizontal in situ compressive stress, therefore a hydraulic stimulation fracture should cut across the fracture sets and help connect them, but should also be subject to branching and screen-outs.

APPENDIX 2.A Workflow and List of Tests, Data

The flow chart presented here (Figure 2.A.1) is an after-the-fact reconstruction of the processes and tests that were conducted at the U.S. Department of Energy's Multiwelll Experiment to understand fluid flow and production in naturally fractured, stress-sensitive, low-permeability sandstones in western Colorado (Lorenz et al., 1993). The workflow indicates the importance of understanding both the fracture-permeability system and the in situ stress system.

Few projects have the luxury of obtaining similar volumes of data or of conducting a similar, comprehensive suite of tests, but the flow chart shows the types of tests that can be carried out, and the types of data that can be collected, to characterize a naturally-fractured reservoir to determine the best practices for stimulation and production. Smaller fracture-characterization projects can be carried out using sub-sets of the listed tests and data-collection techniques.

APPENDIX 2.B Core-Handling, Marking, Sampling, and Analysis Protocol for Core Studies Focused on Natural Fractures

Core offers a one-dimensional and miniscule sample of a reservoir, but it is invaluable because it provides the only actual samples of a reservoir and the permeability-controlling fractures. For that reason, core must be processed and assessed carefully and intelligently so that all possible data are collected and so that data losses are minimized during coring, handling, and processing.

We offer the following protocol as a best-practices outline for processing core to make the most efficient use of the subtle yet abundant fracture data that a core can provide if it is properly assessed. This protocol is a version of the protocol that was developed by CER Corporation and Sandia National Laboratories in the 1980s for processing 4,200 ft (1,280 m) of core on a production-line basis, when 90-ft, back-to-back cores were being cut as part of the US DOE-funded Multiwell Experiment (MWX), and it has served well since then for smaller core projects.

The protocol starts after the core has been recovered from the rig site, removed from the inner barrel, locked together, marked for depth, and marked for the uphole direction with a red-black line pair, but before the core has been sampled, plugged, or slabbed.

1. With the core laid out and locked together as closely as possible, mark the core with a very straight, green line (a "Master Orientation line") along its length, even if the core is not oriented. If the core is not oriented, the location of the MOL on the core surface is arbitrary. If the core is oriented, it should be positioned horizontally on the layout table so that the Principal Scribe Line faces up and the MOL should start coincident with that scribe line at the top of the core. The scribe will rotate, deviating from the MOL. The MOL should be marked with a straightedge.
2. Note on a core log or on the core surface all discontinuities in the core (rubble zones, spin-offs, connections, etc.) where the core cannot be locked together. The green MOL must indicate all non-locking discontinuities in the core. Such discontinuities can be marked by arrowheads on the MOL or recorded in a written log; both is better. If the core is oriented, the core should be repositioned at each discontinuity so that the Principal Scribe Line starts anew facing upward on the layout table, and the MOL again starts at the position of that scribe. Fractures, both induced and natural, can now be oriented relative to each other and to the MOL along each continuous interval of core even if samples are removed.

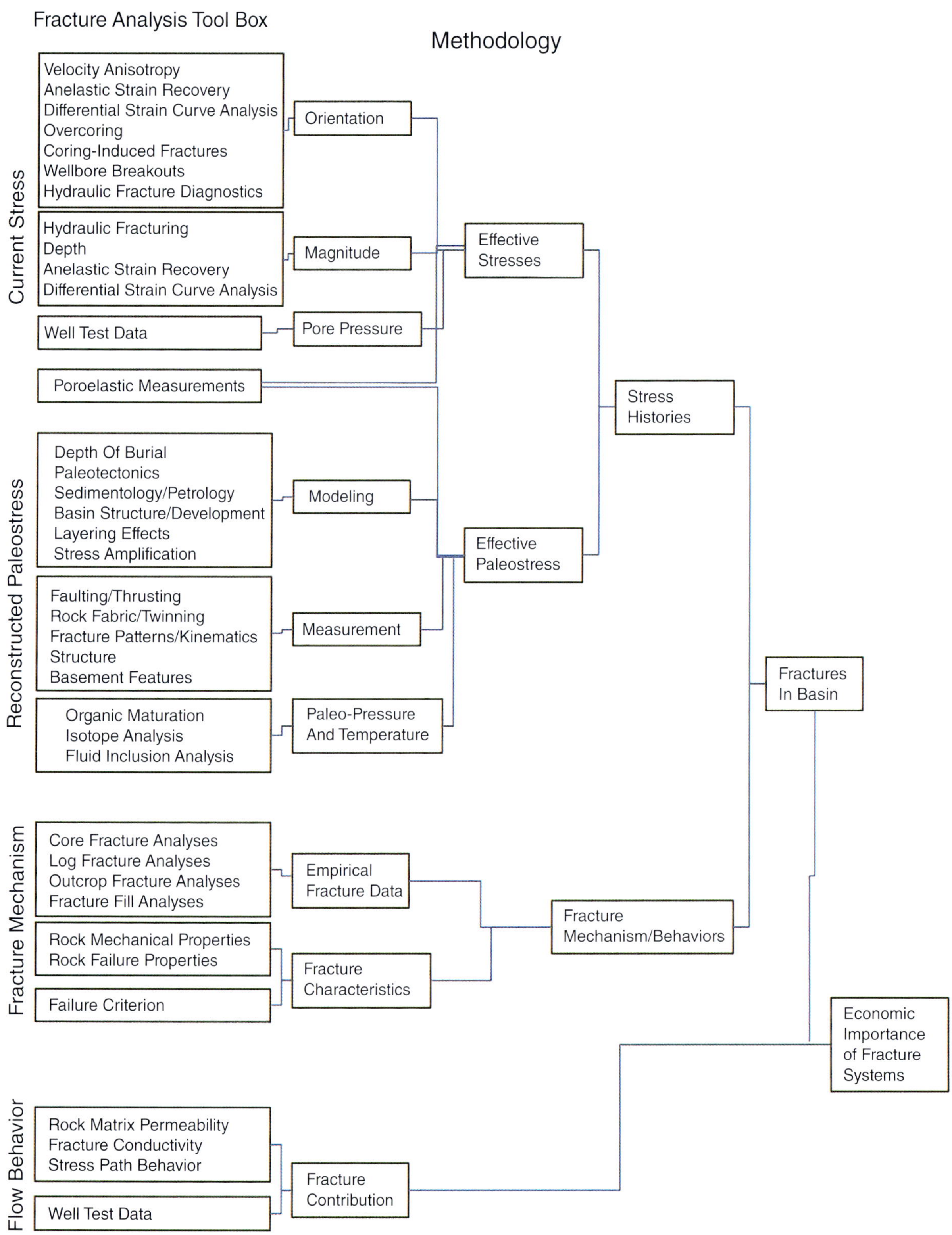

Figure 2.A.1 Flowchart of data collection for understanding naturally fractured reservoirs (Lorenz et al., 1993).

3. With the core continuity and the locations of its discontinuities recorded, any sampling that is time-critical (i.e. sampling for fluid saturations, strain relaxation, etc.) can begin. However, fracture logs will record more and better data if most sampling can wait until after an initial fracture log is made.
4. Fractures are logged as described earlier in this chapter. Logging should include numerous photographs of any cored fractures, taken from different angles. Depths written on the core to be photographed (written on a surface other than the slab face) facilitate correlations between notes and photos during later analyses.
5. Fractures are sampled as desired. Sample types can include:
   a) Thin section samples (chips for thin sections should be marked for their orientation: up direction, horizontal vs. vertical, relative to north or to other fractures, etc.)
   b) Fracture mineralization samples (for isotope and fluid inclusion data)
   c) Plugs for fracture-permeability tests (the orientation of each plug to be picked and marked by the fracture analyst)
6. Normal slabbing and sampling can begin. For fracture studies, the core should usually be slabbed normal to the dominant fracture orientation if it can be determined at this stage of processing, and a desired slabbing plane should be marked for the slabbing technician by the fracture analyst.
7. Additional fracture information can be obtained from most cores if the core is examined and logged again after it is slabbed, since smaller fractures often do not show on the rough exterior of a core, and since additional fracture exposures are often created on the ends of the cores. Although two fracture-logging passes are rarely feasible, they are rewarding.

APPENDIX 2.C Slabbing Recommendations for Horizontal Core

Slabs in horizontal cores should be cut perpendicular to bedding since slabs cut parallel to bedding offer little information to either a sedimentologist or a fracture analyst. Bedding is more readily determined in some horizontal cores from views of the ends of the core, either sawn or broken, than on the cylindrical core surface.

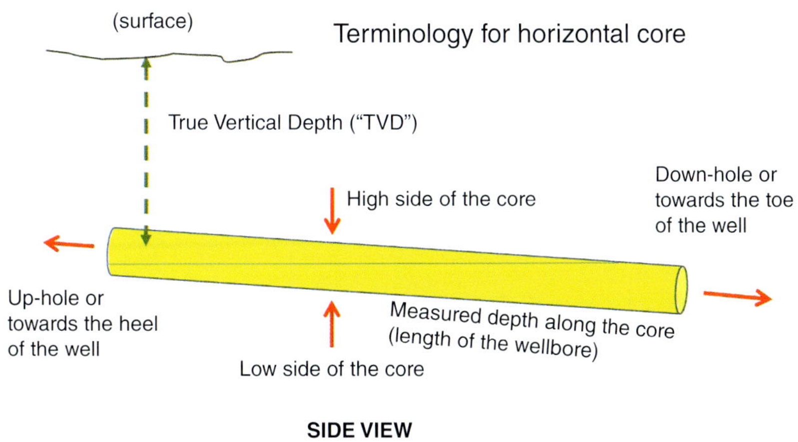

Figure 2.C.1 Terminology for a horizontal core, viewed from the side and parallel to bedding. This sketch shows the core cutting down-section at a shallow angle to bedding (the line drawn along the core oblique to the core axis). If the high side of the core cannot be distinguished in the core, then the core could be inverted and bedding could mistakenly be taken to suggest that the core cut up-section. Image logs are invaluable for orienting horizontal cores.

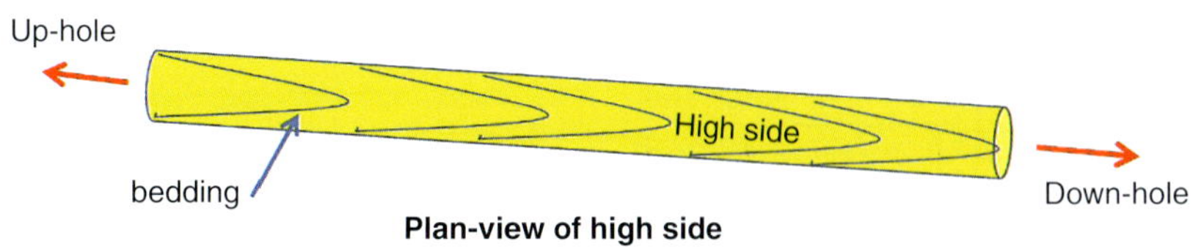

Figure 2.C.2 Horizontal core in plan view, with the high side of the core towards the viewer. The bedding "V's" point downhole indicating that the core cuts down-section as shown in Figure 2.C.1. When the core is rotated 180° around its long axis so that the view is of the bottom or low side of the core, the bedding "V's" point uphole. It can be difficult to distinguish the high side from the low side of a core that cuts un-bedded strata, or across undulating bedding, or where bedding dip relative to hole deviation is unknown.

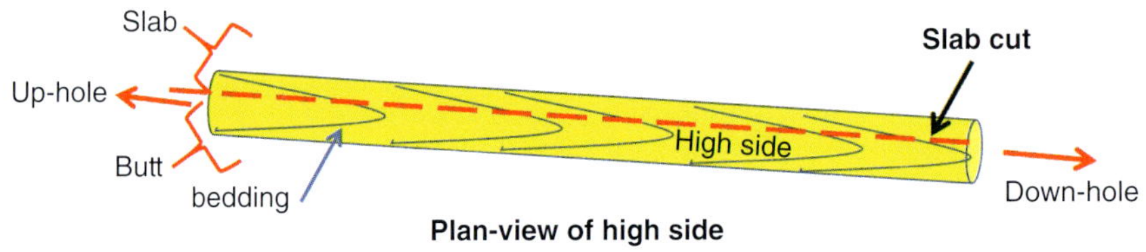

Figure 2.C.3 A horizontal core should be slabbed normal to bedding since this offers the best view of both fractures and sedimentary/stratigraphic features. The slabs that go into the slab boxes should always come from the same side of the core (i.e. the left or right side looking downhole towards the toe of the well with the slab plane vertical); which half of the core the slab comes from should be recorded and the record should stay with the core since this indicates whether the core cut up- or down-section.

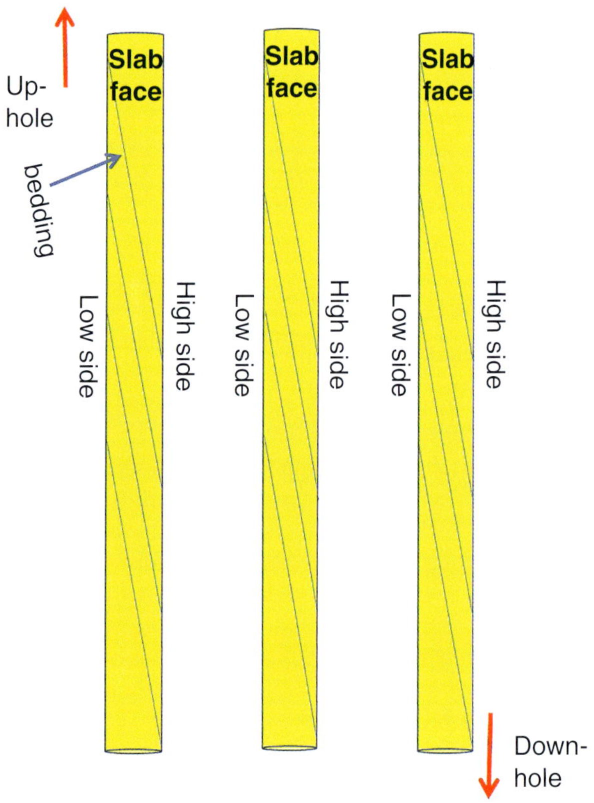

Figure 2.C.4 A sketch illustrating the consistent bedding inclines that will be displayed by slabs of horizontal core in a core box if they are consistently taken as the left half of the core shown in the preceding figure. If the slab faces had been cut with a random orientation relative to bedding, or if alternating left and right halves of the core had been put into the slab boxes, both bedding and fractures would be displayed at inconsistent angles.

If the core is slabbed into halves, either the right or left half (as viewed looking downhole with the bed-normal slab plane being vertical) should be consistently put into the slab boxes. If the core is being slabbed in the more common $^1/_3$–$^2/_3$ sections, the slab should consistently come from either the left or right side of the core. Slabs cut normal to bedding and consistently on the left or right side of the core will provide optimum exposure of the fractures and will present a consistent view of bedding dips relative to the core axis. Figures 2.C.1 and 2.C.2 highlight horizontal-core terminology, and Figures 2.C.3 and 2.C.4 show the recommended slabbing orientation relative to bedding and the high side of a core.

It requires more attention to the slabbing process and more input from the supervising geologist, but a consistent slab orientation is more important for a horizontal core than for a vertical core. If the left or right side of a core consistently becomes the slab, then inclined bedding should maintain a consistent relationship to the core axis when viewed in the slab boxes. If it is not, then the apparent bedding dip alternates along the core in the slab box, with a jarring impact on the viewer and complicating both fracture and stratigraphic interpretations.

PART 3

Effects of Natural Fractures on Reservoirs

3.1 Introduction

Understandings of the effects of natural fractures on hydrocarbon reservoirs have come from laboratory experiments, core measurements, well-tests, and theory. In this section, we present some of those understandings, as well as supporting case studies that offer insights into the behavior of fractured reservoirs, always being mindful of the variability that suggests each naturally fractured reservoir "must be considered as a research project by itself" (Aguilera, 2003). We will not cover engineering calculations, nor do we offer details on how to manage or stimulate fractured reservoirs: for those we refer the reader to industry standards such as van Golf-Racht (1982), Aguilera (1995, 2003), Heinemann and Mittermeir (2015), and numerous engineering treatments of fractured reservoirs in the literature of the Society of Petroleum Engineers. However, we do wish to acquaint the reader with some of the effects of natural fractures on fluid flow in reservoirs including drainage anisotropy, fracture behavior relative to in situ stress systems, interactions between natural and hydraulic fractures, natural-fracture volumetrics, and the difference between individual and system fracture permeabilities.

The default conceptual model of fractures in reservoirs is typically one of open, interconnected slots that enhance reservoir permeability, and in this section, we examine how, why, and to what degree this reflects reality. We will also show that some natural-fracture systems are barriers, and that other systems can be damaged by stimulations and/or by the rapid fluid-pressure decreases in a fracture network related to high production rates. The shape of drainage zones around wellbores can vary depending on fracture type as well as on fracture strikes relative to the in situ stress orientations, and it will become apparent both that in situ stresses are an integral part of the fracture-permeability system, and that neither fracture permeability nor the in situ stresses are static.

In many cases, decisions on well location, horizontal wellbore azimuth, and completion techniques are dictated by partners, lease lines, and economics, but a data-based understanding of the reservoir fracture system still allows for informed choices after these drivers are accommodated. In other cases, the decisions are based on analogy to similar fractured reservoirs, but to know what is and isn't similar one still needs to know the fracture types, distributions, intensities, and their relationship to the in situ stresses. There are commonalities among fracture systems, but each fractured reservoir is built from the common components in different proportions and arrangements, so, as Nelson (2001) points out, an early investment in data pays dividends.

The advantages of investing in data are not always obvious because in the absence of subsurface fracture data the mere fact that a well produces can be defined as success, and it may not be apparent that production rates could have been higher and/or that production decline curves could have been flatter if the development and production strategies were built on better data. Similarly, non-productive wells are commonly plugged and abandoned without knowing why they did not produce.

3.2 Classification

Nelson (2001) offers a four-fold fractured-reservoir classification based on the degree of fracturing and the effects of those fractures (Table 3.1). Reservoirs of course

Table 3.1 Nelson's Fractured-reservoir Classifications.

Type IV
Fractures are baffles: they create anisotropy and degrade the reservoir, adding no permeability or porosity.
Type III
Fractures somewhat enhance permeability in a reservoir that would nevertheless produce without fractures.
Type II
Fractures create the essential permeability in the reservoir, but the matrix provides the reservoir porosity.
Type I
Fractures provide most of the reservoir porosity *and* permeability since there is little of either in the matrix.

Applied Concepts in Fractured Reservoirs, First Edition. John C. Lorenz and Scott P. Cooper.

do not fall into four discrete bins, and as Nelson notes, fractured-reservoir characteristics form a spectrum in terms of fracture intensity and permeability relative to the matrix rock. Nevertheless, if fractures can be characterized well enough to recognize where a reservoir fits in the spectrum then engineers and geologists together can begin to develop strategies for well placement, lateral azimuths, stimulation treatments, production schedules, etc., for efficiently producing hydrocarbons.

Fractures can significantly enhance permeability and the potential for fluid flow in a reservoir, 1) if the individual fracture planes are open and can accommodate both fracture-parallel flow and flow from the matrix into the fracture aperture, and 2) if the pathways are interconnected with other fracture pathways, either natural or artificial, such that the fluids can flow from one fracture into another and ultimately to a wellbore. Fractures that are not open or well-enough developed to enhance system permeability much above matrix values (Nelson's Type III reservoir) are generally inconsequential, whereas other fracture types such as deformation-band shear fractures or tightly mineralized fractures in high-permeability reservoirs may create low-permeability baffles and barriers, compartmentalizing the reservoir and making it difficult to extract hydrocarbons (Nelson's Type IV reservoir). For the most part, this book has been concerned with Nelson's Type I and Type II reservoirs, where open fractures contribute significantly to reservoir permeability. They rarely create significant bulk porosity, although in the case of Type I reservoirs, the *only* bulk reservoir porosity consists of the minimal volume provided by fracture apertures.

Fracture volumes are commonly small relative to matrix porosity because although they have volume, fractures are essentially local planar structures whereas matrix porosity is three-dimensional and pervasive in a reservoir. Production from fractured reservoirs often falls off rapidly because fluids flow more readily through fractures than through the matrix, allowing the small volumes of fluid in the fractures to be rapidly depleted. Fracture aperture closure ("compressibility") can also contribute to rapid production decline. Depletion of hydrocarbons from fracture apertures causes a pressure disequilibrium that can push oil or gas from the matrix to refill the fracture apertures, but it can also bring in an influx of water from below an oil–water contact to fill the depleted fracture apertures. Water influx is unwanted in many reservoirs since it can cause fingering and damage, but it can be a positive development in other reservoirs where water imbibition across fracture faces can displace oil from the matrix, enhancing production.

In evaluating the effects of fractures on fluid flow, it must be kept in mind that although fluid may flow more easily along the direction of open fracture planes, the actual direction of flow is still controlled by pressure gradient, not fracture orientation. It is also important to remember that the effect of a given fracture on fluid flow varies not only with fracture width and interconnectivity but also with the type and viscosity of the reservoir fluids, as well as with the in situ temperature that affects viscosity and the pressure differential that drives fluid flow.

In this section, we will first describe the control individual fractures can have on fluid flow, and then we will describe how the individual fractures form interconnected systems in reservoirs. Later we will discuss the interplay between fracture permeability and changes in the in situ stresses during production, as well as the nature of interactions between natural fractures and hydraulic stimulation fractures. We will also address fracture volumetrics, and, briefly, describe some of the effects fractures have on drilling and coring operations.

3.3 The Permeability Behavior of Individual Fractures

3.3.1 Introduction

Individual fractures do not all have the same effect on fluid flow in a reservoir since, as shown in Part 1, even fractures within the same set typically have a range of heights, lengths, and widths. Two aspects of individual fracture permeability must be considered: flow through the fracture aperture parallel to the fracture plane, and flow across the fracture faces. Flow capacity along a fracture increases exponentially as fracture width increases, but even fracture widths of only a fraction a millimeter can offer important avenues for flow, especially for low-viscosity fluids under high pressures in nano-darcy matrix rock. Some fractures are open and have high-permeability apertures whereas others are occluded with mineralization, blocking or at least retarding fluid flow relative to that offered by open fractures. The permeability across fracture faces can be affected by slickenside development, by the degree/type of mineralization obscuring the faces, and by dissolution along the fracture face.

3.3.2 Three Categories of Fracture Effects

Three basic types of fracture-controlled flow can be considered (Figure 3.1), although many fractures include elements of two or even all three conditions along their length:

1. Fractures may have open, highly permeable apertures as well as surfaces that allow unrestricted flow from the matrix into the fracture aperture, enhancing reservoir quality.

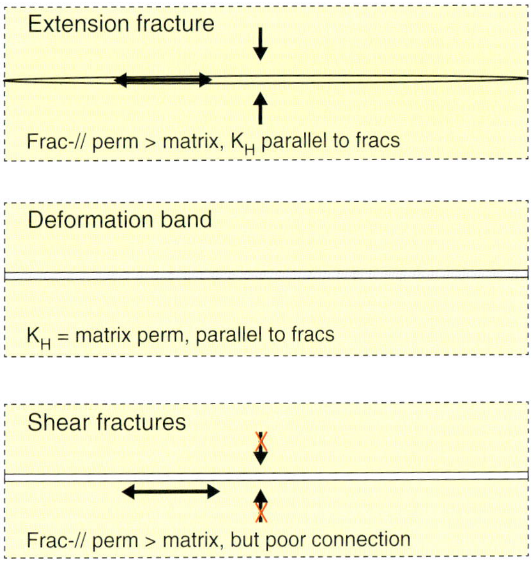

Figure 3.1 Three possible conditions for fracture apertures and faces, depending on the openness of the fracture aperture and the degree of alteration of the fracture faces.

2. Fractures may consist of relatively low-permeability planes that impede flow both across and along the fracture and that degrade reservoir quality.
3. Fractures may have open, high-permeable apertures but low-permeability surfaces, with variable effects on the reservoir.

The first condition is typically the default conceptual model for fractures in a reservoir. Fluids flow readily along open fractures, and can flow from matrix into the open fracture aperture if the fracture faces consist of unaltered matrix rock (Figure 3.2). In simple cases like these, typical of unmineralized extension fractures, system permeability is enhanced above matrix values within and parallel to fractures, but it is limited to the matrix permeability in the direction normal to the fracture planes.

Figure 3.2 Viscous oil seeping from fractures in tilted strata of the Eocene Pila Spi Formation (limestone) in northern Iraq illustrates the preferential flow pathways offered by unmineralized extension fractures.

Deformation-band shear fractures and some completely mineralized fractures have no open aperture, providing no permeability enhancement parallel to the fracture plane while inhibiting flow across the fracture plane. Other types of shear fractures have altered faces that inhibit flow normal to the fracture plane but at the same time have relatively open apertures that enhance flow along the fracture. Vertical permeability depends on fracture height, which, for extension fractures, is commonly limited by the sedimentary heterogeneity of the reservoir (e.g. Ameen et al., 2012). In contrast, shear fractures are more likely to cut across minor bedding planes and other sedimentary heterogeneities, and commonly provide better vertically interconnected permeability within a reservoir (e.g. Loosveld and Franssen, 1992; Nelson et al., 2000; Wennberg et al., 2016).

The geometries of unmineralized shear-fracture apertures differ significantly from the typical apertures of extension fractures. Early on, fracture permeability was modeled as ideal flow in a slot between parallel plates, where permeability is proportional to the cube of the distance between the fracture faces (i.e. Warren and Root, 1963; Kazemi, 1969), and in fact unmineralized extension fracture planes are somewhat planar and the width between opposing fracture walls is relatively uniform along a given fracture. However, mineralization and dissolution are more common than not in fractures (Figure 3.3), adding considerations of tortuous pathways, and turbulent and channeled flow. Even when unmineralized, shear fractures typically have rough,

Figure 3.3 The face of a cored extension fracture (parallel to the plane of the photograph) has been subjected to dissolution so that the fracture aperture does not consist of a parallel-plate slot. Minor lithology changes show as differentially dissolved bedding, with ridges of chert standing out from the embayed chalk matrix. A later druze of small calcite crystals was precipitated onto the etched fracture face, covering the embayed surfaces.

Figure 3.4 *Left*: irregular, intersecting, high-angle, strike-slip shear fractures. *Right*: the surfaces of these fractures are marked by accretionary, congruent steps and horizontal slickenlines. Core from a well-cemented quartzose sandstone in front of the active Andean thrust system in Colombia. The slickensided fracture faces should have low permeability, which would inhibit flow from the matrix rock into the open, unmineralized fracture apertures. (Used with the permission of Equion Energia Ltd.)

irregular, and often stepped surfaces, commonly displaying pinch and swell patterns in cross section since the asperities that are offset during shear leave irregular voids between points of contact across the fracture.

It is now generally recognized that flow within geological fractures is not governed by a width-cubed relationship (e.g. Long et al., 1982; Berkowitz, 2002), and even where it is, it is less applicable to narrower fractures than to wider fractures (Cook, 1992). In fact, just measuring the "width" of a shear fracture or a partially mineralized extension fracture is like trying to specify the width of the void space between a shark's upper and lower jaw with its teeth clenched: there is void space and it allows flow along the fracture, and "widths" are even measurable at specific points, but even an average of such measurements is meaningless in terms of calculating flow capacity. Effective, "hydraulic fracture apertures" must be calculated using data such as the percentage of the fracture-face area that is in contact across the fracture width, and the distribution of that area (e.g. Zimmerman and Bodvarsson, 1996). Although they offer a starting place, even these sophisticated approaches have limitations that do not reflect all the realities of fracture aperture variations.

The second fracture condition listed, where fractures have open, permeable apertures but low permeabilities across the fracture faces, is common since relatively few fractures in reservoirs have faces that consist of freshly-broken host rock. Most reservoir fractures, both shear and extension, are at least partially mineralized, and mineralization alters not only the uniformity and width of fracture apertures but also the permeability of the fracture face.

Shear fractures with enough offset offer an additional complication (Figure 3.4) in that slickensided faces created by shear may consist of nearly impermeable glassy material formed when the rock along the fracture face melts under the locally high temperatures created by shear (Nelson, 2001). Testing whole-core samples containing shear fractures, Nelson found that permeability in the eolian Nugget Sandstone was 1,512 millidarcies in the horizontal direction parallel to an inclined, slickensided shear fracture, but only 18 millidarcies in the horizontal direction across the fracture plane. Shear fractures in carbonates, even without slickensides, may have reduced fracture-normal permeability since the fracture faces are commonly covered with a low-permeability layer of insoluble clay residue left by the pressure dissolution that can accompany shear (see Figure 2.27).

Mineralization may or may not inhibit the ability of fluid to flow across a fracture face, depending on the permeability of the mineralization relative to matrix permeability, and the type of fluid. Insights are offered by laboratory permeability tests conducted on one-inch plugs and containing calcite-mineralized extension fractures several tenths of a millimeter to a few millimeters wide, cut from cores from natural-gas reservoir sandstones with microdarcy-scale matrix permeabilities in the Mesaverde (Lorenz et al., 1989; Morrow et al., 1990) and Frontier (Lorenz et al., 2005) formations. The plug permeabilities, measured at restored-state water saturations and confining pressures, show that these narrow, apparently tightly mineralized fractures do *not* inhibit permeability measured transverse to the fracture planes, and in fact the mineralized fractures enhance permeability parallel to the fracture planes by a factor of two or three over matrix values, presumably due to intercrystalline pathways within the calcite that fills the fracture widths (Figure 3.5). Significantly, since these measurements could only be made on the smaller fractures that remained intact while being plugged, the larger fractures, especially those with macroscopic remnant aperture within the mineralized fracture widths, can be inferred to have much higher fracture-parallel permeabilities.

Calcite-mineralized fractures with similar widths and crystal habit to those tested (Figure 3.5) are common in resource-play mudrocks. Although similar plug-permeability measurements have not been published, it is likely that these fractures have similar microdarcy-scale, fracture-parallel permeabilities, and that they should therefore enhance the reservoir

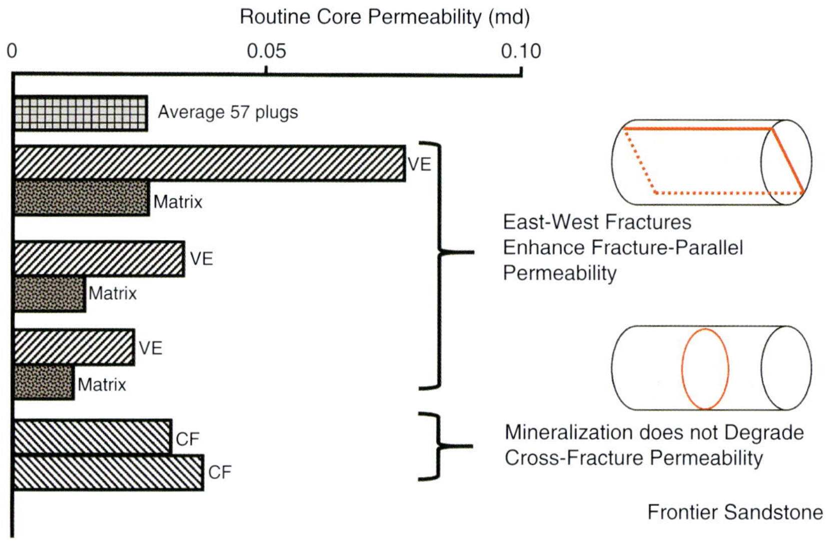

Figure 3.5 One-inch plugs, containing natural, calcite-mineralized vertical-extension fractures ("VE") oriented parallel to the plug axis (as shown in the upper sketch), cut from core taken in sandstones of the Frontier Formation in Wyoming. Permeability tests show that the tightly mineralized fractures enhance permeability in the direction parallel to the fracture plane compared to control, matrix-only plugs ("Matrix") cut adjacent to the fracture. Plugs were also cut with the mineralized fracture plane normal to the plug axis (cross-fracture plugs, "CF," as shown in the lower sketch), and these do not show degraded permeability relative to matrix values. (Adapted from Lorenz et al., 2005).

system permeability over the common background, nanodarcy-scale matrix permeabilities. Landry et al. (2017) have suggested that this is so, based on scanning electron microscope observations of intercrystalline porosity within mineralized fractures and calculations using the dimensions and characteristics of those pores.

Deformation-band shear fractures, and the rarer compaction-band "anti-crack" fractures, consist of planar zones of crushed grains and collapsed porosity. They are most common in sandstones that have high porosities and that were poorly cemented at the time of fracturing (e.g. Jamison and Stearns, 1982; Nelson, 1985a; Antonellini et al., 1994), but they have also been recognized in high-porosity chalks and grainstone limestones (e.g. Rath et al., 2011; Marchegiani et al., 2006). These structures do not enhance permeability in the direction parallel to the fracture planes; rather, they significantly degrade the local permeability normal to those planes (e.g. Fossen and Bale, 2007; Holcomb et al., 2007; Solum et al., 2010). Laboratory measurements indicate that permeabilities normal to deformation-band shear planes can be diminished by several orders of magnitude (e.g. Antonellini and Aydin, 1995; Nelson, 2001; Sternlof et al., 2004) (Figure 3.6), although Fossen and Bale (2007)

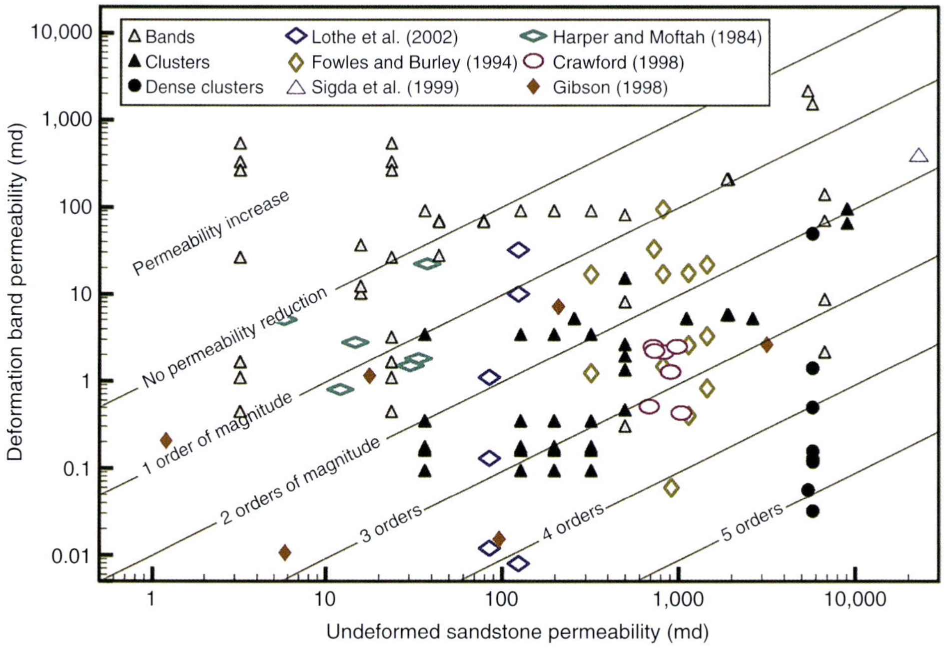

Figure 3.6 Cross plot of matrix permeability vs. deformation-band permeability, showing that most deformation bands degrade local reservoir permeability (reprinted from Fossen and Bale, 2007, with permission from AAPG, whose permission is required for further use). Those few measurements that suggest enhanced permeability may come from the rarer examples of dilation bands (e.g. Du Bernard et al., 2002).

Figure 3.7 *Left*: a stylolite with associated short extension fractures in a limestone may form a permeability streak in the reservoir. In one example from the Permian Basin a plug containing a stylolite that was inadvertently sampled during routine core plugging tested at 1.67 microdarcies whereas the two adjacent plugs that sampled the matrix tested at 0.0189 and 0.0155 microdarcies. *Right*: a horizontal slot lined with scattered, large white dolomite rhombs, in the Arbuckle Dolomite of Kansas. This partially mineralized slot may represent preferential fluid movement, dissolution, and remineralization along a stylolite. Teufel et al. (1993) suggest that such stylolite-related extension fractures are locally well interconnected and form bed-parallel zones of enhanced permeability in chalk reservoirs in the North Sea.

suggest that deformation bands that are discontinuous may be less effective in compartmentalizing fluid flow.

Nevertheless, where deformation bands are well-developed, maximum reservoir permeability is the matrix permeability and it will be channeled parallel to the fracture planes; permeability normal to the planes will be less than matrix-only permeability. Tightly mineralized fractures in high-porosity matrix reservoir rock should create the same type of anisotropic, degraded flow system if fracture permeability is less than matrix permeability.

3.3.3 Stylolites

The seams of ductile, insoluble residue found along stylolites suggest that stylolites inhibit flow normal to stylolite planes, especially since narrow low-porosity zones are commonly present adjacent to a stylolite. However, Heap et al. (2014) and Toussaint et al. (2018) suggest that because stylolite planes are limited in extent they form "perforated layers" that do not significantly inhibit overall fluid flow in a reservoir, and that at a gross scale the potential for fluid flow normal to stylolite planes is not significantly reduced below that of the matrix.

A stylolite zone must have had enough permeability to allow the removal of a significant amount of material as solute in the formation fluids when the stylolite formed, but that is not an indicator of the present permeability of the zone. However, the short extension fractures that commonly form on one or both sides of stylolites can be closely spaced and sufficiently open and interconnected to create local permeability streaks in a reservoir (Figure 3.7). Nelson (1981) measured the permeabilities of core samples containing stylolites and associated fractures, finding permeabilities that are higher by nearly two orders of magnitude than associated samples without stylolites and fractures.

3.3.4 Microfractures

The term "microfracture" generally applies to those fractures that are small enough to be best observed with a microscope (e.g. Gale, et al., 2007; Milliken and Land, 1994; Milliken and Laubach, 2000). Microfractures within sand grains in clastic rocks may form as extension fractures parallel to the maximum compressive stress, or they may originate as cracks radiating away from point contacts between grains. Some microfractures cut across multiple grains; others form in the cement between grains. Continued deformation can connect microfractures to form macrofractures, or can shatter grains as microfractures develop into gouge-filled cracks.

Since they are easily overlooked without the aid of thin sections, the potential contribution of microfractures to fluid flow is not always considered. Also, as shown below, a consensus has not yet formed regarding the role of microfractures in reservoirs. Narrow microfracture apertures are highly sensitive to stress, being prone to closure when the confining stresses increase (Kranz, 1983), as they do during production. Paterson (1978) also reported that in the laboratory, in "compact" rock (i.e. very low-porosity rock such as granite), permeability initially decreases with the application of stress as existing microfractures close. However, Paterson also noted that the permeability of a sample increases "very markedly…even under confining pressure" with continued strain, due to the formation of numerous new microfractures. Fewer microfractures form during strain within higher porosity rock, since some of the initial strain is accommodated by pore collapse rather than microfracturing.

Microfracturing may not be significant in reservoir rock that is not being actively strained, since Anders et al. (2014) suggest that most natural microfractures in sandstones are completely healed. Even where open microfractures exist, those that are confined to individual grains may not form an interconnected, effective permeability network. Loucks and Reed (2016) suggest that microfractures in mudrocks are also poorly interconnected and therefore likely to be ineffective in enhancing permeability. Nevertheless, Perez et al. (1993) offered theoretical considerations suggesting that microfractures can act to increase oil mobility because capillary forces can force water into microfractures, driving oil out under the right conditions, while Geiser and Sansone (1981) suggested that microfractures together with macrofractures can provide conduits that transport dissolved carbonate away from developing cleavage planes in deformed limestones; microfractures cannot be dismissed as potential contributors to reservoir quality.

Other published studies report that microfractures have measurably enhanced reservoir quality. Ameen and Hailwood (2008) describe microfractures in well-cemented, low-permeability/low-porosity sandstone reservoirs in Saudi Arabia, suggesting that these microfractures increase effective reservoir porosity by up to 50% and increase permeability by up to 75%. Moreover, they report that reservoirs containing connected microfractures have as much as 14 times greater productivity than wells drilled in non-microfractured sections of the reservoir. Likewise, Zeng and Li (2009) and Zeng (2010), using an extensive database of core and core tests, report that numerous open intra-grain and grain-edge microfractures in low-permeability feldspathic reservoirs in the Ordos Basin in China provide as much as 1% of the total reservoir volume (which have an average 10% porosity) and 25% of the reservoir permeability (which ranges from 0.1 to 10 millidarcies). Samples containing microfractures have laboratory-test permeabilities on the order of ten times greater than unfractured control samples of the matrix (Figure 3.8), and the microfracture permeability is reported to be highly stress-sensitive. The dataset of nearly 10,000 samples demonstrates the potential for a set of microfractures, if natural and well developed, to influence reservoir permeability.

Liu et al. (2013) have also suggested that microfractures can be important, contributing to reservoir quality in low-permeability Cretaceous sandstone reservoirs at a depth of 6,000 m (19,686 ft) in China. These microfractures are short and can be filled with bitumen and calcite. Liu et al. (2013) report that half of the bulk porosity, which ranges from 0.5% to 3.0% in these tight reservoirs, is due to microfracturing, presumably based on point counts from the several hundred thin sections made for the study. The permeabilities of samples without microfractures are less than 0.1 millidarcy, whereas samples with microfractures have permeabilities than can exceed 1 millidarcy: the authors suggest that microfractures contribute 99% of the stress-sensitive reservoir permeability.

Microfractures that are unmineralized must be assessed critically, since many are artifacts. Induced microfractures are common where a core has been cut

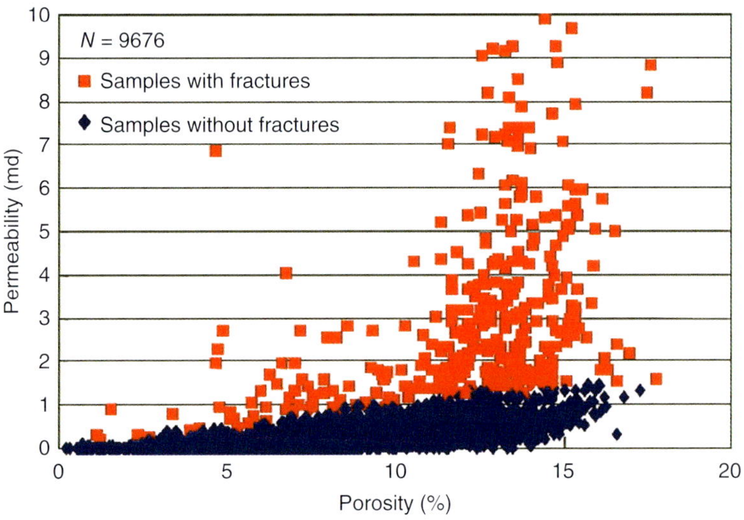

Figure 3.8 A cross plot of porosity vs. permeability, distinguishing microfractured samples (red) from those without microfractures (blue), shows that samples containing microfractures have significantly higher permeabilities than unfractured samples, but that the two types of samples have similar porosities (n = 9676). Before attributing fluid flow in a reservoir to microfractures, they must be demostrated to be natural rather than artifacts. (Reprinted from Zeng and Li, 2009, with permission from AAPG, whose permission is required for further use).

and released from the in situ confining stresses (e.g. Teufel, 1983; Lin et al., 2006). Induced microfractures can also form during processing and sample preparation, and they may form as a mudrock core dehydrates (e.g. Grover, 2011; Milliken and Land, 1994.) Loucks and Reed (2016) found that natural microfractures are in fact relatively rare in mudrocks, and that the common microcrack artifacts are easily misidentified as natural.

Landry et al. (2014) suggest that mineralized microfractures in the mudrock cores they studied have permeabilities that are similar to those of the host rock, on the order of a hundred nanodarcies, and that the importance of microfractures lies not so much in their inherent permeability as in the weak adhesion of the microfracture mineralization to the host rock and the potential for microfracture reactivation during hydraulic stimulations.

3.4 The Effects of Fracture Systems

3.4.1 Introduction

Fractures, no matter how permeable, cannot influence reservoir permeability unless they are sufficiently numerous and have enough intersecting strikes to form an integrated, interconnected fracture system. The flow capacities of individual fractures must be considered together with information about fracture strikes, lengths, spacings, intensities, etc., in order to assess the full effect of a fracture system on a reservoir.

Many fracture systems consist of intersecting sets, but few if any consist of the three sets of ideal, mutually orthogonal fractures that resemble the faces of stacked sugar cubes (e.g. PetroWiki, 2019; Heinemann, and Mittermeir, 2015). Sugar-cube engineering models are necessary due to the limitations of the mathematics used to create them, so engineers and geologists must work together to iteratively create the models that most nearly approximate the effects of the fracture system being modeled. Guerriero et al. (2013) recognized that fracture populations consist of numerous small fractures and fewer larger fractures and proposed a four-tier permeability model where fluids in the matrix flow into pervasive small fractures, which in turn feed larger strata-bound fractures, and then flow into the largest but less common faults. This type of model accounts for the spectra of fracture sizes, and can be built to include criteria for fracture intersections and interconnections.

Fracture system permeabilities cannot be isolated and measured directly, but the effects of a fracture system can be estimated using measurements of fracture characteristics in cores and image logs, and they can be more quantitatively assessed by comparing 1) laboratory measurements of restored-state matrix permeabilities from core plugs, to 2) pre-stimulation well production tests that measure the combined matrix-fracture system permeability. The difference between them, which can be several orders of magnitude (e.g. Warpinski and Lorenz, 2008), can be assumed to be the contribution of the natural-fracture system to the total reservoir permeability (e.g. Belfield, 1988). The difference can be significant in both low-permeability and conventional reservoirs (Figure 2.72; Figure 3.9). For example, Teufel and Farrell (1992) and Teufel et al. (1993) report that the chalk reservoirs of the Ekofisk field have system permeabilities, measured from well tests, that are as high as 150 md, whereas the matrix permeabilities measured from core plugs in the laboratory are on the order of 1 md.

Fractures contribute directly to production rates, but typically only slightly to reservoir volumes: contributions of a fracture system to the Estimated Ultimate Recovery (EUR) of a well usually depend on whether or not the fracture system connects a wellbore vertically or laterally to remote parts of a reservoir that might not otherwise be drained, and not on the volume provided by fracture apertures. Elkins (1953), describing intensely fractured reservoirs of the Spraberry Formation in West Texas, reported that wells were capable of producing up to 1,000

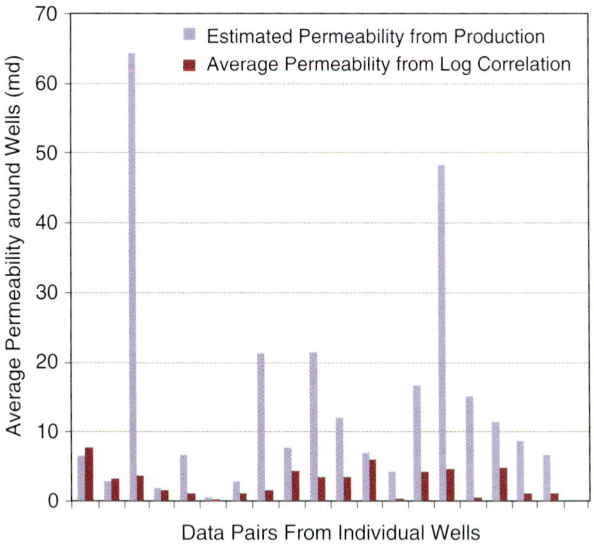

Figure 3.9 Fractures can significantly enhance system permeability in conventional reservoirs, as shown by this chart of data from eolian Tensleep Sandstone reservoirs in Wyoming. The figure compares 1) system permeabilities calculated from pre-stimulation well tests (blue bars) to 2) matrix permeabilities calculated from wireline log measurements that were calibrated with laboratory tests on core plugs (purple bars). System permeabilities are greater, by a factor of up to 20, than associated matrix permeabilities. (S. Wo, personal communication, as published in Lorenz and Cooper, 2013).

barrels of oil per day, after treatment, from sandstones that laboratory tests suggested should have the capacity of only five to ten barrels of oil per day. However, he also calculated from core data that the volume of the fracture system is "negligible," on the order of 110 barrels per acre.

3.4.2 Fracture-Controlled Permeability Anisotropy

Production tests from single wells provide assessments of the delivery capacity of a fracture-plus-matrix permeability system, but they do not indicate the orientation of the fracture-controlled permeability anisotropy. Nevertheless, simple systems of parallel extension fractures can create significant permeability and drainage anisotropies in a reservoir. That information must be derived from well-interference tests and/or knowledge of fracture strike. Anisotropy information can be used to plan the placement and azimuths of development wells, while knowledge of the angular relationship between natural-fracture strikes and the in situ stresses can be used to develop completion and production strategies.

Bell and Babcock (1986) documented significant reservoir permeability anisotropies in Alberta, with the long axes oriented normal to the fold and thrust belt in the strata filling the adjacent basin, related to the presence of well-developed, parallel extension fractures created by the local thrust-related stress anisotropy. Several other well-documented examples of fracture-controlled permeability anisotropy are presented below. These examples are from relatively simple fracture systems where the effects of the fractures in enhancing permeability in one direction are obvious. The effects of fracture systems consisting of multiple sets of intersecting fractures can be more difficult to interpret and predict, but the simple systems presented here can be used as building blocks for more complex systems.

3.4.2.1 Case Study: The Midale Field

The Midale oil field, located in Saskatchewan in the northern part of the Williston Basin, illustrates the effects that a well-developed but relatively simple fracture system can have on a reservoir. As described by Beliveau (1987, 1989) and Beliveau et al. (1993), the Mississippian-age Midale reservoir is composite, consisting of an upper, predominantly dolomite unit and a lower, predominantly limestone unit. The average gross thickness of the two units together is about 54 ft (16.5 m), and they are under- and overlain by unfractured anhydrite seals. The Midale reservoir, at a depth of about 4,600 ft (1,400 m), is normally pressured and the unfolded strata have a gentle, uniform dip. Average matrix porosities are 20–35% in the dolomite and 10–15% in the limestone. Laboratory-measured, unstressed air permeabilities of the matrix range from 1 to 150 millidarcies, and permeability is generally higher, by a factor of two, in the upper dolomitic zone than in the underlying limestone.

The Midale natural-fracture system has been characterized using data from core (from both vertical and horizontal wells), observations of preferred injection breakthrough directions, and well tests such as tracer tests, well-interference tests, and pulse tests. Beliveau et al. (1993) emphasize that "…a multidisciplinary team approach was *absolutely necessary* [their italics] to characterize the Midale reservoir…" The data and analyses show that the Midale reservoirs are dominated by a system of vertical, parallel, NE-SW striking extension fractures. The fracture system creates a strong NE-SW trending, horizontal permeability anisotropy parallel to fracture strike. The NE-SW fracture trend is also reflected in the long, highly asymmetric axes of the ellipses formed by contour maps of the percentage of water ("water cut") in the produced fluids across the field, created by the preferred along-strike migration direction of injected water (Figure 3.10).

Cores show that spacings of the natural fractures average about two feet (0.6 m) in the dolomite but only one foot (0.3 m) in the underlying limestone. The fracture spacings measured in core are consistent with the effective fracture spacings calculated from draw-down/interference tests. Some of the fractures are narrow and are expected to contribute little to the system permeability, but more of them are open, with wide, irregular apertures. Fracture heights in the vertical cores range from a few inches to 15 ft (5 m).

During water flooding, water broke through rapidly between injectors and those producing wells that are aligned along the NE-SW trend. In contrast, breakthroughs to producers that are oblique to the NE-SW trend are slower and more prolonged, and the water cuts are lower. Some 70 measurements from pulse tests and interference tests in the field show that the ratio of maximum to minimum horizontal permeability ranges from 2:1 to 150:1 and averages 25:1. Differences in the performance between wells that are on trend and those that are "just a few degrees" off trend are noticeable: for example a controlled 850 psi (5,860 kPa) pressure reduction in one test well created a 51 psi (352 kPa) pressure deviation in an on-trend monitor well, a 23 psi (159 kPa) deviation in an oblique monitor well, and a "slight but discernable" pressure response in an off-trend monitor well.

Fractures dominate the Midale system permeability, but even these closely spaced fractures are reported to contribute negligibly to the bulk reservoir porosity, and most of the oil is in the matrix rock between the fractures. Rapid depletion of the fracture volume creates

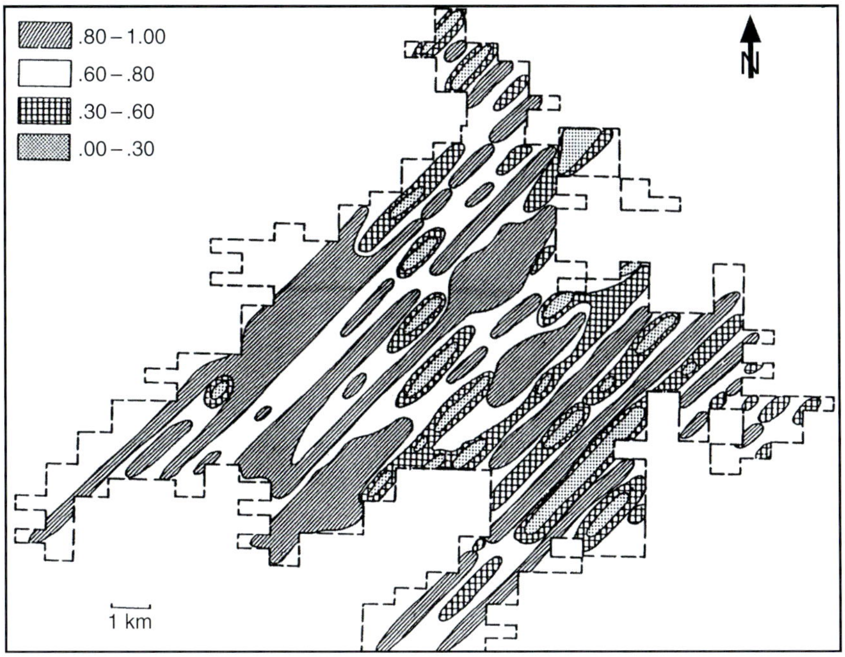

Figure 3.10 An anisotropic fracture-permeability system in the Midale carbonate reservoirs of Saskatchewan is recorded by a map of the distribution of percent water cut in the produced fluids. Wells located on-trend relative to water injectors have early and abrupt water break-throughs, as well as continued high-percent water cuts over their production histories (from Beliveau, 1987).

a pressure gradient between the interior of the matrix blocks and the fracture faces, which causes fluids to flow from the matrix into the fracture system.

This primary flow of oil from the matrix is relatively inefficient, and knowledge of fracture characteristics has been used in several ways to improve the effectiveness of fluid extraction from the matrix. For example, horizontal wells drilled across the fracture trend intersect significantly greater numbers of fractures and on average produce six times more fluid than vertical wells. The natural-fracture characteristics are also used to advantage in Enhanced Oil Recovery (EOR) designs. Pilot projects showed the benefit of injecting CO_2 into the more highly fractured but lower-porosity limestone zone, where the density difference between oil and CO_2 forces the CO_2 upward and into the overlying, higher-porosity, oil-bearing dolomite. CO_2 is miscible with oil, lowering the oil viscosity and allowing it to be more easily swept and produced. The pilot test results led to a successful CO_2 flood across the field. The importance of the fracture surface area to EOR design is discussed in Section 3.6.5.

3.4.2.2 Case Study: The Rulison Field

A dominant set of parallel-striking, variably mineralized extension fractures is also present in Cretaceous sandstones of the Mesaverde Group in western Colorado, where fractures create horizontal drainage anisotropies of between 10:1 and 100:1, documented by a research project called the "Multiwell Experiment." (Multiwell Experiment Project Groups Final Reports; see also a summary in Warpinski and Lorenz, 2008). This research project consisted of three closely spaced wells arranged in a triangle with legs on the order of 150–200 ft (46–61 m) long. Drainage anisotropy was documented by well-interference tests: the reservoir pressure in two of the wells was drawn down, then allowed to build up while at the same time natural gas production was pulsed in the third well. The WNW-ESE trend of the natural fractures, documented in oriented core and associated outcrops, creates a WNW-ENE drainage anisotropy, and interference was not observed in the two monitor wells over the course of a month-long test despite the close well spacing (Figure 3.11) because the long axis of high-aspect-ratio, fracture-controlled drainage ellipse runs between the two monitor wells and does not link them (Branagan et al., 1984).

3.4.2.3 Case Study: The Spraberry Formation

A similar, single set of parallel, open, vertical extension fractures, creating drainage anisotropies of up to 1:1000, was postulated to exist in the Permian sandstones of the Spraberry Formation in West Texas as early as the mid-20th century (Elkins, 1953, 1963; Elkins and Skov, 1960). Vertical core had shown the presence of natural fractures in the formation, and its generally NE-SW strike had been documented by inter-well communication patterns (Figure 3.12), but the evidence was contradictory, never providing an entirely consistent picture of the subsurface permeability system.

More recent fracture data from image logs and cored horizontal wells have confirmed the presence of a dominant set of parallel, variably mineralized, NE-SW

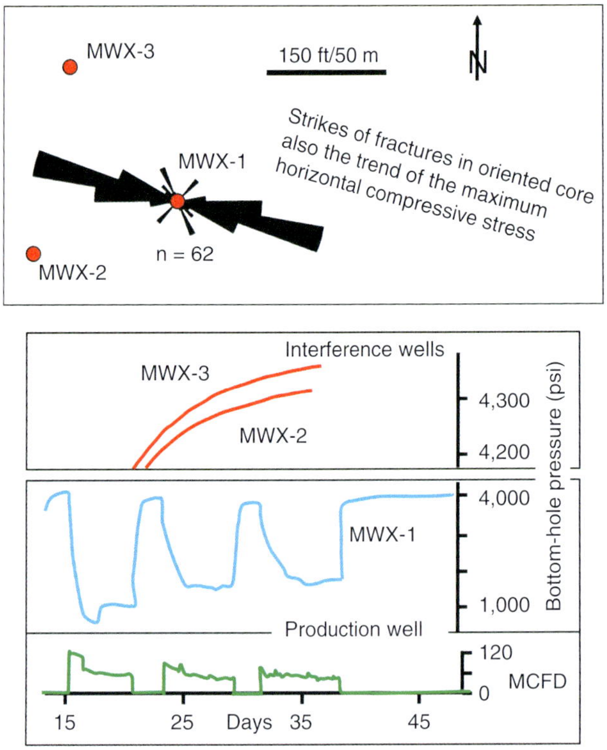

Figure 3.11 *Top*: a map showing the map-view locations (red dots) of the three closely spaced MWX wells, drilled into low-permeability sandstone reservoirs of the Mesaverde Formation in Colorado, with a superimposed rose plot of the strikes of 62 fractures measured in oriented core cut from these wells. *Bottom*: interference tests demonstrated elliptical drainage, trending WNW-ESE, parallel to the natural-fracture strikes at this site. During the course of the illustrated test, the reservoir pressure was drawn down in wells MWX-2 and MWX-3, then shut in and allowed to build up back to initial pressures as production from the same sandstone was pulsed in well MWX-1. The pressure in MWX-1 varied from 1,000 to 4,000 psi (7–28 MPa) as the well was alternately flowed and shut in over the course of the test. No interference was measured in the two observation wells since drainage was from a high-aspect-ratio ellipse between MWX-2 and MWX-3. (Adapted from Lorenz et al., 1989).

striking extension fractures in some layers of the formation, but they also record the presence of an entirely different fracture system in other layers, explaining some of the anomalies and providing an opportunity to compare the effects of two fracture systems in otherwise similar reservoirs. The cores and logs, as well as detailed engineering tests, were part of a research project called the "Advanced Reservoir Characterization CO_2 Gravity Drainage in the Naturally Fractured Spraberry Trend Area" (Schechter, 2002). Fracture measurements from the oriented horizontal cores revealed two distinct fracture systems in sandstones separated vertically by 140 ft (43 m) (Lorenz et al., 2002). A set of parallel, NE-SW striking, partially mineralized extension fractures is

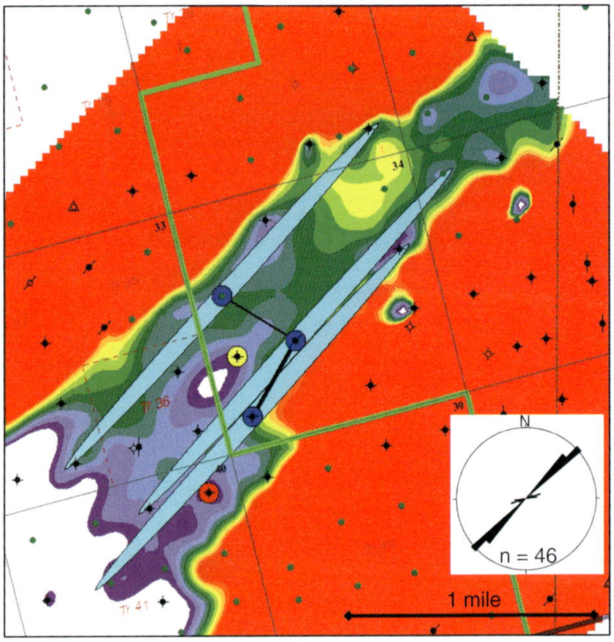

Figure 3.12 A single set of well-developed extension fractures creates a definitive NE-SW permeability anisotropy in sandstone reservoirs of the Permian Spraberry Formation, West Texas. The anisotropy is documented by an area of reduced Gas-Oil Ratios (cool colors) where water injected into three wells (blue dots) has repressurized the reservoir and dissolved gas *back into* solution along a well-defined, fracture-controlled azimuth. (Courtesy D. Schechter, personal communication, February 2019.) The lower-right inset shows fracture strikes measured in horizontal core from the same formation in a nearby well (see Lorenz et al., 2002).

present in one reservoir, designated Bed A here, and a set of intersecting, unmineralized conjugate shear fractures, striking NNE-SSW and ENE-WSW, is present in the similar sandstone of Bed B (Figure 3.13).

Tracer tests conducted between two wells positioned along the NE-SW trend illustrate the effects of the difference in fracture systems between the two reservoir units. Tracers were injected simultaneously into Bed A and Bed B, and production from the two units was comingled at the monitor well. The tracers had a distinct, bimodal breakthrough pattern: sharp tracer spikes showed in the monitor well three to six days after injection started, and broader peaks of the same tracers showed up again in the same well 28–32 days after injection (Figure 3.14). The most plausible explanation for the bimodal breakthrough is that tracer, following the short, direct pathway through extension fractures in layer A, broke through early, whereas tracer injected into the conjugate fracture system in layer B had to follow a more circuitous route along the intersecting fractures and through a larger volume of the reservoir, delaying arrival at the monitor well.

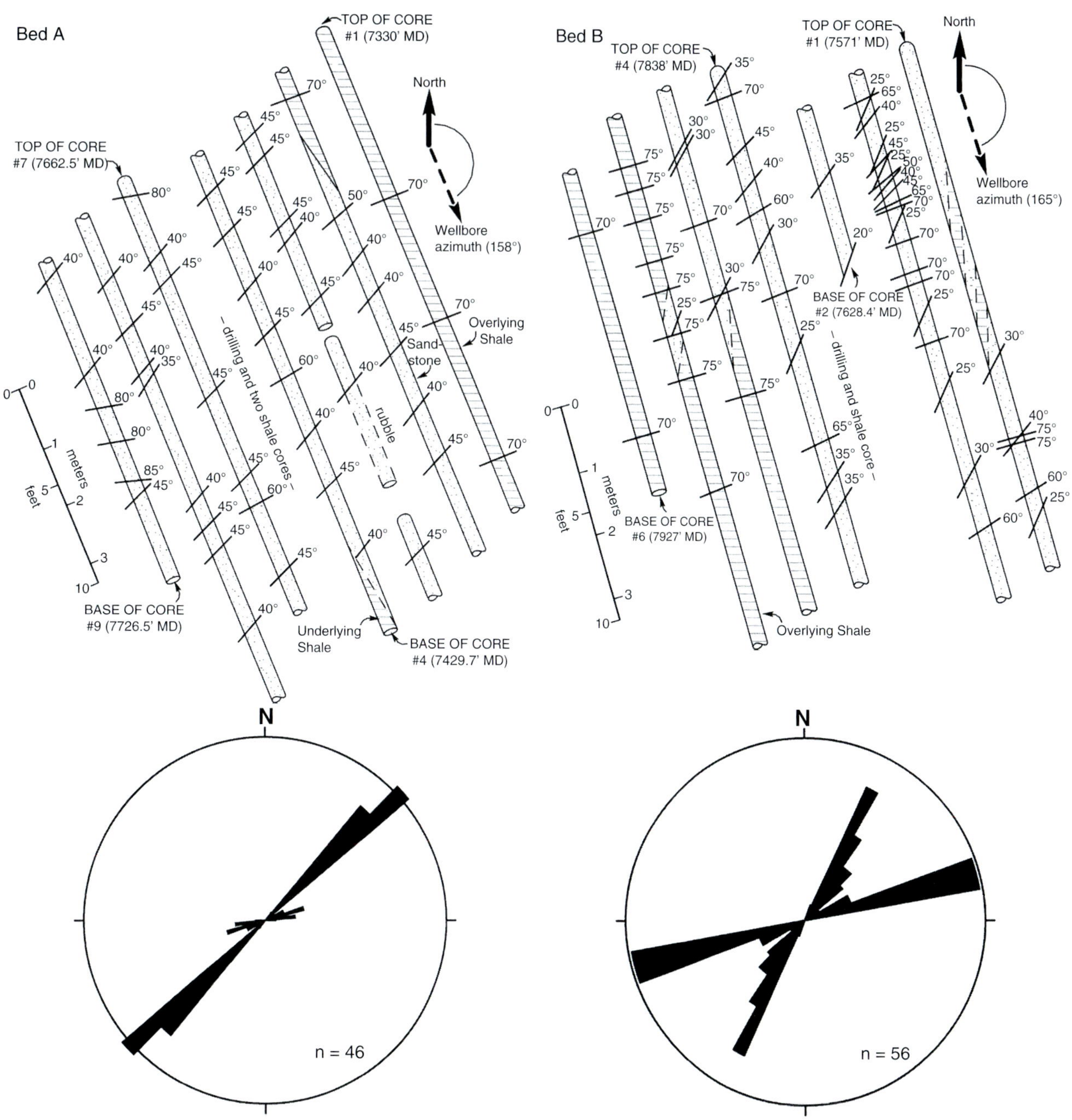

Figure 3.13 *Top*: map views of fractures captured by near-horizontal cores cut from Beds A and B at a depth of about 7,000 ft (2,130 m) in marine sandstones of the Permian Spraberry Formation, West Texas. Most of the illustrated core sections are connected end-to-end but they are displayed side-by-side for convenience. True fracture strikes are shown by the lines draw across the cores. The cores were cut across bedding at a shallow angle, intersecting both the reservoir sandstones and the bounding mudrocks. *Lower left*: core from Bed A captured a systematic set of parallel extension fractures. These incompletely mineralized fracture planes create highly anisotropic permeability and drainage patterns in the reservoir. *Lower right*: in contrast, the core from Bed B 140 ft (43 m) deeper in the section captured a pair of conjugate, strike-slip shear fractures. (Adapted from Lorenz et al., 2002, with permission from AAPG, whose permission is required for further use.)

The presence of dissimilar fracture systems in adjacent reservoirs is an example of "dynamically compatible" systems, where different fracture sets can form in adjacent beds under the same stress conditions due to dissimilarities in the basic mechanical properties of the layers (Hancock, 1986). The NE-SW bisector of the acute angle between the strikes of the two sets of the conjugate strike-slip shear-fracture system in Bed B is parallel to the average strike of the extension fractures in Bed A (Figure 3.15), thus the simplest explanation is that both fracture systems formed when the maximum in situ compressive stress was horizontal and oriented

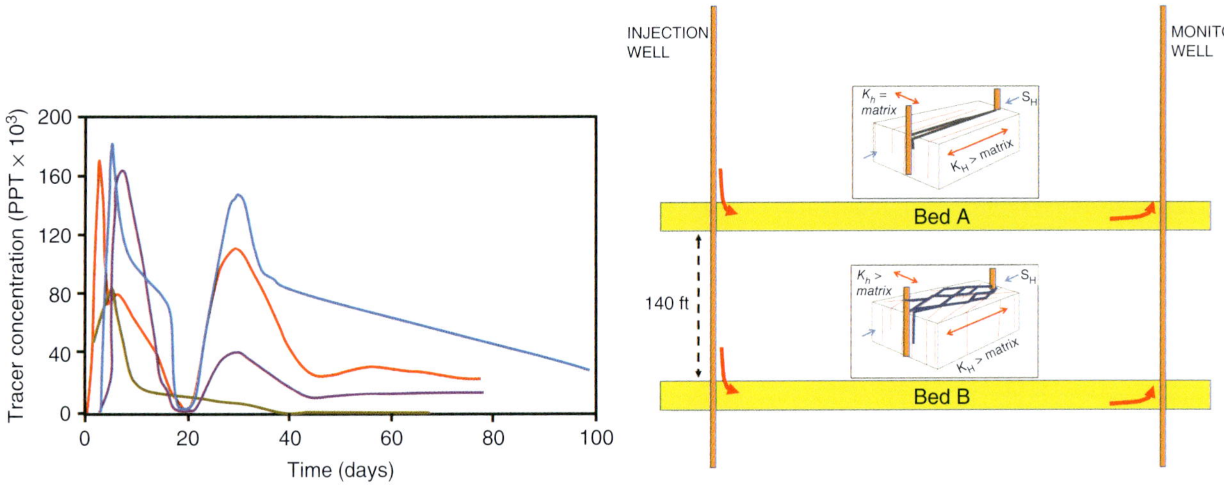

Figure 3.14 *Left*: the pattern of tracer breakthrough between injection and observation wells aligned along the NE-SW main fracture trend was bimodal, showing early, abrupt spikes and later, broader peaks (courtesy D. Schechter, personal communication, February 2019). *Right*: the tracers were injected simultaneously into Beds A and B, and production was comingled in the observation/monitor well. The most plausible interpretation for the bimodal tracer-breakthrough pattern is that the early-arrival tracers followed a direct pathway along the parallel extension fractures in Bed A, whereas tracers injected into Bed B had to follow more circuitous routes through the intersecting conjugate shear fractures.

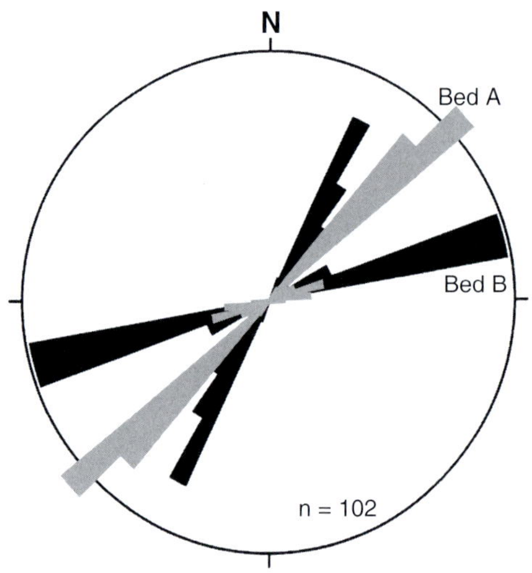

Figure 3.15 Dynamically compatible fracture sets, consisting of parallel, uniformly striking extension fractures in Bed A and a pair of strike-slip shear fractures in Bed B that are symmetrically arrayed clockwise and counterclockwise from the strike of the extension fractures. The two fracture sets formed in different layers only 140 ft (43 m) apart vertically. Bed B contains more detrital clay and less quartz cement, and is both weaker and more ductile than Bed A. The difference in mechanical properties between the two beds, caused by variations in deposition and diagenesis, controlled the different modes of fracturing (reprinted from Lorenz et al., 2002, with permission from AAPG, whose permission is required for further use).

NE-SW, the minimum stress was horizontal and oriented NW-SE, and the weight of the overburden provided the intermediate compressive stress. The present-day maximum horizontal in situ compressive stress at this location still has a NE-SW trend and is probably a residual from a more anisotropic system that existed when the fracture sets formed. Other recent vertical and horizontal cores cut from the Spraberry Formation show that although there is a strong NE-SW fracture fabric in the reservoirs, these specific fracture types are not universally characteristic of specific reservoirs. The local fracture response of the different layers is controlled by the sedimentary and diagenetic heterogeneity of the individual sandstones as well as by proximity to larger structures.

Observations of off-trend inter-well communication in the Spraberry are explained by the secondary NNE-SSW and ENE-WSW fracture sets. Moreover, patterns where communication between two on-trend wells skips over an intervening, on-trend well can be understood in the context of the irregular drainage patterns that can occur in such conjugate shear-fracture systems. Heffer et al. (2010) reported similar maximum-permeability trends in North Sea Chalk reservoirs that are oblique to the maximum horizontal in situ compressive stress, attributing this to preferential flow along conjugate faults and fractures in the formation.

The differences between a conjugate shear-fracture system vs. a system consisting of one or multiple sets of extension fractures affect reservoir drainage, permeability anisotropy, and vertical connectivity, as well as drawdown, production rates, and stimulations (e.g. Loosveld and Franssen 1992), but differences usually cannot be observed in single-well tests or production histories since decline curves and well-test results can also be affected by variables including the in situ stress magnitudes and orientations, drilling and completion techniques, and sedimentary heterogeneities in a reservoir. Tests using multiple wells, and subsurface information about the fracture system from image logs and/or cores, are needed to constrain interpretations of well-test results.

Both the Spraberry and Mesaverde sandstones contain systems of regional fractures in almost flat-lying strata, and the fractures are widespread and well-developed despite the absence of major structures. The in situ stress orientations in both formations have not changed orientation since the time of fracturing, thus hydraulic stimulation fractures propagate parallel to the extension-fracture strikes in both cases, and oblique to the strikes of the conjugate fractures in the Spraberry field.

Drainage patterns in fractured reservoirs can be much less anisotropic where two or more sets of superimposed, oblique-striking extension fractures exist, provided that all fracture sets are equally well developed and equally permeable. If one of two fracture sets is better developed and more closely spaced it may dominate the reservoir, creating drainage anisotropy despite the presence of intersecting fracture sets. However, if the numerically dominant fracture set is also more completely occluded by mineralization, drainage and system permeability anisotropy may reflect the less-well-developed but more open fracture set (Figure 3.16). Likewise, a less-well-developed fracture set may dominate if the better-developed set strikes normal the maximum compressive stress and is therefore more susceptible to closure during production-related in situ stress changes.

In contrast to the parallel and poorly interconnected planes that comprise a set of extension fractures, shear-fracture sets commonly consist of intersecting conjugate pairs that create drainage ellipses with much smaller aspect ratios (Figure 3.17). The two subsets that form the pair may not be equally developed, i.e. one subset may consist of a few large faults and the other of a more pervasive series of antithetic shear fractures, but where they are equally well developed and where the subsets have similar spacings and degrees of mineralization, they should have equal effects on system permeability. Unlike extension fractures, shear fractures strike oblique to the in situ stresses at the time of fracturing, which becomes important when considering their interactions with hydraulic stimulation fractures as well as their sensitivity to changes in the in situ stress during production.

3.4.3 Fracture-Controlled Sweet Spots

Some fracture systems are developed regionally, but others are restricted to, or better-developed within, the localized fracture domains that create "sweet spots" of high production rates within a reservoir. Such fracture corridors or fracture swarms can be related to extension

Figure 3.16 Plan view of a sandstone bedding surface in the Cretaceous Frontier Formation in Wyoming, showing two unequally developed sets of bed-normal extension fractures. If all fractures are equally open and effective, system permeability and drainage anisotropy are greatest parallel to the more closely spaced fractures of Set 2. However, if the Set 2 fractures are occluded by mineralization and the Set 1 fractures are not, then the less well-developed Set 1 fractures may dominate system permeability. Ameen et al. (2012) describe a reservoir in Saudi Arabia where reservoir permeability is dominated by a set of unmineralized fractures and not the oblique, older, fully mineralized fracture set. Likewise, if the Set 2 fractures strike normal to the maximum horizontal stress, they may close quickly during production, leaving the less-numerous Set 1 fractures to dominate production. Teufel et al. (1993) demonstrated that although fractures with numerous orientations are present in the chalk reservoirs of the Ekofisk oilfield in the North Sea, the most effective fractures strike parallel to the maximum horizontal compressive stress because they are least susceptible to closure. Gale (1990) even suggests that if two fracture sets are present in a reservoir, they may have different aperture distributions and will need to be modeled with different flow laws.

caused by folding at the hinge of an anticline (e.g. Zahm and Hennings, 2009; Gilbertson, 2006), and such hinges may also locally reactivate and enhance regionally developed fracture systems (e.g. Bellahsen et al., 2006). Local domains of intensely developed fractures also commonly form as fracture halos around faults. The width of a fault-related fracture corridor is commonly related to the magnitude of slip along the fault (Figure 3.18) (Robertson, 1983; Marrett and Allmendinger, 1991). The scatter in the data is such that the width of a specific fracture halo can only be estimated from fault offset, and in fact the halo may include a damage zone of gouge and breccia that can reduce rather than enhance permeability, especially if the associated fractures in the zone have been occluded by mineralization.

Faulted Cretaceous sandstones in the San Juan Basin of New Mexico provide an example of a localized fracture domain that has a strong positive influence on production. As documented by Hart (2006), a well drilled in the early 1950s produced steadily until a second well was drilled nearby in the 1970s. Completing the second well

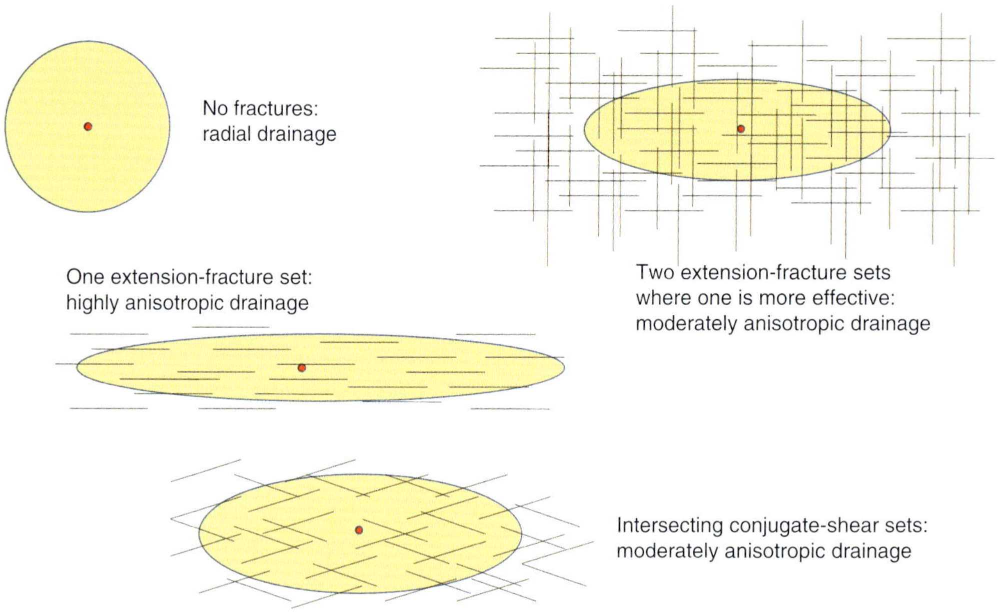

Figure 3.17 Map views showing conceptual drainage ellipses around vertical wellbores (red dots) drilled into reservoirs containing different fracture systems. *Upper left*: a homogeneous reservoir with no fractures is likely to have radial drainage. *Left-middle*: fractured reservoirs with single sets of open extension fractures have drainage ellipses with high aspect ratios. *Lower left*: reservoirs with intersecting conjugate shear fractures should have drainage areas with smaller aspect ratios. *Upper right*: drainage from reservoirs containing intersecting sets of extension fractures depends on the relative development of the two fracture sets and their degrees of effectiveness due to aperture, mineralization, orientation relative to the in situ stresses, etc. They may create near-circular drainage if they are equally developed, but it is more likely that one fracture set is more effective and that the drainage is anisotropic.

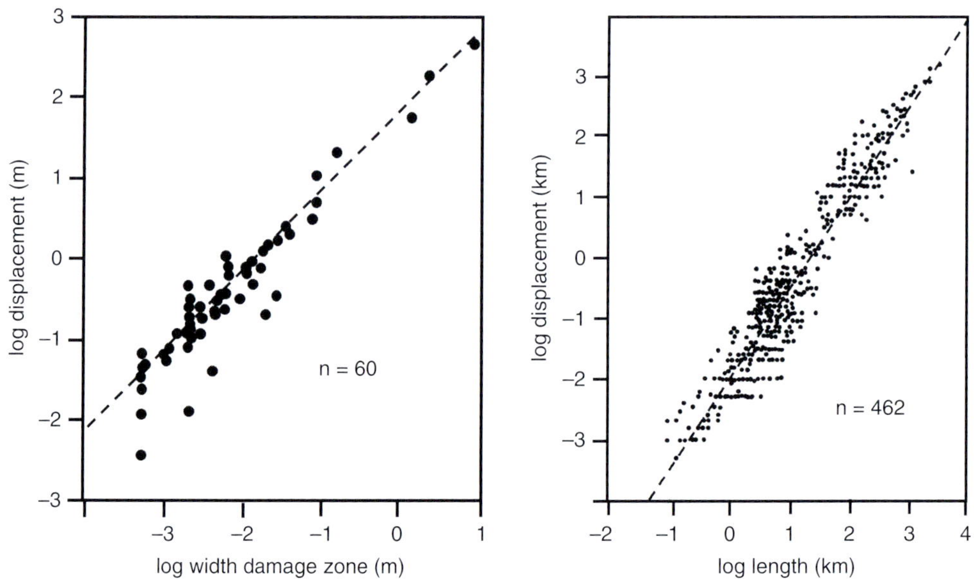

Figure 3.18 *Left*: chart showing the relationship over several orders of magnitude between fault/shear-fracture displacement and the width of the associated fracture/gouge/damage zone (from Marrett and Allmendinger, 1990). The scales are logarithmic meaning that the scatter in the data is wider than desirable for predicting precise widths from offsets, but it at least provides a starting point for an estimate. *Right*: fault length vs. displacement (from Marrett and Allmendinger (1991).

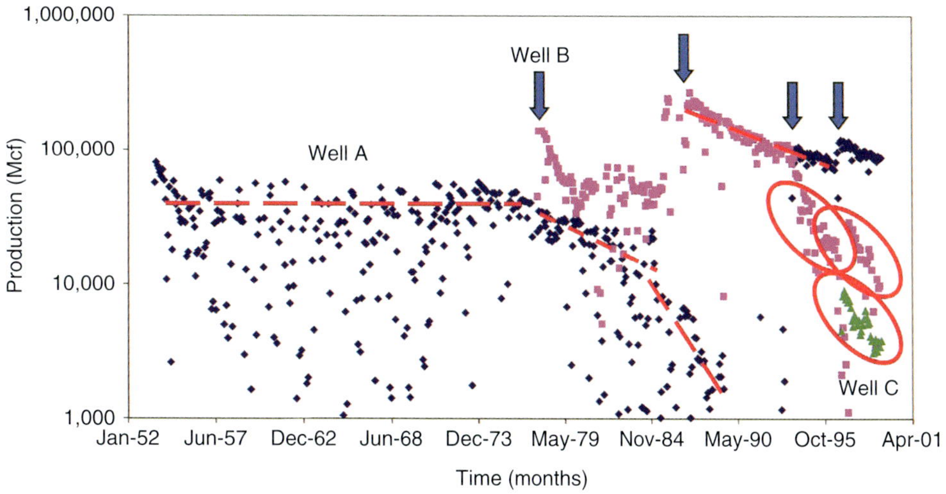

Figure 3.19 *Top*: production curves show interference between wells A (blue) and B (pink), and that well C (green) between them was drilled into a depleted zone. (Modified from Hart, 2006, with permission from AAPG, whose permission is required for further use.) *Bottom*: a fracture corridor in marine sandstones, Wyoming (courtesy R. Billingsley, 2004).

caused an abrupt decline in the production from the first, and the two wells showed definitive interference patterns over the years. A third well, drilled on a line between the first two in the 1990s, penetrated a depleted zone (Figure 3.19).

The three wells were drilled by different companies, and the fact that there was interference between them only became apparent when one company bought the other and compared the production and completion records of the wells. In turn, the reason for that interference only became obvious when a 3D seismic survey was shot over the area. The survey showed a local anomaly that could be mapped using seismic attributes (see Figure 2.68). Outcrops of equivalent strata show that the strata are faulted and the faults have fracture halos, explaining the linear trend of the production sweet spot as well as the observed interference between wells.

3.5 The Sensitivity of Fracture Permeability to Changing Stress

3.5.1 Stress-Sensitive Extension Fractures

Fractures viewed in a core laid out on a laboratory table have the appearance of immutability, but rock, and

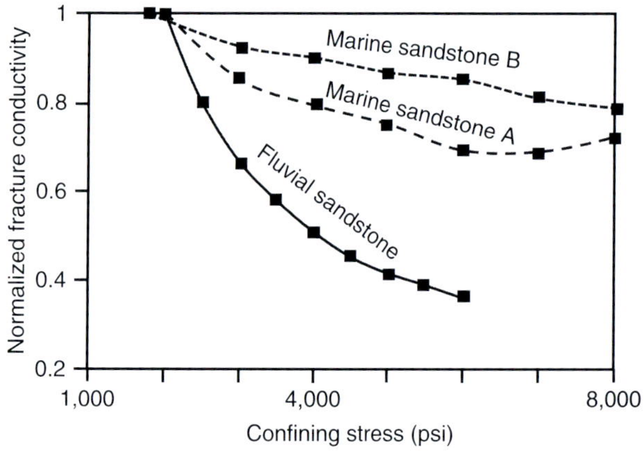

Figure 3.20 Laboratory tests show that mineralized natural fractures in different sandstones of the Mesaverde Group in Colorado have different degrees of closure under the increased confining stresses that approximate decreasing fluid pressures in fracture apertures. Fractures in the immature fluvial sandstones, with a higher percentage of relatively ductile sand grains, close rapidly and lose about 70% of their original conductivity as the confining stress increases from 2,000 to 6000 psi (from 13.7 to 41.4 MPa), whereas the cleaner, more quartz-rich marine sandstones lose less than 30% of their conductivity under the same simulated conditions of reservoir drawdown. (From Lorenz et al., 1993.)

the fractures within rock, are not rigid under the high confining stresses at depth. As shown in this section, laboratory and well tests indicate that fracture apertures, and therefore fracture permeability, can change by varying degrees due to the changing in situ stress conditions associated with production from a reservoir (e.g. Aguilera, 1999; Lorenz, 1999). The closure of fracture apertures and the related damage to fracture permeability (Figure 3.20) has been called both "stress sensitivity" and "compressibility."

The strain accommodated by rock containing natural fractures is typically greater than that of otherwise-similar unfractured rock, i.e. rock containing an open natural fracture compresses more readily and by a greater amount under a given stress increase than rock without an equivalent fracture, since closing an open fracture aperture requires less energy than compressing intact rock. However, in terms of permeability reduction, the *effect* of a given amount of strain in closing narrow pore throats and reducing permeability in the matrix is typically less than the effect of the same amount of strain in rock containing fractures (Figure 3.21). The degree to which a fracture is compressible varies with factors including:

- Fracture type
- Degree and type of fracture mineralization
- The mechanical properties of the host rock
- Percent water saturation in the fracture
- Fracture orientation relative to the in situ stress axes
- In situ stress magnitudes and anisotropy
- The characteristics of the matrix permeability (primarily pore throat size)

As long as a reservoir is under production, fluid pressure in the fracture apertures should be less than the overall pore pressure in the formation, and the effective compressive stress acting normal to the fracture planes is relatively high, narrowing fracture apertures and degrading permeability. In the laboratory, the same effect is achieved by increasing the confining stresses on a fractured sample (e.g., Myer, 1991). Stress sensitivity also operates in the other direction: some natural fractures can widen as fluid pressure in the fracture apertures increases during injection, even when the injection pressure is less than the pressure needed to break rock and create a hydraulic fracture in the formation (e.g. Warpinski and Lorenz, 2008).

Laboratory experiments that have tested fractured rock under compression, with no fluid pressure in either the fracture or the matrix, have shown that the asperities along unmineralized but mismatched fracture faces crush and the apertures slowly close with increasing normal stress despite contact across the fracture walls and bridging at the asperities (e.g. Bandis et al., 1983). Witherspoon et al. (1979) suggested that artificial fractures with mismatched faces in different rock types close but that the contact areas hold open a percentage of the original aperture, "maintaining space for fluids to continue to flow as the fracture aperture decreases," i.e. the apertures are stress-sensitive, but they do not close completely. Likewise, Bandis et al. (1983) reported that the apertures of mismatched fracture faces in the laboratory are initially highly compressible as the stress normal to the fracture plane increases, with 70% of the fracture closure occurring during the imposition of the first 10% of the increase in confining stress that it takes to close the apertures. Closure is defined as the point where the asperities were no longer being crushed, where a high percentage of the area of the opposing fracture faces is in contact across the fracture plane, and where the rock mass begins to respond to stress increases in an elastic fashion. Thus, fracture closure or "compressibility" is a matter of degree, it is not an all-or-nothing effect.

Some authors (e.g. Dyke, 1992) have suggested that fracture permeability is not stress-sensitive because natural-fracture apertures are typically mineralized and the mineralization should act as a natural proppant, keeping a fracture open. Propping by mineralization would seem to be more effective for narrow than for wide fractures, but in fact the same linear magnitude of closure reduces the width of a narrow fracture by a greater percentage than of a wide fracture, so the permeabilities of narrow fracture apertures are typically most

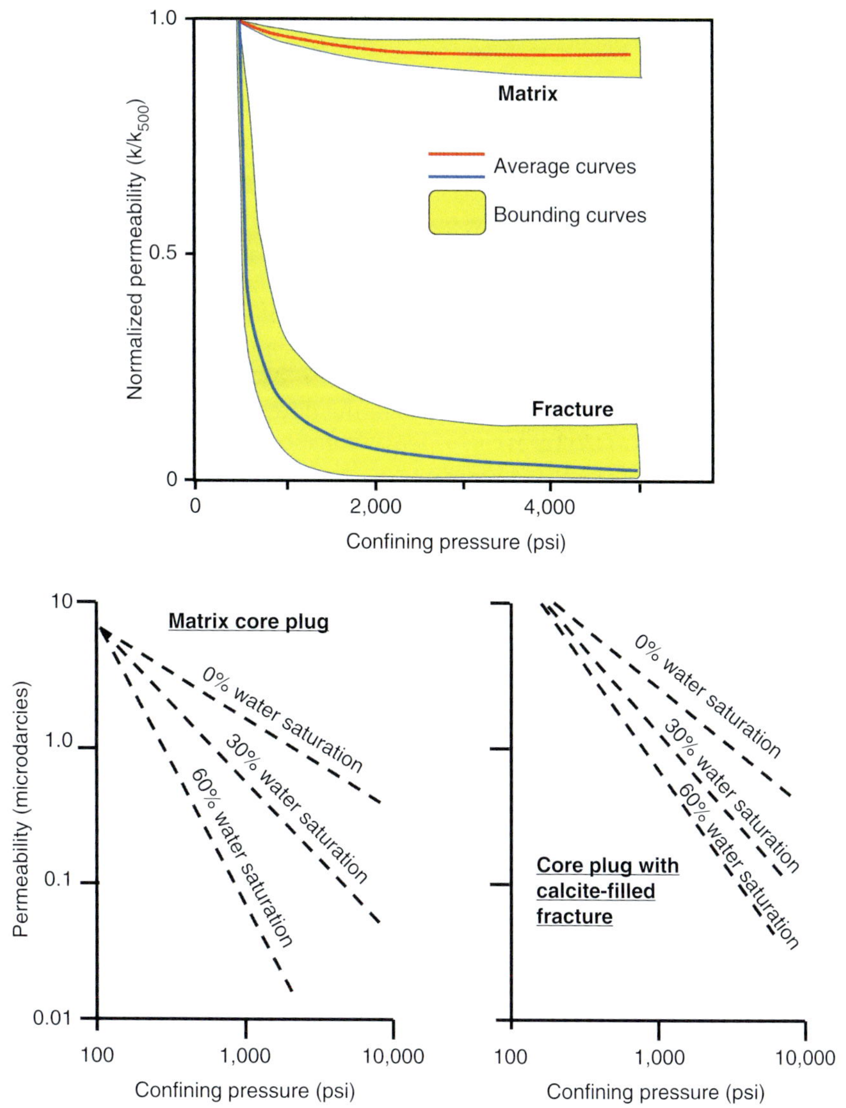

Figure 3.21 Examples of laboratory tests comparing the susceptibility of fracture and matrix permeabilities to changes in the confining stress. *Top*: example of fracture permeability in Triassic sandstone that is more susceptible than matrix permeability (Daw et al., 1974). *Bottom*: tests in a Cretaceous sandstone showing fracture and matrix permeabilites decrease with increasing confining stress and increasing water saturation (Multiwell Experiment Project Groups).

sensitive to changes in the confining stresses (Aguilera, 2003). Regardless, fracture closure should be less where fractures are filled with a strong mineral such as quartz, where the mineralization bridging the fractures is more complete, and where the fractures occur in a strong, mechanically stiff formation.

Mineralization that is incomplete and patchy creates points of stress concentration that are similar to the asperities of mismatched fracture faces and that are equally prone to crushing (e.g. Cook, 1992). Fracture-filling mineralization that is weaker than the host rock, for example a calcite-filled fracture in a siliceous sandstone, is more susceptible to crushing than asperities of the host rock during fracture closure. Aguilera (1999) constructed a chart of idealized fracture compressibility as a function of both the confining stress acting on a fracture and its degree of mineralization. The chart suggests that an unmineralized fracture is almost ten times more compressible than a fracture with 50% mineralization, but that even a fracture that is 50% mineralized compresses by an order of magnitude as the confining stress increases from 0 to 12,000 psi (from 0 to 83 MPa). The loss of fracture-parallel permeability created by closure can be exacerbated by the fines created by crushed mineralization and asperities since the fines can migrate and block the narrower parts of the remaining fracture aperture.

When fractures close, whether due to decreasing fluid pressure in the aperture or increasing confining stresses, the associated permeability degradation may be either temporary or permanent. The Lisburne Limestone reservoir of Alaska provides an example of lasting damage, where laboratory tests on naturally fractured, mineralized samples showed significant permeability reductions under compressive stress increases equivalent to the amount of pressure drawdown in the field (Figure 3.22),

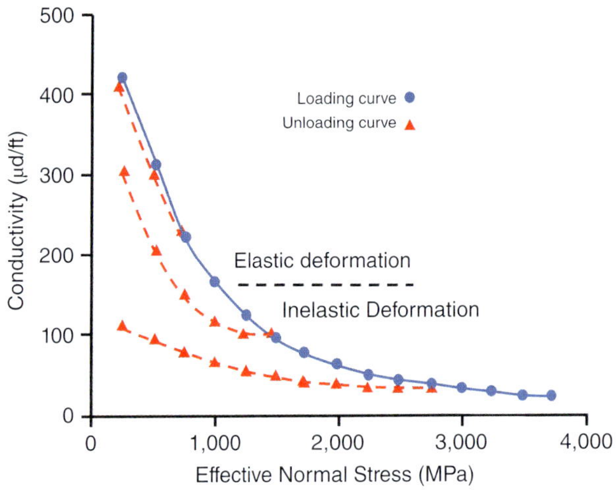

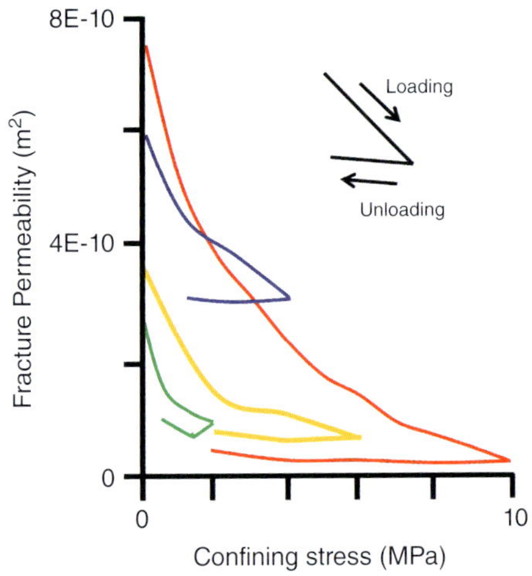

Figure 3.22 *Left*: The conductivity of fractured plugs cut from cores of the Carboniferous Lisburne Limestone in Alaska is elastic and not damaged if the confining pressure in the lab test is increased to about 900 psi (6.2 MPa), but a permanent 20% loss of conductivity occurred when the confining pressured was raised to 1,400 psi (9.7 MPa). The fractures had a permanent 75% loss when the confining pressure was raised to over 2,700 psi (18.6 MPa). (Data by L. Teufel, reprinted from Hanks et al., 1997, with permission from AAPG, whose permission is required for further use.) *Right*: the results of four laboratory tests on fractured samples of the Jurassic Kimmeridge Clay showed that permeability decreased as the normal stress across the demineralized fracture plane was increased, and that the permeability reduction was permanent, remaining low even after the confining stress was released. (Adapted from Gutierrez et al., 2000.)

and where release of the compressive stresses did not restore the samples to their original permeability (Hanks et al., 1997). Similarly, laboratory tests have shown that fractures in the Kimmeridge Clay, a Jurassic mudrock in southern England, can be significantly and permanently damaged by increasing confining pressures (Gutierrez et al., 2000).

The tested Kimmeridge fractures were narrow, 0.10 mm wide, and were originally mineralized with calcite. The calcite was removed by dissolution for the experiment. Permeability along demineralized fractures was reduced by 50%–90% under a closure stresses of up to 1,450 psi (10 MPa) normal to the fracture planes, and the fracture-parallel permeability remained at the low, damaged level even after the confining stress was released (Figure 3.22). Nevertheless, permeability along the damaged factures was still orders of magnitude greater than that of the host shale.

Other fracture systems are equally stress-sensitive but they are more elastic. The fracture-controlled permeability of these systems is not permanently damaged but can be restored by shutting in production and allowing fluid pressures in the fracture apertures to equalize to formation pressures. For example, laboratory and well tests of extension fractures in the natural-gas reservoirs of Cretaceous sandstones of the Mesaverde Group of Colorado show this type of elastic behavior (Figure 3.23). As measured by flow rate at the well head, subsurface fracture-system permeability decreased to nearly zero when the wells were opened to atmospheric pressure (depleting gas from the insignificant fracture volumes near the wellbore and closing the fracture apertures), but the fracture system in this reservoir is not permanently damaged. When the wells were shut in, allowing the formation pressure to rebuild to initial conditions, the natural fracture system regained its initial permeability.

3.5.1.1 Case Study: The Bulo Bulo Field

Vega Navarro (2014) described another example of a stress-sensitive fracture system in gas-condensate reservoirs in the Bulo Bulo field in front of the Andean thrust system in Bolivia. The reservoirs in this field consist of fine-grained, shallow-marine Devonian sandstones with an average porosity of 6% at a depth of about 13,000 ft (4,000 m). The laboratory-measured matrix permeabilities in the field range from 0.001 to 0.01 md, but the permeabilities measured from drill-stem tests are orders of magnitude higher, ranging from 50–500 md. The sandstones are overlain by shale which forms the reservoir seal, and formation pressures in three target zones range from normally pressured to over-pressured (from 0.49 to 0.75 psi/ft.) Detailed core analyses show that the Bulo Bulo fracture system consists of partially mineralized fractures ranging from microscopic to centimeters in size. Open fracture apertures are calculated to comprise 0.6% of the bulk rock volume, or a tenth of the total porosity in the reservoirs.

Rapid declines in production are common in the over-pressured reservoirs, but not in the normally pressured reservoirs. In one reservoir where the original pressure gradient was 0.75 psi/ft, the initial production of 22 MMCF/D declined rapidly to 6 MMCF/D and then

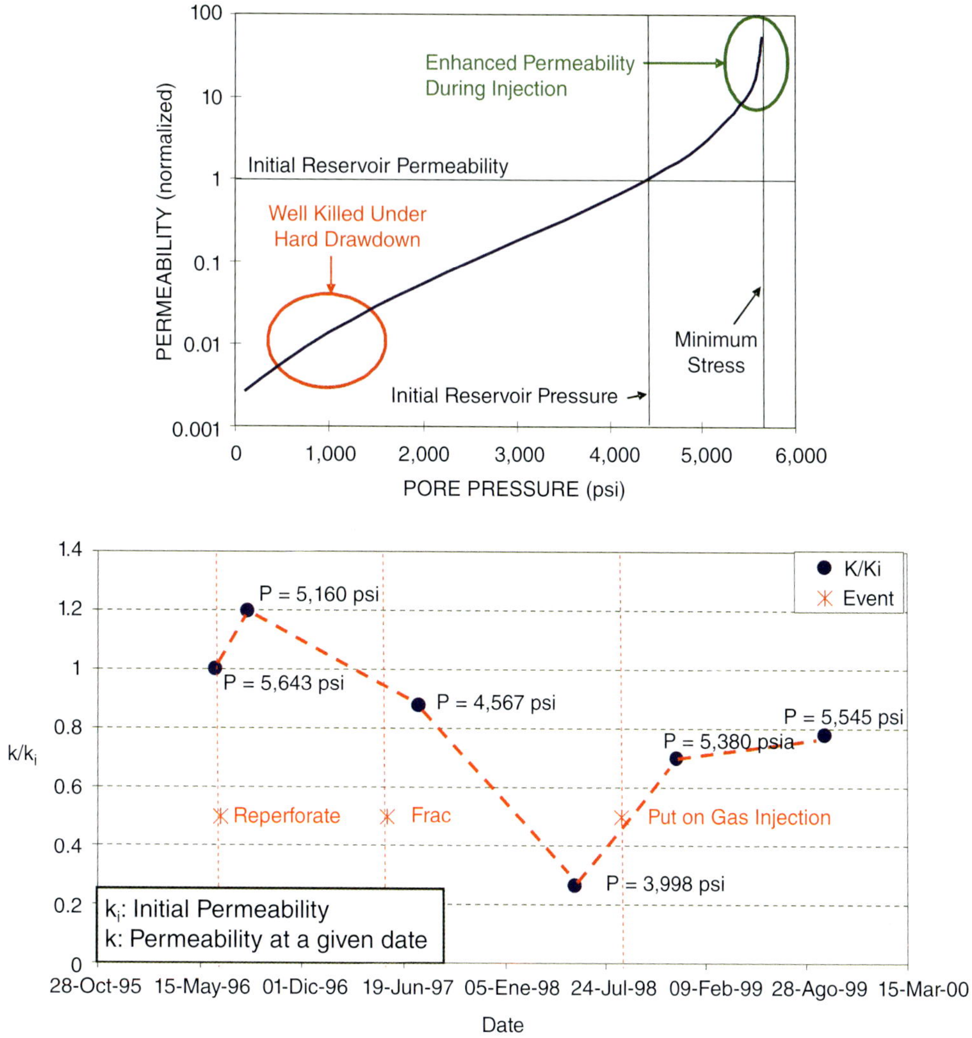

Figure 3.23 Stress-sensitive, elastic fracture systems. *Top*: well tests in the naturally fractured Mesaverde sandstones in Colorado showed that, unlike the fracture systems of the Lisburne and Kimmeridge strata, this stress-sensitive fracture system is not permanently damaged. The system permeability effectively decreased to 0 when the wells were opened to flow at atmospheric pressure, but the system permeability recovered once the wells were shut in, reducing the effective stresses on the fractures. This cycle was repeated several times without obvious damage to the system. The natural fracture system also became more permeable when fluids were injected at greater than reservoir pressure (but at pressures less than minimum stress so that hydraulic fractures were not created). (Reprinted from Warpinski and Lorenz, 2008, with permission from AAPG, whose permission is required for further use.) *Bottom*: the fracture-system permeability of reservoirs in the deeply buried, highly over-pressured, tectonically active strata of the Cupiagua field immediately east of the active Andes thrust belt in Columbia are stress-sensitive, with permeability changing as a function of pore pressure changes during production and injection as determined from transient testing analysis (courtesy J. Gildardo Osorio, personal communication, January 2019.)

to 1 MMCF/D (from 600 to 170 to 28 m^3 per day) in less than two months, followed by a sudden natural shut off of the well. This was attributed to increases in the effective confining stresses acting on the fractures during production, which resulted in closure of the natural-fracture apertures. Fracture collapse starts as soon as the wells are put on production, ultimately turning the reservoirs into simple, matrix-only, low-permeability systems with very low production rates and resulting in significant losses in the volume of recoverable hydrocarbons. The system is elastic since when a reservoir unit is shut in for a few years, the pressure builds back up as gas migrates from the matrix to the fractures to restore pressure equilibrium between the fractures and the matrix, and production rates recover; subsequent production again undergoes a rapid decline.

Vega Navarro recognized three production behaviors depending on the magnitude of the calculated effective in situ confining stress. Natural fractures are permeable and dominate production where the effective confining pressure ranges from about 1,500 to 4,000 psi. Fracture compressibility takes its toll on permeability, narrowing fracture apertures, in a transition range between about 4,000 and 6,000 psi. Fractures are closed, fracture permeability is negligible, and production rates are non-commercial at effective stresses greater than 6,000 psi (at and below the normal gradient of 0.45 psi/ft),

Interestingly, attempts to hydraulically fracture the reservoirs with fracture-pressure gradients of up to 1.5 psi/ft never reached a recognizable formation-breakdown pressure. The hydraulic-fracture energy is suggested to have been dissipated into the natural fractures despite injection pressures up to 18,000 psi. Nevertheless, the stimulation was deemed to be successful, breaking rock and opening the natural fracture system, improving production rates to levels on the order of 10 MMCF/D (Vega Navarro, personal communication, January 2019).

3.5.2 Stress-Sensitive Shear Fractures

Shear fractures (Figure 3.24) are also subject to damage induced by changes in a reservoir stress system. Most subsurface strata are suggested to be in a near-equilibrium stress state where perturbing the stresses easily reactivates shear fractures and faults (e.g. Hubbert and Willis, 1957; Heffer et al., 2010; Zoback 2007). Shear fractures and faults can become reactivated by either increasing or decreasing the pore pressure in a stress system (e.g. Molina and Zeidouni, 2017): *increasing* pore pressure decreases the component of the effective stress that acts normal to a shear-fracture surface, reducing the critical value of the shear stress required to initiate offset and allowing the existing effective-stress anisotropy to cause offset along the fracture plane (e.g., Hubbert and Rubey, 1959; Healy et al., 1968). *Decreasing* pore pressure increases the effective stresses as well as the stress anisotropy, and the latter can grow to exceed the frictional strength of a shear surface regardless of the compressive stress normal to the shear plane. Here we deal with the second case, where the matrix pore pressures are reduced during production. Teufel et al. (1993) describe a chalk reservoir where fractures not only shear during production, but where new shear fractures form due to the production-related increase in the differential between the maximum and minimum effective stresses.

Shear fractures, which are typically oblique to the in situ compressive stress axes and therefore prone to reactivation when the stresses change, may both shear and close during production. Laboratory experiments (e.g. Durham and Bonner, 1994; Kassis and Sondergeld, 2010) and models (e.g. Bisdom et al., 2016) suggest that a few millimeters of shear offset can dramatically increase permeability along a fracture since offsets of the asperities, the highs and lows along a fracture face, require the fracture to open. However, greater degrees of shear, as well as shear that occurs under high stress magnitudes normal to the fracture plane, can quickly turn asperities into gouge that plugs the aperture and damages fracture permeability.

Olsson (1992) and Olsson and Brown (1993) conducted laboratory tests on different lithologies including limestone and welded tuff, their results showing that the permeability along artificial fractures that were cut across small cylinders of the rock can increase by 20%

Figure 3.24 Two views of an inclined, dip-slip shear fracture with an irregular pinch-and-swell aperture that is partially occluded by calcite. *Left*: a view parallel to fracture strike. *Right*: a view of one of the fracture faces. The dashed line drawn on the core below the shear fracture in the photo on the left outlines a drilling-induced petal fracture, the strike of which is parallel to the maximum in situ horizontal compressive stress, showing that the natural shear fracture is oblique to that stress and subject to further shear during production. Core from the Pennsylvanian-age eolian Tensleep Sandstone, Wyoming.

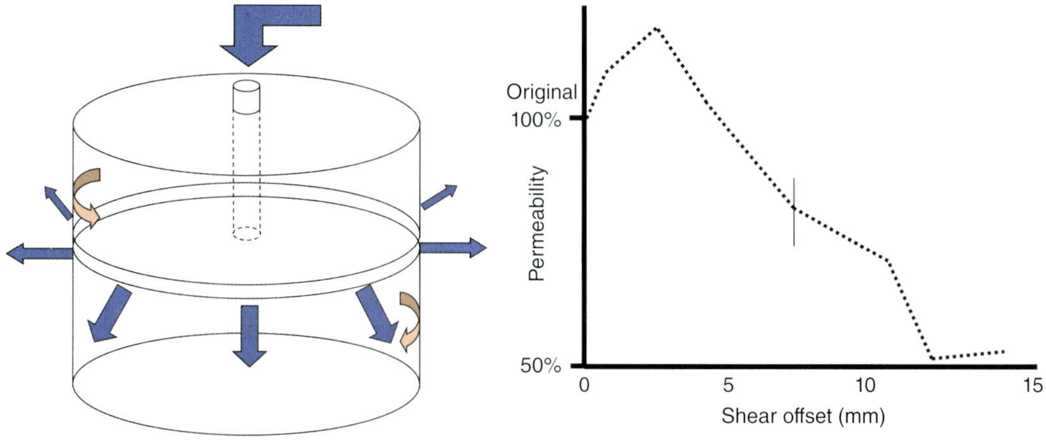

Figure 3.25 *Left*: Olsson (1992) and Olsson and Brown (1993) measured the changes in fracture permeability due to shear along an artificial fracture in the laboratory. The fracture was created normal to the axis of a cylinder of rock, and silicone oil was injected into the fracture from a hole in the top of the cylinder. The rate of fluid flow out of the fracture was measured as the top part of the cylinder was rotated and sheared against the lower part of the cylinder under different stress magnitudes normal to the fracture plane. *Right*: the permeability along this artificial fracture increased to a maximum of about 120% of the original permeability due to offset asperities for the first few millimeters of shear, but then decreased with continued offset as the asperities were sheared off and the fracture aperture was damaged. The direction of rotation was reversed at about 7 mm of offset (the black vertical bar), but damage continued, stabilizing at about half of the original permeability after 12 mm of cumulative offset.

after a few millimeters of shear offset. However, in at least one sample, the permeability then systematically and irrevocably decreased in proportion to the magnitude of shear offset as shear continued. Damage to the flow capacity of the fracture continued to decline even though an attempt was made to reverse the damage by reversing the direction of shear (Figure 3.25). All damage had been done by the time shear offset, as measured along the outside of the circumference of the sample, reached 11–12 mm. At that point the fracture permeability stabilized at about 50% of its original value. Shear-related damage in other samples was immediate and permanent.

Gutierrez et al. (2000) extended their tests on the fractured Kimmeridge Clay samples to measure permeability when the fractures were sheared under different levels of stress normal to the fracture planes (Figure 3.26). They found, in parallel with Olsson, that permeability *in*creased ten-fold when a fracture was sheared with a low confining stress (145 psi/1 MPa) acting normal to the fracture plane, but that the fracture-parallel permeability *de*creased by seven orders of magnitude when the fracture was sheared while a normal stress of 870 psi (6 MPa) was applied.

In addition, the amount of force required to shear the fracture increased only up to the point where 1–2 mm of shear had taken place, after which the force remained constant. This indicates that the asperities had been sheared and the damage to the permeability had been done with only 1–2 mm of shear offset. It is also worth noting that although permeability was reduced dramatically when the fracture was sheared while compressed under a high normal stress, the permeability of the most damaged fracture was still a thousand times greater than the 110 nD/(10^{-19} m^2) permeability of the shale host rock.

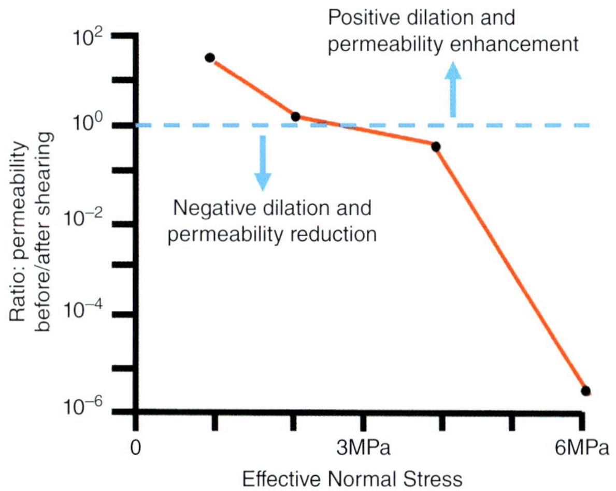

Figure 3.26 Gutierrez et al. (2000) used a laboratory setup similar to that shown in Figure 3.25 to measure permeability as they rotated the upper block against the lower block of a naturally fractured mudrock sample, shearing the fracture under different normal-stress conditions. Permeability along the fracture increased if the fracture was sheared while under low normal stresses, but permeability was significantly damaged when the fracture was sheared under high normal stress. (Adapted from Gutierrez et al., 2000.) Tests by Teufel et al. (1993) also documented the strong dependence of shear-fracture permeability damage on the normal stress acting across the fracture plane during shear, for naturally fractured chalk samples from the North Sea.

The rapid decline curves displayed by the production histories of naturally fractured reservoirs are often, rightly, assumed to reflect the rapid depletion of fracture-aperture volumes and the transition to matrix production. However, fracture sensitivity to changes in stress during production can also play a role in rapid production declines, and it is worth distinguishing between the two mechanisms because of the permanent damage to the fracture-delivery system that can be associated with the latter.

There are several possible ways to assess the relative roles of fracture depletion vs. fracture closure. Laboratory measurements of the sensitivity to stress changes, measured directly on natural fractures cored from a reservoir, can help in the evaluation. Comparisons between the fluid volume produced during initial decline and estimates of the in situ fracture volume can also help address the question: the fractures may be closing if less fluid is produced than could be supported by the estimated in situ fracture volume.

If fracture-related permeability is found to be stress sensitive it can be managed through pressure maintenance; either by fluid injection or by restricting production. Pressure maintenance will keep the natural-fracture production pathways open longer, increasing the ultimate recovery from a well or field. Although it takes more time to amortize the cost of a field with low production rates, the ultimate fluid recovery and return on investment can be greater over the long run. Rapid declines in production following attempts to produce at high rates with no pressure maintenance, especially if rates do not recover after a well is shut in, suggest stress sensitivity and fracture-system damage, but that assessment may be too late to be of use in remediating the situation. High in situ stress magnitudes and anisotropies, and high pore pressures, increase the likelihood of fracture closure during drawdown, especially if the fractures are shear fractures and oriented oblique to the in situ stress axes (e.g. Hillis, 1998).

Stress sensitivity and fracture closure are most likely to occur in reservoirs with low matrix permeabilities and narrow fractures, i.e. where fractures dominate permeability and where even small increases in compressive stress close the fracture apertures by an appreciable percentage. Extension fractures oriented normal to the maximum compressive stress, i.e. where the stress system has changed since the fractures formed, are more likely to close than those striking parallel to the maximum stress, but even the later fractures can close significantly (Hillis, 1998). Most fracture systems consist of numerous small fractures and fewer larger fractures, and the large fractures in the system should not be as susceptible to closure. However, flow through these few conduits must be supported by contributions from the more common and more stress-sensitive smaller fractures. Stress-sensitive fracture permeability is probably more common than not, and the data required to recognize it should be collected from a reservoir early in the life of the field.

3.5.3 Damage Due to Production-Related Scale

Reservoir fracture apertures have the potential to be actively mineralized during the short time-span of the production history of a reservoir. The same changes in pressure and temperature that cause geochemical instability and the precipitation of "scale" in production tubing can also occur within subsurface fracture apertures since they are part of the production plumbing system.

Examples of tubing scale are common (Figure 3.27), but production-related mineralization in natural fractures has not been widely reported, probably because fewer cores are taken after a reservoir comes online and because scale mineralization may not be easily differentiated from natural mineralization. There are several laboratory reports of fracture mineralization that occurs on a relatively short time scale: for example, Gale and Reardon (1984) described a laboratory test where calcite was precipitated within a

Figure 3.27 Calcium carbonate scale buildup reduces the diameter of oilfield tubing. Wikipedia. Similar mineralization can occur both in the matrix porosity and within the apertures of natural fractures during production since the reservoir fluids are less capable of carrying mineral in solution as fluid pressures and temperatures are reduced during production. Photo by Александр Юрьевич Лебедев - Ранее нигде не публиковалась, Public Domain, https://en.wikipedia.org/wiki/Oilfield_scale_inhibition#Calcium_carbonate_scale accessed 18 April, 2019.

grouted natural fracture in a granite core sample over the course of several months. The precipitated calcite, which measurably decreased the permeability of the fracture, formed due to the interaction between the fracture-filling grout and the calcium bicarbonate in the fluid that flowed through the fracture. Meredith (2013, 2015) reported that fractures created in granite samples in the laboratory become mineralized with locally derived quartz when "cooked" at high temperatures and pressures in a chemically closed system over time intervals on the order of days to a few months. The common oilfield practice of acidizing during recompletions can remediate scale buildup not only in the production tubing but also within the natural-fracture system.

3.6 Fracture Volumetrics

3.6.1 Introduction

"Volumetrics" is a loosely defined term referring to measurements of those fracture dimensions and characteristics that can be used to estimate fracture volumes, permeabilities, surface areas, and other effects on a reservoir. Volumetrics starts with measurements of the numbers and sizes of fractures, but parameters such as fracture interconnectivity and flow capacities within and across fractures should also be considered in the calculations. Since assumptions are involved, and since most of the subsurface fracture data that are used in volumetric calculations are derived from limited samples and incomplete datasets, "estimations" is perhaps a better term than "calculations." It is important to be as quantitative as possible, but the numbers derived from volumetric estimations are subject to the same caveats for quantifying fracture systems discussed in Part 2.

Dershowitz and Herda (1992) and Dershowitz and LaPointe (1994) laid out a basic suite of intensity and density parameters that are used in some models of fractured reservoirs. These parameters use a "P" nomenclature (for Persistence) with subscripts for the designated dimension for the fractures (number, length, area, or volume) and the sample type (line, area, volume). As used by Dershowitz and Herda, fracture "density" is measured in terms of simple *number* of fractures per length, area, or volume of the host rock, whereas fracture "intensity" is a measure of fracture *area* relative to the area or volume of the host rock. Some authors reverse the definitions, using density for area/volume and intensity for count. Fracture "porosity" is the ratio of fracture volume to the total volume of the host rock. (Note that this is different from our designation of "remnant fracture porosity" as the percentage of void space in a partially mineralized fracture.) Since the "P" nomenclature is used by the engineering and modeling community, it worth describing some of the specifics.

$\mathbf{P_{10}}$ is a linear fracture density defined by the number of fractures per unit length. Presumably this originated as a measure of the number of fractures intersected by a tape measure or scan line laid out across a vertical outcrop or on a bedding-plane pavement, and equivalent measures can be derived from horizontal core. By definition (number per length), a similar number can be obtained in the more common case where core and fractures are both vertical, but where the number has a different significance since the axis of the linear sample (the wellbore) is parallel to the fracture planes. In the latter case, the sample provides a poorly constrained measure of the probability of intersection rather than a representation of a fracture population having measurable distributions and spacings.

The measurements obtained from core also vary significantly depending on whether whole core, core butts, or just the core slabs are used, since slabs comprise less than a third of the core volume and are often cut to intentionally miss cored fractures. The P_{10} value for vertical cores is usually best used in a relative sense, comparing fracture intensities between wellbores where the fracture/wellbore geometries are similar and where the same formation was cored.

$\mathbf{P_{20}}$ is similar to the P_{10} measure of fracture density in that the fractures are still being measured only as a frequency, but here the frequency is related to a unit area sample rather than a line, offering a better definition of the fracture/host geometry. Both fractures and their host rocks are three-dimensional, and each higher level of P classification offers a more complete characterization of both the fractures and the host sample. Measurements involving a sample area are useful for outcrop pavements, less so for image logs and cores.

$\mathbf{P_{30}}$ is a volumetric fracture density, designating the number of fractures per unit volume, and providing a still better representation of the host rock, i.e. volume rather than just its area or length, but it does not improve on the number-only characterization of the fractures.

$\mathbf{P_{21}}$ is an intensity measurement that offers an improved characterization of the fractures, designating fractures as their measured unit lengths, relative to samples measured in terms of unit area. Measurements of fracture trace lengths on a bedding-plane pavement, or heights in a vertical outcrop, begin to be fit into this type of intensity measurement. The measure has different significance depending on the orientation of the fracture plane relative to the sample area, i.e. fracture heights can often be obtained from cores and logs, but lengths cannot. Nevertheless, the truncated fracture heights and lengths measurable in the limited volume of a core should be useful since the sample volume is similarly constrained.

P_{32} is an intensity measurement that treats fractures as surface areas and compares them to samples of rock volume. This is a useful measure where fracture heights and lengths but not fracture widths can be determined.

P_{33} designates fracture porosity, i.e. the volume of fracture apertures per total volume of the host rock. Most models assume that fractures are open, but the percent mineralization in the fracture widths must be considered if fractures are mineralized. Measurements or designations of the P_{33} fracture intensity parameter require knowledge of fracture height, aperture, and length as well as constraints on the dimensions of the rock volume within which the fracture population occurs. Core offers the opportunity to make this type of detailed measurement, as well as to measure a small but commensurate volume of the host rock. Since it is a 3D measurement and the samplings of both fractures and the host rock are three dimensional, the relative orientations of the fracture plane and core axis are less important for the P_{33} measurements than for other P measurements. Image logs may also provide useful P_{33} data once the fracture signatures for width have been calibrated with core.

3.6.2 Fracture Volume/Fracture Porosity

Fracture porosity values are relatively easy to calculate given the heights, lengths, and open widths of a fracture population, but fracture porosities must be used carefully. Capillary forces can inhibit the recovery from narrow fractures so a smaller percentage of the fluids can be recovered from narrower fractures than from wide ones during primary production. Even though all fractures contribute to porosity according to their size, they do not all contribute to fluid recovery in the same proportional manner. For the same reason, fluid recovery is more efficient from fractures than from the matrix, although that factor is usually rendered moot by the small fracture volume in most formations.

P_{33} is ultimately the desired intensity designator for fracture porosity, and it is typically small, less and often much less than 1%. Calculating fracture porosity is simple for hypothetical cases, for example, where a cubic meter of rock contains an open, one-centimeter-wide by one-meter tall and one-meter long fracture (Figure 3.28). The open fracture volume (H x W x L) of 10,000 cm^3 seems large but in fact it comprises only one percent of the one-million-cubic-centimeter total volume of the cube. Similarly, if the same 1 m^3 cube of rock is more intensely fractured but with narrower fractures, for example containing 10 open, 1-mm wide fractures, the fracture volume still comprises only 1% of the total

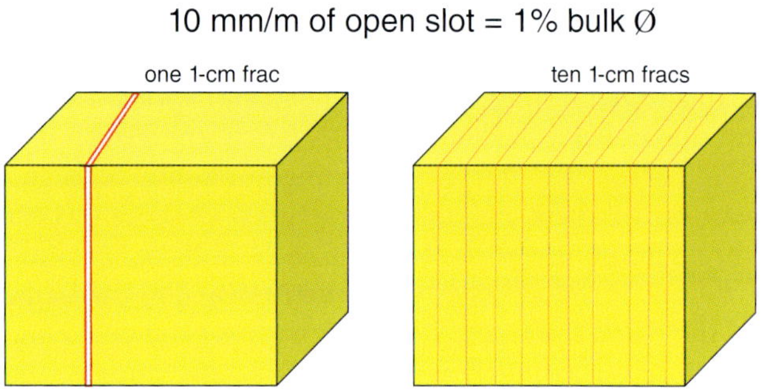

Figure 3.28 A fractured 1 m^3 block of rock that contains either a one-centimeter wide open, unmineralized fracture (*left*) or 10 one-millimeter-wide open, unmineralized fractures (*right*) has a P_{33} fracture porosity of 1%.

volume of the rock. (A smaller percentage of the fluid in the fractures may, however, be recovered from the system consisting of narrow fractures.) Actual fracture spacing is not critical to this calculation where the average spacing is adequate, but actual spacings become important when fracture permeability is considered.

Core studies suggest that subsurface fracture spacings are commonly wider, and fractures apertures are commonly narrower, than the dimensions used for this hypothetical case, so if we use half-millimeter average widths that are 50% occluded by mineralization, and an average spacing of 25 cm, the bulk fracture porosity in a reservoir is one tenth of one percent (0.10%). An actual example is the horizontal core cut from the naturally fractured Cretaceous marine sandstone in Colorado (Lorenz and Hill, 1994). The core sampled a population of 37 partially-mineralized, parallel-striking, vertical extension fractures (Figure 3.29). The average fracture spacing is very close to 1 m and the average remnant fracture porosity is 50% of the average 1-millimeter-wide extension fractures. From these numbers, the contribution of the fracture void space to the bulk porosity in this section of the reservoir can be calculated as ([0.5 x 1.0] mm/m divided by 1,000 mm/m x 100), or 0.05%. The average matrix porosity of this sandstone, as measured from unfractured core plugs in the laboratory, is 7%, so the bulk reservoir porosity with the added fracture

Figure 3.29 Subsurface data from 115 ft (35 m) of horizontal core that was cut through a fractured marine sandstone at a depth of 8,000 ft (2,400 m): 37 partially mineralized extension fractures strike nearly normal to the core axis. The average matrix porosity is 7%, the average fracture width is one millimeter, and calcite occludes an average of 50% of the fracture widths. The average fracture spacing is 3 ft (1 m). P_{33}, the fracture porosity, is 0.05% of the total rock volume.

volume is 7.05%. The contribution of the fracture volume to the bulk reservoir porosity is negligible.

In another example (unpublished data) a wellbore was unintentionally deviated by 10°–12° from vertical and cored across an interval containing horizontal beds and vertical, partially mineralized fractures. The core was cut through a poorly bedded marine shale, and the deviation azimuth of the wellbore was nearly normal to the strike of the cored fractures. The tall fractures extended along the core for as much as eight feet (2.4 m), most of the fractures exiting the sides of the core before terminating. Conceptually, since no terminations were observed, the fractures could be extended vertically beyond the core volume so that an average fracture spacing of 1.4 feet (0.43 m, or 2.3 fractures per meter) could be calculated from the basic geometry provided by vertical fracture spacings and the intersection angle between the core axis and the fracture planes. Average fracture width was 1 mm, so there were 2.3 millimeters of fracture width per meter, but the fractures were on average 50% occluded with calcite, leaving only 1.15 millimeters of open fracture porosity per meter, or 0.115% fracture porosity.

As pointed out by Nelson (2001) fractures typically contribute minimally to reservoir porosity. This is why, before the advent of image logs, attempts to identify fractured intervals in a reservoir from porosity logs by identifying the local increase in porosity due to the presence of natural fractures had marginal success. Nevertheless, Aguilera (2003) suggests that fractures can contribute significantly to the bulk porosity of some reservoirs, particularly vuggy fractures with wide, irregular aperture. Likewise, Ameen et al. (2012) report that fractures and microfractures create "a few percent" bulk porosity in some unconventional reservoirs in Saudi Arabia.

For reference, wide, open, closely spaced fractures that that create 2% of the reservoir bulk volume in a rock where the matrix porosity is only 4% would increase the total reservoir porosity by 50%, but the total porosity in the reservoir would still only be 6%. To continue the thought experiment, a 2%-volume fracture system might consist of one open, 2-cm wide fracture, one 4-cm wide fracture that is 50% mineralized, or perhaps 20 open 1-mm wide fractures (with an average 5 cm spacing), per cubic meter of rock.

Fracture porosity can also be estimated from image logs if the fracture-signature widths are calibrated with core. Sampling bias should be considered however, since closely spaced fractures striking parallel to the azimuth of a horizontal wellbore will be less well represented, if at all, than widely spaced fractures that strike normal to the wellbore azimuth. If a core or image log intersects two fracture sets, the data for both sets must be geometrically corrected to true spacings.

3.6.3 Fracture Permeability

All fractures large and small contribute equally to reservoir porosity, but they do not contribute equally to permeability since capillary and viscous forces have greater effects and can restrict flow within the narrower fractures of a population. Nevertheless, fracture permeability is an important parameter and should be addressed to the extent possible in a volumetrics assessment. Permeability can be measured directly in laboratory tests performed on whole cores, or more commonly on one-inch plugs cut from cores, but only the smaller fractures in the system, those fractures that are small enough or tightly enough cemented so that they do not fall apart during the coring and plugging process, can be tested.

Plugs from cores intended for permeability testing should be cut both parallel and normal to the fracture planes to test permeability along and across the fractures, and they should be tested at restored-state water saturations and confining pressures. The sensitivity of fracture permeability to changes in confining pressures can be tested at the same time. Control plugs cut from the adjacent matrix are taken for comparison to the fractured-plug permeabilities. Laboratory tests of matrix-only permeabilities provide the baseline for assessing the full system permeabilities measured by pre-stimulation well tests. Where the former are less than the latter, the difference must be due to the enhanced permeability provided by the reservoir fracture system.

If intentional fracture-specific laboratory testing is not an option, the routine porosity and permeability sampling programs that often plug fractures inadvertently may offer insights. The reports of such plugging programs typically consider the anomalously high permeabilities measured from fractured plugs to be invalid, and plugs that tested induced fractures are irrelevant while some of the naturally fractured plugs may have parted along the fracture planes during testing. However, other fractured plugs can give ballpark values for the permeabilities for natural fractures in a reservoir (Figure 3.30). The plugs themselves are not always available for examination, but the holes in the core left by plugging can and should be examined for fractures at those depths where anomalously high permeabilities were measured.

Since they were not intentionally sampled, the fractures captured by routine plugging programs may have been cut oblique to the axis of the plugs and so may not entirely traverse the long plug axis. Also, the samples are usually tested at two or three standard confining pressures rather than true restored-state conditions. Nevertheless, but such plugs offer at least approximations of the potential for natural-fracture permeability enhancement in a reservoir.

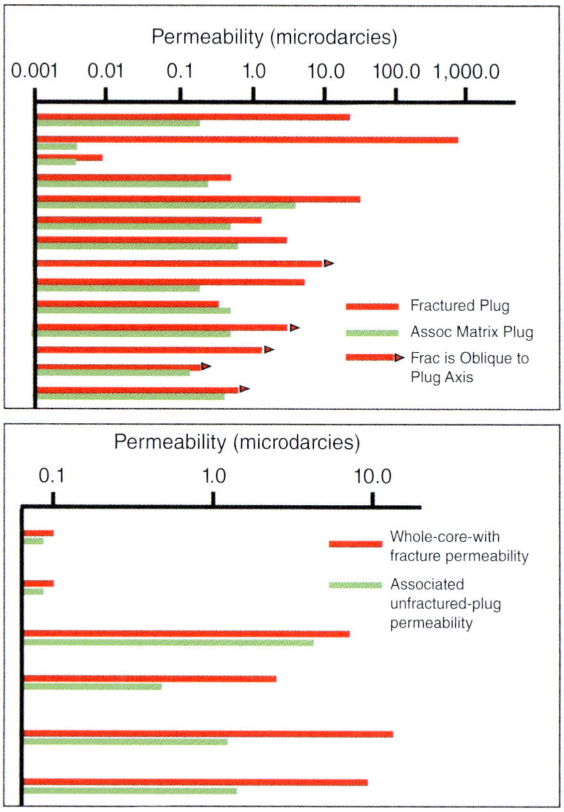

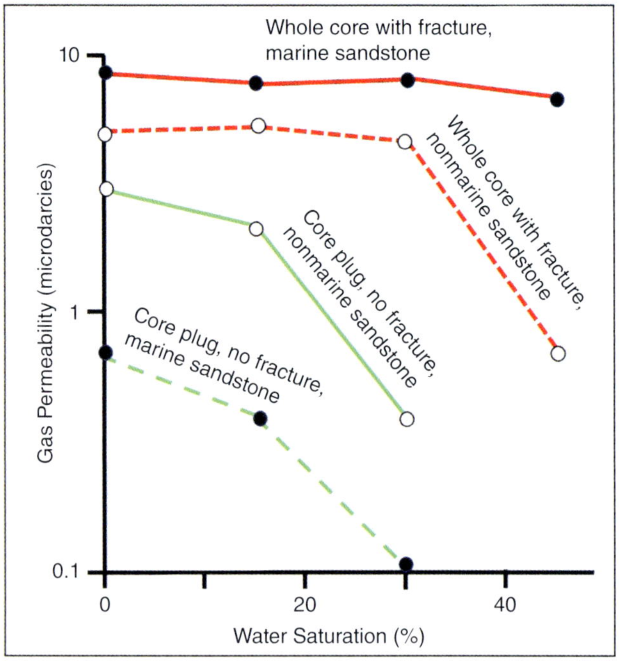

Figure 3.30 *Upper left*: routine plugging of a core often captures natural fractures. These calcite-mineralized fracture permeabilities measured from one-inch plugs cut from low-permeability sandstone reservoirs in Colorado show that fractured plugs consistently have greater permeabilities than adjacent matrix-only plugs, despite complete fracture mineralization. Some of the fractures exited the side of the plug before transiting the entire plug length since the plugs were not specifically cut to follow the fracture planes, and for these plugs the fracture permeability is probably higher than measured (dry Klinkenberg corrected permeabilities). *Lower left*: fracture permeabilities measured in whole-core samples of the same formation, compared to the permeabilities of nearby matrix plugs. *Right*: increasing water saturation has a greater effect on matrix permeability than on fracture permeability (samples intentionally taken to test fracture permeability), in both a clean marine sandstone and in an immature nonmarine sandstone in low-permeability reservoirs. (Adapted from Sattler, 1989.)

There are plausible arguments that fracture apertures measured from core are invalid because the apertures have expanded after being released from the in situ stresses, (e.g. Wennberg et al., 2016; Nelson, 2020). However, many fractures recovered from the subsurface are still bridged by undeformed mineralization indicating that minimal expansion has occurred, and we have never had trouble fitting circular protractors over a fractured core even though the inside diameter of the protractors are only $^{1}/_{16}$ inch (1.6 mm) larger than the inside diameter of the core bit. In fact, measurements of the expansion of core show that the recovery strain is on the order of 0.01%. (e.g. Teufel, 1983, Warpinski et al., 1993b; Lin et al., 2006). Moreover, the thermal contraction of a core, due to cooling after being brought to the surface, can be greater than its expansion due to release from the confining stresses. The two offsetting processes occur at the same time but the thermal effect is isotropic whereas the strain recovery is not. Measurements taken to determine the direction and magnitude of the maximum anelastic expansion of a newly cut core (made to determine stress orientation and magnitude) often end up measuring the direction of minimum contraction instead (L. Teufel, personal communication, 1994). With few limitations, fracture-aperture data measured in core are the best hands-on physical data for the calibration of image log fracture-aperture data, for modeling, and for calculating fracture porosity and effectiveness.

Fracture-controlled permeability anisotropy can also be assessed with well interference tests, tracer tests, and from anecdotal information from the oilfield personnel familiar with the pumping and gathering systems they are charged with maintaining. The in situ permeability anisotropy of a fracture system can likewise be estimated qualitatively by measuring fracture strike(s), the absolute and relative development of different fracture sets (spacing, intensity/density), the relative orientations of fracture set(s) and the in situ stresses, and the degree of mineralization of the fractures. Without data, however, there is little foundation for making such assessments.

3.6.4 Transfer Function

The ability of fluid to flow from the matrix across a fracture face and into a fracture aperture, the "transfer function" of engineering use, is a parameter used to tweak fractured-reservoir models so that their results more closely conform to actual production patterns. This parameter is usually assigned empirically in a model and modified as necessary to history-match production, but preliminary estimates can be made and constrained by examining cored fracture faces for features such as slickensides or clay residues, and refined with thin sections oriented normal to the fracture planes and cut carefully to avoid losing the fracture face during preparation. The transfer function can also be assessed with plugs cut from core normal to the fracture planes and tested for flow across the fracture surfaces.

3.6.5 Fracture Surface Areas

In addition to the permeability across fracture faces, the sizes of those surfaces are important. The larger the fracture the more surface area there is across which fluid can flow from the matrix and into fracture apertures. If the average fracture spacing is on the order of 1.0 m (3.2 ft) as in the hypothetical cubic meter of reservoir rock, and if each fracture extends the full dimensions of the block, then each fracture has an area of 1 m^2, but since each fracture consists of two opposing faces, this cubic meter of rock has a total of 2.0 m^2 of surface area available for the transfer of fluid from the matrix into the fracture aperture (Figure 3.31). This calculation is independent of fracture width, but width is related to spacing (i.e. wider fractures accommodate more strain so fewer are required for any given amount of strain). Using the cubic meter of rock, one wide fracture provides 2 m^2 of surface areas whereas 10 narrower fractures provide 20 m^2 of surface area. Many narrow fractures can be more efficient in extracting fluids from the matrix than fewer wide ones, and imbibition EOR techniques are more effective when the fracture surface area in a reservoir is large.

Another way to assess the significance of surface area provided by a fracture system is to consider a hypothetical vertical fracture 1 m tall by 3 m long intersected by a standard 8-inch diameter vertical wellbore. Fracture lengths are typically controlled by percent strain whereas heights, at least for extension fractures, are controlled by bedding, and since the two controls are unrelated the ratio between height and length is not constant, but for this example, we will assume, based on outcrop observations, that meter-tall fractures are on the order of three meters in length. The surface area provided by the cylindrical wellbore and available for fluid flow into the well across the vertical interval spanned by the fracture, assuming an open-hole completion, is some 6,340 cm^2 (0.634 m^2; height x circumference). In contrast, the two faces of the natural fracture provide 60,000 cm^2 (6.0 m^2) of surface area, an order of magnitude more surface than the wellbore wall by itself (Figure 3.32).

A horizontal wellbore cutting normal to the fracture planes should intersect more fractures than the vertical wellbore and therefore the ratio of fracture surface to wellbore surface area should be commensurately greater. If the well is cased and production is through

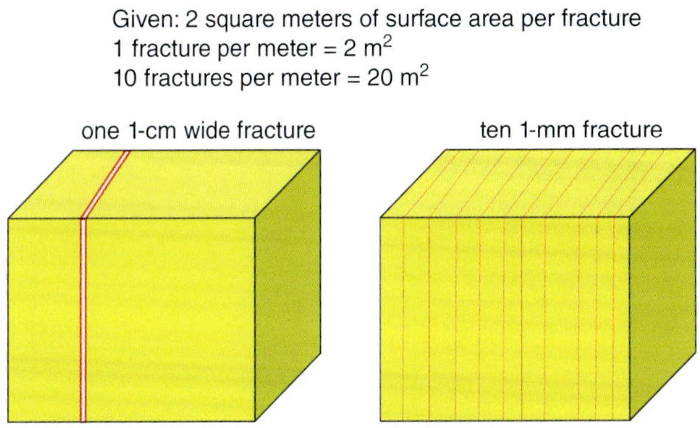

Figure 3.31 Calculating fracture surface area per volume of rock. Fracture faces are the connection from the matrix and the fracture apertures. They provide surfaces for drainage, and for the imbibition of water or other EOR fluids into the matrix, driving oil out of the matrix pores. The size and number of the fracture faces as well as the permeability across the faces (along with pressure differentials and fluid viscosities) govern the rates of matrix drainage and imbibition.

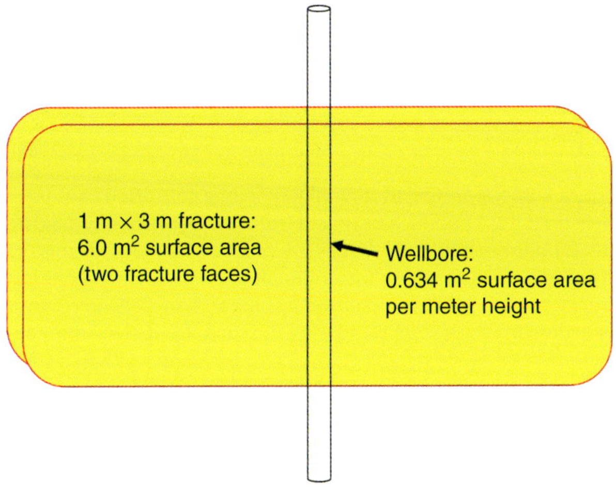

Figure 3.32 The two fracture faces of a 1 m x 3 m fracture provide significantly more surface area for drainage from the matrix than the wellbore surface alone.

perforations as is more commonly the case, the wellbore itself provides essentially no surface area for such transfer.

The additional surface area provided by fractures is important not only for drainage from the matrix into the fracture slot and ultimately into the wellbore, but also for imbibition and related EOR processes. Wettability and capillary pressures, which control the ability of water to imbibe into matrix rock across any face and thereby expel oil, are more important in fractured reservoirs than in conventional reservoirs (D. Schechter, personal communication, 2019). The more fractures there are and the larger the fractures are in a reservoir, the greater the surface area where imbibition can act to exchange water for oil in the matrix, and the more significant imbibition becomes to oil recovery.

The importance of the large surface areas provided by fracture faces, and the significance of the reservoir wetting state, are illustrated by the intensely fractured reservoirs at Midale, where imbibition and fluid exchange across fracture faces are important to both water-flooding and CO_2-flooding EOR projects (Beliveau et al., 1993). The matrix system at Midale is predominantly a water-wet system, where water coats the surfaces of pores and is in direct contact with the rock. Oil in the pores in such systems is surrounded by water. In such water-wet systems, the capillary and viscous forces work together, expelling oil from the matrix and into the fracture system, as well as retarding water breakthrough since much of the injected water initially imbibes into the matrix. Water-floods are less efficient in oil-wet systems, where injected water imbibes into the matrix more slowly so that in turn the oil is expelled more gradually. Since the injected water stays in the fractures longer, breakthroughs between on-trend injectors and producers are quicker and more abrupt in oil-wet systems.

Wetting state is one of degree rather than being an either-or (water or oil) condition. The limestone unit at Midale is both more water-wet and more highly fractured than the overlying dolomite, thus the limestone is efficiently swept by water flooding and has better displacement efficiency, and it therefore has a relatively low residual oil saturation. In contrast, the overlying dolomite unit has a relatively high residual oil saturation, and a CO_2 sweep is the more efficient process in this unit.

Midale has undergone three basic stages of production: 1) primary (depletion); 2) secondary (water injection); and 3) tertiary (EOR with gas injection). The considerations in modeling these processes include the transfer mechanisms of aqueous phase imbibition or gas injection, and involve multi-contact miscibility, wettability, and capillary pressure, but those are subjects for a book by experts in the petroleum engineering field.

3.7 Effects of Fractures on Drilling and Coring

This section is a non sequitur of sorts in that it describes the effects of fractures on phenomena unrelated to fluid flow. However, drilling and coring are important to the industry, so we have included it. If the fractures are tightly cemented and have little remnant fracture porosity, their effects on drilling and coring are minimal, but larger and more open fractures can cause core jamming and wellbore rugosity. One of the more obvious effects, especially if mud weight overbalances formation pressure, can be lost circulation, i.e. the loss of drilling mud into fractures as they are encountered, as evidenced by pump pressure variations and falling mud levels in the mud pit. Lost circulation during coring is often apparent in archived core where various types of lost circulation material can be found jammed into fracture apertures. Fluid flow in fractured reservoirs can also go the other way, from the formation into the mud pits, if mud weights are less than formation pressures. The encounter between a bit and a highly fractured interval can be dangerous, especially in low-permeability reservoirs where capillary pressures acting on the small-aperture fractures and narrow matrix pore throats allow a well to inadvertently be drilled underbalanced. Capillary forces are not as effective on wider apertures of fractures, so if the bit penetrates a zone of wider, more open fractures it is possible that a previously adequate mud weight will become insufficient to control a well even though the mud weight and reservoir pressure have not changed. Though dangerous during drilling, such intervals commonly produce as sweet spots in a reservoir.

One of the old-timer tricks for detecting fracture zones was to watch the turning Kelly Bushing that drove the rotating drill string on the rig floor. Fractures that caused wellbore spalling instead of the gradual chipping while drilling hypothetically created a temporary jam at the bottom of the hole, causing the bit to stop turning and winding up torque energy in the elastic drill string until the jam was cleared by the extra torque. This manifested itself at the surface by an erratic rotation speed of the Kelly Bushing. The presence and severity of the effect depends on any number of things including the size of the rock spalls at the bottom of the hole, the length of the drill string, and the size and strength of the fractures.

Natural fractures are one of the structures, along with inclined bedding planes, vuggy zones, etc., that can disrupt the integrity of a core and cause it to jam inside a core barrel. A jam occurs when the core tries to expand laterally within the core barrel due to wedging action of a weakness plane in the core inclined to the core axis. In a vertical core, the weight of the core above an inclined break pushes the top of the core down and past the lower core section, effectively expanding the core diameter so that it becomes stuck inside the barrel.

When a core jams it becomes impossible for additional core to be cut since the new core cannot push the jam upward in the barrel. The weight of the drill string rests on the core stump at the bottom of the hole instead of on the teeth of the hollow core bit. Since the core bit no longer seats on the bottom of the hole, the circulating mud pressure decreases. Lost pressure and decreased rate of penetration during coring are classic signs of a jam in the core barrel. If they are not recognized, the core bit can continue to make hole, usually at a reduced rate, while grinding up the formation rather than capturing additional core.

A fracture-related jam in the core barrel, if it hasn't been turned into rubble during core retrieval and slabbing, is usually released when the core is removed from the core barrel, so the wedge itself cannot always be recognized when the core is laid out in boxes in the laboratory. Torque-related induced fractures with characteristic helical surfaces are often created in a core below a jam (see Lorenz and Cooper, 2018a, Chapter 2F) recording the jam if there is no other evidence.

Poorly cemented, well-developed fractures in a formation can degrade the integrity of a formation, leading to enlargement and rugosity of the wellbore. This in turn causes poor contact between an image-log tool and the wellbore wall, and it can increase the volume of cement required for casing a well. The shadowy, irregular shapes of fracture-related wellbore enlargements, sometimes referred to as "washouts," seen in image logs can usually be differentiated from the more symmetric wellbore breakouts created by stress anisotropy in the formation, but the characteristics of the fractures that cause such enlargements are difficult to determine from the image-log shadows.

3.8 Completions: The Interaction Between Natural and Hydraulic Fractures

3.8.1 Early Conceptual Models

Descriptions of the interactions between natural fractures and hydraulic stimulation fractures need to be prefaced by a brief review of the early concepts of hydraulic fracturing to appreciate the strengths and weaknesses of our present-day understandings. No one had seen the in situ characteristics of a stimulation fracture until the late 1970s, so early concepts of hydraulic-fracture propagation directions and features were based on an amalgam of presumably analogous geologic structures, simple laboratory tests, and geomechanical theory. Moreover, the paucity of natural fractures in the existing, incomplete subsurface datasets was implicitly and incorrectly equated to a general absence of natural fractures in reservoirs. If natural fractures were considered at all, the laboratory tests and analytical models seemed to show that they had no influence on propagating hydraulic fractures (e.g. Lamont and Jessen, 1963; Daneshy 1974).

In fact, during the early history of hydraulic fracturing, subsurface stresses were commonly considered to be isotropic by the industry, even though Anderson (1905) had shown that the faulting that is common in geology requires an anisotropic stress system. With assumed isotropic stresses, most hydraulic fractures were thought to form along horizontal planes, lifting the overburden towards the unconstrained free surface of the earth. Hubbert and Willis (1957) were among the first to discuss why in situ stresses at depth should be *an*isotropic, suggesting that the minimum in situ compressive stress lies in the horizontal plane in the absence of structural/tectonic complications, and that hydraulic stimulation fractures should therefore be vertical. Moreover, Hubbert and Willis suggested that vertical igneous dikes are the natural analogs to hydraulic fractures. The conceptual model of a hydraulic fracture therefore evolved into that of a vertical, single-plane fracture where proppant was comparable to the magma that fills dikes. Dikes, driven by the enormous energy of geologic processes, generally cut indiscriminately across layered formations, therefore it was also considered that the stratified geologic media into which industrial stimulations are injected behave in a mechanically homogeneous fashion despite layering.

The principles that stress systems at depth are anisotropic and that a hydraulic fracture should be vertical were accepted slowly (see Reynolds and Coffer's 1957 discussion of the Hubbert and Willis paper). By consensus, however, the new conceptual model slowly became that of a vertical, single-plane hydraulic fracture, extending outward in opposite directions as "wings" from a wellbore. The fractures were generally considered to be unbounded vertically and therefore circular or "penny-shaped" when viewed from the side, normal to the plane of the fracture.

With these assumptions about stimulation-fracture geometry, both fracture width and the distance that one should propagate from the wellbore, the "wing length," could be calculated based on the mechanical properties of the rock and the volumes and pressures of the injected fluid (e.g. Perkins and Kern, 1961) and an ideal penny-shaped, single-plane, bi-winged, vertical hydraulic fracture replaced the horizontal pancake fracture as the accepted conceptual model. The potential for interactions between hydraulic stimulation fractures and natural fractures was not even considered since there was, as yet, little evidence for the natural fracture systems that are pervasive in the subsurface.

3.8.2 Direct Evidence of the Characteristics of Hydraulic Fractures

This concept of injected hydraulic fractures was challenged, and the importance of geologic heterogeneities such as bedding and natural fractures began to be recognized, during a project designed to physically examine injection fractures. The "Mineback" experiment was conducted at the Nevada Test Site in the western U.S., where relatively small experimental stimulation fractures consisting of a few hundred barrels of color-dyed grouts were injected into a formation at a depth of 1,400 ft (427 m) (see Northrop et al., 1978; Warpinski et al., 1981). The extents and characteristics of the injected grouts were then mapped by incrementally mining into and through the hydraulically fractured zone of the formation, providing the first direct, hands-on view of an in situ hydraulic fracture.

The fractured rock at the Nevada Test Site consists of layers of variably welded igneous tuffs, but the layering and the anisotropic in situ stress system even at the relatively shallow depth of the tests provided enough similarities to reservoir conditions that the unexpected observations of fracture complexity caught the industry's attention. Industry was surprised by the fact that the vertical hydraulic fractures exposed in the walls, ceiling, and floor of the mine were considerably shorter laterally than anticipated. It was also apparent that the injected fractures were vertically bounded by the stress differences associated with lithologic changes and the related mechanical-property differences of the layered rock (Figure 3.33). Equally important were the observations that the injected fractures were multi-stranded and branching, and that they had been diverted or even terminated by geologic heterogeneities such as natural fractures, bedding, and faults.

Concurrently, detailed measurements of in situ stresses began to suggest heterogeneous stress systems within layered formations. Since hydraulic fractures are controlled by stress, the limitations on height growth provided by layered stress systems began to be incorporated into models, modifying the concept of penny-shaped fractures. Although they are not universal, height limitations have been recorded by numerous microseismic surveys which listen to the small seismic events created in the rock in and near a hydraulic fracture as the rock mass adjusts to the added stresses imposed by an injection. The heights of hydraulic fractures are limited where high stress contrasts are created by reservoir-scale interlayering of ductile and brittle strata. It is the stress difference created by varying ductilities, not the change in ductility, that limits height growth.

Given the Mineback observations of hydraulic and natural fracture interactions, and related laboratory experiments, Blanton (1982, 1986), and Warpinski and Teufel (1987a) proposed that natural fractures create important mechanical discontinuities capable of diverting a propagating hydraulic fracture and creating branching, multi-stranded, and tortuous hydraulic fractures. Conceptual models for hydraulic fractures began to include interactions with natural fractures and bedding planes, leading to branching, tortuosity, reduced wing lengths, and diminished capacities for carrying proppant, and sophisticated models of complex hydraulic fractures have been constructed (e.g. Potluri et al., 2005; Dahi-Taleghani and Olson, 2011).

Figure 3.33 Observed complexity of blue-dyed grout injections pumped into welded tuff, Nevada. *Left*: a view on a vertical mine face showing that the injections are roughly vertical, offsetting at bedding planes (yellow arrow). This injection follows a natural fracture below the bedding plane but parallels the fracture (red arrow) in unbroken rock to the left of the fracture above the bedding plane. The fabric of vertical striations on the wall was produced by teeth on the drum of a Continuous Miner. *Right*: a view of the roof of the mine showing irregular planes of the blue-dyed injection emanating from the vertical wellbore (white circle) and interacting with open natural fractures in the welded tuff. (From Warpinski et al., 1981.)

These models allowed engineers to consider that the weakness planes provided by natural fractures could and should capture and divert stimulation fractures, especially where natural fractures are aligned with or less than 30° oblique to the trend of the maximum in situ horizontal compressive stress. In contrast, if the fracture planes are normal to or within 30° of normal to that compressive stress, the stimulation would be more likely cut across the natural-fracture fabric, although the natural fracture planes might still cause branching and act as thief zones where the advancing stimulation fluids could leak off into the formation, and where jogs in the developing hydraulic fracture planes would cause interruptions in the flow and allow proppant to settle out of the stream (Figure 3.34). These characteristics are similar to the complex injection fractures observed in coal mines (see Vincent, 2009).

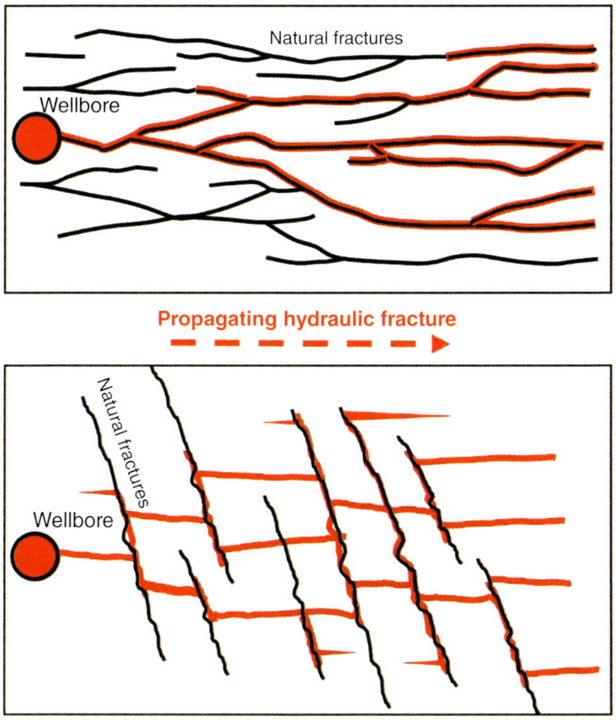

Figure 3.34 Map-view conceptual models of the interaction between natural and hydraulic fractures at depth, made prior to observations of actual interactions in core. *Top*: visualization of the interaction between natural and hydraulic fractures where the in situ stress anisotropy directs a hydraulic fracture *parallel* to a natural-fracture fabric. Later core across this system showed that the hydraulic fractures did *not* in fact exclusively exploit the mineralized natural-fracture planes as anticipated, but rather that many of them propagated parallel to the natural fractures in the adjacent unbroken rock. (Unpublished figure from the MWX experiment, late 1980s.) *Bottom*: "authors' visualization" of the interaction between a hydraulic fracture and a natural-fracture fabric, where the in situ stress anisotropy controls hydraulic fracture propagation *across* a natural-fracture fabric. (From Warpinski and Teufel, 1987a.)

This concept of interaction between hydraulic and natural fractures seemed to explain the observation that injecting stimulation fluids into some fractured formations was just as likely to decrease production as it was to improve it, especially if the stimulation fluids consisted of high-viscosity gels. In the case of low-permeability Mesaverde sandstone reservoirs in Colorado, the damage was ascribed to a combination of plugging of the essential natural-fracture apertures by high-viscosity stimulation fluids, and the narrowing of those apertures by the lateral stress increases created by the injection of wide, propped fractures along planes parallel to the natural fractures (Branagan et al., 1987, Lorenz et al., 1989; Multiwell Experiment Project Groups Final Reports, various dates).

At the experimental MWX site in Colorado, where such pre- and post-stimulation production tests were conducted, one horizon produced natural gas at a rate of 200 MCFG/D (5,660 m^3/day) unstimulated, and only 160 MCFG/D (4,530 m^3/day) after being hydraulically fractured. However, that same zone produced at a rate of 400 MCFG/D (11,330 m^3/day) when it was re-entered after being shut in for 18 months. The damage had been remediated by a combination of viscous stress relaxation of the stresses imposed by injection on the formation, imbibition of the stimulation fluid (moving it from the natural-fracture apertures into the adjacent matrix rock), and thermal degradation of the injected gels. All three factors contributed to a reopening of the natural-fracture passageways, allowing the positive aspects of the stimulation to finally take effect.

Such damaging interactions between natural and stimulation fractures in hindsight also explained the puzzling results of the Massive Hydraulic Fracture experiments of the 1970s (Table 3.2), where injections that were large for the time were pumped into formations yet the production enhancements were not as large as anticipated (Randolph, 1974; van Poollen et al., 1977; Medlin and Fitch, 1988).

Nevertheless, there was still room for improvement in the model of hydraulic fracture interactions with natural fractures. This was provided by the first hands-on view of a hydraulic fracture in a naturally fractured, low-permeability, natural-gas reservoir *at depth*. Warpinski et al. (1993b) described this hydraulic fracture, as revealed by core cut from nonmarine sandstones of the Mesaverde Group in Colorado. Where the core crossed the plane of the stimulation fracture it contained about 30 sub-parallel, sub-vertical, hydraulic-fracture planes containing residues of the injected gels, but only rare crushed remnants of the injected proppant. The WNW-ESE-striking fracture strands span a lateral interval of about three feet (1 m)

Table 3.2 The injected volumes of hydraulic fracture fluid, and the weights of proppant in four intervals of the Mesaverde Group in Colorado during the "Massive Hydraulic Fracture" (MHF) experiment. Although small by present-day standards, at the time these were huge fracture jobs. The four isolated, stimulated intervals were relatively thin, and, based on the injected volumes, the hydraulic stimulation fractures had design lengths up to 1,700 ft (520 m). In contrast, the effective fracture lengths calculated from post-stimulation production profiles were only a few tens of feet. More importantly, the resulting production enhancement was much less than anticipated. These factors suggest that the stimulations damaged the essential natural-fracture permeability system which controls production in the reservoirs, which was not known to exist at the time of the experiment (e.g. van Poollen et al., 1997).

	Interval 1	Interval 2	Interval 3	Interval 4
Fluid volume (gallons)	117,500	285,000	344,000	228,000
Sand Proppant (pounds)	400,000	880,000	809,000	448,000
Frac Interval (ft)	40	80	112	20
Anticipated Frac Length (ft)	840	920	860	1648
Post-Test Frac Length (ft)	110-170	150	100	20–30
(from type-curve match)				
Pre-Frac Flow (MCFG/D)	60	57	43	55
Post-Frac Flow (MCFG/D)	61	137	160	69

and strike parallel to the measured WNW-ESE trend of the local maximum horizontal compressive stress. A narrower, secondary zone composed of eight WNW-ESE striking stimulation-fracture planes is also present in the core, offset from the main hydraulic-fracture zone in the direction normal to the fracture planes by about 60 ft (18 m) laterally.

The descriptions of complex, multi-stranded hydraulic fractures were not immediately or universally accepted (see Discussion, Nolte, 1993; and Reply, Warpinski et al., 1993c). Moreover, the fact that the stimulation fractures strike parallel to two narrow, calcite-filled natural fractures in the same interval of the core but did *not* open and exploit them was largely overlooked. Nevertheless, this suggested that all natural fractures do not create mechanical weakness planes in a rock, and/or that the in situ regional stresses and the local stresses driving propagation of a stimulation fracture can have more control over hydraulic-fracture propagation than some natural-fracture fabrics. In the Mesaverde core, some of the stimulation-fracture planes terminate vertically against bedding planes and shale partings in the cored sandstone; others jog laterally a few centimeters but then continue across these discontinuities.

Since the publication of this core description in 1993, other reports of similar cored hydraulic fractures have been published in locations ranging from California to Michigan to Texas, and in reservoirs ranging from chalks to sandstones to phyllite (e.g. Fast et al. 1994; GRI, 1999; Peterson et al., 2001; Hopkins et al., 1998; Jeffrey et al., 2010; Raterman et al., 2017; Roggenthen and Doe, 2018). The natural fractures in these cores have strikes ranging from normal to oblique to parallel to the in situ maximum horizontal compressive stress, but the multi-stranded nature of the cored stimulations are similar (Figure 3.35). Interactions between the stimulation fractures and natural fractures have not been prominent in the descriptions, in part because such intersections have a low probability of being sampled with cores that represent such a small volume of a reservoir, but the few cored intersections suggest that even where hydraulic fractures do exploit natural-fracture planes they only do so for a short distance before cutting back into the matrix rock to follow planes controlled by the in situ stresses.

Observations from cored hydraulic fracture contrast significantly with the stimulation-fracture/natural-fracture interactions observed at the Mineback test at the Nevada Test Site, where natural fractures played a major role in diverting and disrupting the stimulation fluids. This can be explained by the low stress magnitudes, the absence of pore pressure, the small injection volumes, and the low injection rates at the Mineback site. Because of these factors, lithologic heterogeneities played a larger role in controlling stimulation-fracture propagation at this shallow test site than they do in reservoirs at greater depths where the in situ compressive stresses and pore pressures are greater, where the driving injection pressures are higher, and where the natural fractures are typically more tightly cemented.

A series of cores has recently been cut across a complex system of hydraulic fractures in the Permian Wolfcamp shales of Texas at the Hydraulic Fracture Test Site in the Midland Basin. Preliminary published reports

Figure 3.35 Cored multi-stranded vertical hydraulic fractures captured by deviated core at the Mounds Drill Cuttings Injection Project in Oklahoma. *Left*: three closely spaced hydraulic-fracture strands completely separate the core. (From Peterson et al., 2001.) *Right*: in another section of the core three of the four closely spaced hydraulic-fracture strands do not completely break the rock even though they are stained and filled with the injected cuttings slurry. (From GRI, 1999.)

(e.g. Gale et al., 2018; Ciezobka et al., 2018) suggest that the hydraulic fractures at this research site have characteristics comparable to those found in earlier studies, although full descriptions and the datasets have not been released as of this writing.

A complementary hands-on look at the interaction between natural fractures and hydraulic fractures at a smaller scale is afforded by core samples where induced centerline fractures interact with natural fractures. Most centerline fractures are small-scale hydraulic fractures that form immediately below an advancing core bit due to the weight of the drilling mud combined with pressure pulses as drilling mud jets from nozzles in the rotating bit. As with other hydraulic fractures, these induced fractures strike normal to the minimum compressive in situ stress. Examples from core (e.g. Lorenz and Cooper, 2018a) show that hydraulic centerline fractures can form in unbroken rock immediately adjacent to natural fractures rather than exploiting a natural-fracture plane, whether the natural fractures strike parallel or oblique to the centerline fracture. In other examples centerline fractures enter and follow weakly mineralized natural fractures, exploiting that weakness plane for short distances before breaking back into the matrix rock.

3.8.3 The Developing Hydraulic-Fracture Model

The limitation that can be imposed on the vertical growth of hydraulic fractures by heterogeneous, layered strata is now widely accepted, as is the typically vertical orientation of fractures except at shallow depths or in complex geologic provinces where the maximum compressive stress is other than vertical. However, the developing hydraulic-fracture model has two other components that are still under discussion: the nature of interactions between hydraulic fractures and natural fractures, and the origin of the cored, multi-stranded fractures that contain little or no proppant but which are obviously related to injection.

In regard to the first, the answer to the question "Do stimulation fractures follow or cut across natural fractures?" is still: "It depends." This interaction is controlled by several independent and highly variable factors including the natural fracture system (strength of mineralization, degree of mineralization, fracture spacing, etc.), the in situ stress magnitudes and anisotropy, the angle between fracture strike and the maximum compressive stress, and the local stresses created by the injection pressures and rates.

Service companies quantify these parameters for use in predictive equations (e.g. Aimene et al., 2018; Cruz et al., 2018), but in general, in situ stresses should control hydraulic-fracture characteristics where stress magnitudes are high and where natural fractures are poorly developed, whereas a well-developed fabric of incompletely mineralized fractures should have significant influence, especially if the effective stress magnitudes and stress differentials are low. Higher-pressure injections of more viscous fluids create local injection-related stresses and have a greater probability of cutting across a natural-fracture fabric, especially if the fractures are widely spaced and tightly mineralized.

Recent modeling (e.g. Aimene et al., 2018) has provided better tools with which to model and understand the interactions between hydraulic and natural fractures. The problem is not yet solved for all fracture types and stress conditions (A. Ouenes, personal communication, 2018), but we have come a long way from the concept of penny-shaped fractures formed in homogeneous media where the in situ stresses are isotropic.

In formations filled with narrow, tightly mineralized fractures, the high-volume/high-pressure stimulation technology currently in use would seem to render moot much of the question of natural-fracture/hydraulic-fracture interactions. The large volumes of injected fluids and proppant create local stress fields which overwhelm the mechanical inhomogeneities associated with relatively subtle fracture fabrics (e.g. Raterman et al., 2017; Gale et al., 2018). Regardless, natural fractures are pervasive in most reservoirs and, if they are open and not damaged, they should add significantly to production from fracture-connected areas outside of the stimulated rock volumes.

The other developing component of the hydraulic-fracture model, that of an explanation for the numerous cored stimulation-related fractures that, unexpectedly, contain little or no proppant, is still under debate. The ubiquitous presence of remnant stimulation fluids in the fractures is usually assumed to indicate that they were forced open by the injection of fluids, but this does not explain the common paucity and absence of the proppant that was injected together with those fluids.

The concept of complex, multi-stranded hydraulic fractures with truncated wing lengths met with resistance when it was first described (see Nolte, 1987; Warpinski and Teufel, 1987b), and in fact the basic question "Are hydraulic fractures simple planes?" is still under lively discussion (e.g. Vincent, 2009; McClure, personal communication, 2019). The increasing number of descriptions of cored, multi-stranded hydraulic fractures would seem to suggest that the answer is "no," yet it is still possible that the cores are misleading, and there may yet be room for a single-plane conceptual model for hydraulic stimulations based on the Hubbert and Willis (1957) igneous-dike analog. Numerous zones of parallel to sub-parallel, irregular fractures and cracks are undeniably associated with a stimulation, but they may only be a secondary result of an injection; there may yet be evidence for a more limited, main propped fracture, which, like an igneous dike, may be relatively localized.

It is possible that cores are preferentially capturing the fractures of process zones that form in front of and adjacent to a few main injections, because the former are more numerous and more widely distributed than the latter. Fracture process zones are a response to the altered and enhanced stresses in rock around an injection, and they are common adjacent to igneous dikes (Figure 3.36) (e.g. Delaney et al., 1986; Baer and Reches, 1991).

If so, the numerous smaller fractures would not have formed as minor injection planes, even though they are filled with stimulation fluids. Rather, they may be the result of a two-step process consisting first of fracturing due to altered stresses adjacent to a main injection during pumping, and second, were later filled by stimulation fluids leaking off into the formation, when the flow of fluids was insufficient to carry proppant along with the fluids.

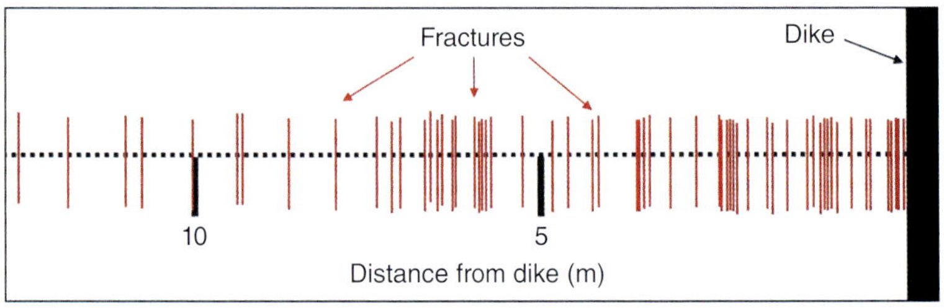

Figure 3.36 *Left*: dike-parallel vertical extension fractures in sandstone are irregularly spaced, but spacing decreases irregularly with proximity to a vertical igneous dike in the Raton Basin of Colorado-New Mexico, suggesting that the fractures are part of a process zone that formed in front of or parallel to the dike during intrusion. (Adapted from Lorenz et al., 2004.) *Right*: one of the many igneous dikes in the Raton Basin, cutting through a light-colored sandstone and an overlying shale in a road cut south of Trinidad, Colorado. The irregularly spaced, brown-stained vertical fractures parallel and adjacent to the dike have characteristics that are different from the more widespread calcite-mineralized fractures in the formation, and define the process zone associated with dike propagation.

Similar processes create hydrothermal mineralization in the closely spaced process-zone cracks adjacent to igneous dikes. If this is the case, the process-zone fractures would dominate cores cut through a stimulation since they are numerous and relatively widespread, while an apparent absence of wide, propped fractures would be a function of the minute volume of the reservoir sampled by cores combined with the limited distribution of these main, propped hydraulic fractures.

Widespread process-zone fractures can nevertheless be important contributors to a stimulation even though they are not propped since, as described in Section 3.5, the studies of Gutierrez et al. (2000) show that a planar break in a mudrock, even when it is compressed or sheared, retains fracture-parallel permeability that is orders of magnitude greater than that of the adjacent matrix rock. The main propped fracture may consist of only a few strands and be relatively planar, but the act of driving it through the rock creates numerous, adjacent, process-zone fractures that contribute significantly to the local permeability enhancement.

Regardless of the interpretation of the cored stimulation fractures, stimulations are essential in making resource-play reservoirs produce. However, the Stimulated Rock Volume created by a hydraulic fracture is still significantly less than the total volume of naturally fractured rock in reservoirs if the natural fracture system is well developed, and care should be taken to design stimulations that do not damage a natural-fracture permeability system.

3.8.4 Nuclear Stimulations

Before leaving the subject of hydraulic fractures it is worth taking a quick look at the attempts of the 1960s and 1970s to stimulate reservoirs with the nuclear technology that was originally developed for weapons. These tests were undertaken with the idea that if enough energy could be put into a reservoir the rock would shatter, providing significant surface area for fluid drainage into a greatly enlarged wellbore. Following Project Gnome, a proof of concept test in 1961, three such tests were carried out in low-permeability, natural-gas-bearing sandstones in Rocky Mountain basins of the U.S., and three were undertaken in oil-bearing limestones in the former Soviet Union (Table 3.3) (Lorenz, 2001). The first test in the United States was Project Gasbuggy in the Carson National Forest of New Mexico (Figure 3.37).

In the U.S., the sandstone reservoirs had not been well-enough characterized to recognize the presence and importance of the pervasive natural-fracture systems that turn microdarcy matrix rock into millidarcy reservoirs, thus it was not apparent that the nuclear blasts damaged the natural-fracture production system. Only a few of the results from the Soviet tests have been published.

Based on calculations of the measured energy release, the reservoir rock around the subsurface nuclear explosions was expected to be shattered within a few hundred feet of the wellbore and for several hundred feet vertically, creating breccia "chimneys." The fact that

Table 3.3 List and characteristics of the six reported attempts to stimulate oil and gas reservoirs in the U.S. and the USSR in the 1960s and 1970s.

USA		USSR	
GASBUGGY		**FIELD A**	
	New Mexico		Kuybyshev
	1967		1965
	29 kt		(2) 2.3 kt plus (1) 8 kt
RULISON		**FIELD B**	
	Colorado		Prem
	1969		1969
	40 kt		(2) 8 kt
RIO BLANCO		**UNDESIGNATED FIELD**	
	Colorado		Ob
	1973		1979
	(3) 30 kt		21 kt

Figure 3.37 Site of the first U.S. nuclear stimulation experiment, named "Project Gasbuggy," in the Carson National Forest east of Farmington, New Mexico. The sign reads: "This was the site of the first U.S. underground nuclear experiment for the stimulation of low productivity gas resources. On December 10, 1967, a 29-kiloton nuclear explosive was detonated at a depth of 4222'. The Atomic Energy Commission currently monitors water sources in the surrounding area for radioactive traces."

the produced gas was radioactive to some degree, and that it consisted of significant percentages of CO_2 and water vapor created by the blast instead of the previous 99% methane, diminished the value of the produced natural gas. Both problems would probably have been ameliorated over the course of long-term production, but neither the Soviets nor the Americans proceeded from experiment to production mode. The deciding factor was probably largely economic: production in the nuclear-stimulated U.S. wells was enhanced by a factor of between 2 and 5, depending on the wells to which production was compared, but the hoped-for ten-fold increase in production rate and recovery was not achieved. Environmental concerns and liability issues undoubtedly also played a part in discontinuing the programs.

References

Abdullah, R.A., and Amin, M.F.M., 2008, Verification of post failure behavior of rock using closed-circuit servo-controlled testing machine; Bulletin of the Geological Society of Malaysia, v. 54, p. 17–20.

Addis, M.A., 1997, Reservoir depletion and its effects on wellbore stability evaluation; International Journal of Rock Mechanics and Mining Sciences, v. 34, p. 3–4.

Addis, M.A., 2017, The geology of geomechanics: Petroleum geomechanical engineering in field development planning; in, Turner, J.P., Healy, D., Hillis, R.R., and Welch, M.J. eds., Geomechanics and Geology, The Geological Society, London, Special Publication 458, p. 7–29.

Addis, M.A., Last, N.C., and Yassir, N.A., 1996, Estimation of horizontal stresses at depth in faulted regions and their relationship to pore pressure variations; SPE-28140-PA, Society of Petroleum Engineers, Formation Evaluation, March, v. 11, p. 11–18. https://doi.org/10.2118/28140-PA

Aguilera, R., 1995, Naturally Fractured Reservoirs; 2nd ed.; Pennwell, Tulsa, 521 p.

Aguilera, R., 1999, Recovery factors and reserves on naturally fractured reservoirs; Journal of Canadian Petroleum Technology, Distinguished Authors Series, p. 15–18.

Aguilera, R., 2003, Geologic and engineering aspects of naturally fractured reservoirs; Canadian Society of Exploration Geophysicists Recorder, v. 28, no. 2, Calgary, Canada, 12 p.

Aimene, Y., Hammerquist, C., Nairn, J., and Ouenes, A., 2018, 3D Anisotropic damage mechanics for modeling interaction between hydraulic and natural fracture planes in a layered rock – Application to Eagle Ford and Wolfcamp; URTeC paper #2902985, Unconventional Resources Technology Conference, Houston, 23–25 July, 2018, 20 p.

Al-Aasm, I.S., Muir, I., and Morad, S., 1993, Diagenetic conditions of fibrous calcite vein formation in black shales: Petrographic, chemical, and isotopic evidence; Bulletin of Canadian Petroleum Geology, v. 41, p. 46–56.

Amadei, B., and Stephansson, O., 2012, Rock Stress and its Measurement, 2nd ed.; Springer Science and Business Media, Dortrecht, 490 p.

Ameen, M.S., 1995, Fractography: Fracture topography as a tool in fracture mechanics and stress analysis, an Introduction; in, Ameen, M.S., ed., Fractography: Fracture Topography as a Tool in Fracture Mechanics and Stress Analysis; The Geological Society, London, Special Publication 92, p. 1–10.

Ameen, M.S., 2018. Operational Geomechanics – A Rock-Based Science for Environmental, Energy, and Engineering Applications; European Association of Geoscientists and Engineers, Education Tour Series 14, 327 p.

Ameen, M.S., and Hailwood, E.A., 2008, A new technology for the characterization of microfractured reservoirs, test case: Unayzah reservoir, Wudayhi field, Saudi Arabia; American Association of Petroleum Geologists Bulletin, v. 92, p. 31–52.

Ameen, M.S., MacPherson, K., Al-Marhoon, M.I., and Rahim, Z., 2012, Diverse fracture properties and their impact on performance in conventional and tight-gas reservoirs, Saudi Arabia: The Unayzah, South Haradh case study; American Association of Petroleum Geologists Bulletin, v. 96, p. 459–592.

Anders, M.H., Laubach, S.E., and Scholz, C.H., 2014, Microfractures: A review; Journal of Structural Geology, v. 69, p. 377–394.

Anderson, E.M., 1905, The dynamics of faulting; Edinburgh Geological Society, Transactions, v. 8, p. 393–402.

Anderson, E.M., 1951, The Dynamics of Faulting and Dyke Formation with Application to Britain, 2nd ed.; Oliver and Boyd, Edinburgh, 206 p.

Antonellini, M., and Aydin, A., 1994, Effect of faulting on fluid flow in porous sandstones: Petrophysical properties; American Association of Petroleum Geologists Bulletin, v. 78, p. 355–377.

Antonellini, M., and Aydin, A., 1995, Effect of faulting on fluid flow in porous sandstones: Geometry and spatial distribution; American Association of Petroleum Geologists Bulletin, v. 79, 642–671.

Applied Concepts in Fractured Reservoirs, First Edition. John C. Lorenz and Scott P. Cooper.

Antonellini, M.A., Aydin, A., and Pollard, D.D., 1994, Microstructure of deformation bands in porous sandstones at Arches National Park, Grand County, Utah; Journal Structural Geology, v. 16, p. 941–959.

Arp, G.K., 1992, Effusive microseepage: A first approximation model of light hydrocarbon movement in the subsurface; Association of Petroleum Geochemical Explorationists Bulletin, v. 8, p. 1–17.

Ashena, R., and Thonhauser, G., 2018, Coring Methods and Systems; Springer International Publishing AG, 243 p. http://doi.org/10.1007/978-3-319-77733-7

Ashena, R., Thonhauser, G., Ghalambor, A., et al., 2019, Determination of maximum allowable safe-core-retrieval rates; SPE-189480-PA, 17 p. https://doi.org/10.2118/189480-PA

Aydin, A., 1978, Small faults formed as deformation bands in sandstone; Pure and Applied Geophysics, v. 116, p. 913–930.

Baer, G., and Reches, Z., 1991, Mechanics of emplacement and tectonic implications of the Ramon Dike Systems, Israel; Journal of Geophysical Research, v. 96, p. 11,895–11,910.

Bahat, D., 1979, Theoretical considerations on mechanical parameters of joint surfaces based on studies of ceramics; Geological Magazine, v. 116, p. 81–166.

Bai, T., and Pollard, D.D., 2000a, Closely spaced fractures in layered rocks: Initiation mechanism and propagation kinematics; Journal of Structural Geology, v. 22, p. 1409–1325. http://dx.doi.org/10.1016/S0191-8141(00)00062-6

Bai, T., and Pollard D.D., 2000b, Fracture spacing in layered rocks: A new explanation based on the stress transition; Journal of Structural Geology, v. 22, p. 43–57. http://dx.doi.org/10.1016/S0191-8141(99)00137-6

Bandis, S.C., Lumsden, A.C., and Barton, N.R., 1983, Fundamentals of rock joint deformation; International Journal of Rock Mechanics, Mining Sciences & Geomechanics, v. 20, p. 249–268.

Barker, C., 1990, Calculated volume and pressure changes during the thermal cracking of oil to gas in reservoirs; American Association of Petroleum Geologists Bulletin, v. 74, p. 1254–1261.

Barree, R.D., Gilbert, J.V., and Conway, M.W., 2009, Stress and rock property profiling for unconventional reservoir stimulation; SPE-118703-MS, Society of Petroleum Engineers Hydraulic Fracturing Technology Conference, The Woodlands, TX, 19–21 January, 2009, 18 p. https://doi.org/10.2118/118703-MS

Barton, N., 2007, Rock Quality, Seismic Velocity, Attenuation, and Anisotropy; Taylor and Francis Group, London, 729 p.

Baud, P., Zhu, W., and Wong, T.-f., 2000, Failure mode and weakening effect of water on sandstone; Journal of Geophysical Research, v. 105, p. 16,371–16,389.

Bayly, B., 1983, Mechanics in Structural Geology; Springer-Verlag, New York, 253 p.

Belfield, W.C., 1988, Characterization of a naturally fractured carbonate reservoir: Lisburne Field, Prudhoe Bay, Alaska; SPE-18174-MS, Society of Petroleum Engineers Annual Technical Conference and Exhibition, Houston, Texas, 2–5 October, 9 p. https://doi.org/10.2118/18174-MS

Beliveau, D.A., 1987, Midale CO2 flood pilot; PETSOC-87-06-05, Society of Petroleum Engineers, The Journal of Canadian Petroleum Technology, v. 26, p. 66–69. https://doi.org/10.2118/87-06-05

Beliveau, D.A., 1989, Pressure transients characterize fractured Midale Unit; SPE-15635-PA, Society of Petroleum Engineers, Journal of Petroleum Technology, v. 41, p. 1354–1362. https://doi.org/10.2118/15635-PA

Beliveau, D.A., Payne, D.A., and Mundry, M., 1993, Waterflood and CO2 flood of the fractured Midale Field; SPE-22946-PA, Society of Petroleum Engineers, Journal of Petroleum Technology, v. 45, p. 881–887. https://doi.org/10.2118/22946-PA

Bell, J.S., and Babcock, E.A., 1986, The stress regime of the western Canadian basin and implications for hydrocarbon production; Bulletin of Canadian Petroleum Geology, v. 34, p. 364–378.

Bellahsen, N., Fiore, P., and Pollard, D.D., 2006, The role of fractures in the structural interpretation of Sheep Mountain Anticline, Wyoming; Journal of Structural Geology, v. 28, p. 850–867.

Benedicto, A., and Schultz, R.A., 2010, Stylolites in limestone: Magnitude of contractional strain accommodated and scaling relationships; Journal of Structural Geology, v. 32, p. 1250–1256.

Bengtson, C.A., 1981, Statistical curvature analysis techniques for structural interpretation of dip data; American Association of Petroleum Geologists Bulletin, v. 65, p. 312–332.

Bergbauer, S., 2007, Testing the predictive capability of curvature analysis; in, Jolley, S.J., Barr, D., Walsh, J.J., and Knipe, R.J., eds., Structurally Complex Reservoirs; The Geological Society, London, Special Publication 292, p. 185–202.

Berkowitz, B., 2002, Characterizing flow and transport in fractured geological media: A review; Advances in Water Resources, v. 25, p. 861–884.

Berryman, J.G., 1992, Effective stress for transport properties of inhomogeneous porous rock; Journal of Geophysical Research, v. 97, p. 17409–17424.

Billingsley, R.L., 2005, Identifying and Remediating High Water Production Problems in Basin-Centered Formations: Final Report; DOE Award No. DE-FC26-02NT41437, plus Appendix D, Working Model for Reservoir Development, 102 p.

Biot, M.A., 1941, General theory of three-dimensional consolidation; Journal of Applied Physics, v. 12, p. 155–164.

Bisdom, K., Bertotti, G., and Nick, H.M., 2016, A geometrically based method for predicting stress-induced fracture aperture and flow in discrete fracture networks; American Association of Petroleum Geologists Bulletin, v. 100, p. 1075–1097.

Bishop, J.W., Sumner, D.Y., and Huerta, N.J., 2006, Molar tooth structures in the Neoarchean Monteville Formation, Transvaal Supergroup, South Africa II: A wave-induced fluid flow model; Sedimentology, v. 53, p 1069–1082.

Blanton, T.L., 1982, An experimental study of interaction between hydraulically induced and pre-existing fractures; SPE-10847-MS, Society of Petroleum Engineers and United States Department of Energy, Unconventional Gas Recovery Symposium, Pittsburgh, Pennsylvania, 16–18 May, 1982, 13 p. https://doi.org/10.2118/10847-MS

Blanton, T.L., 1986, Propagation of hydraulically and dynamically induced fractures in naturally fractured reservoirs; SPE 15261-MS, Society of Petroleum Engineers, Unconventional Gas Technology Symposium, Louisville, KY, 18–21 May, 1986, 15 p. https://doi.org/10.2118/15261-MS

Branagan, P., Cotner, G., and Lee, S.J., 1984, Interference testing of the naturally fractured Cozzette Sandstone: A case study at the DOE MWX site; SPE-12869-MS, Society of Petroleum Engineers, Unconventional Gas Recovery Symposium, Pittsburgh, Pennsylvania, 13–15 May, 1984, 8 p. https://doi.org/10.2118/12869-MS

Branagan, P., Cipolla, C.L., Lee, S.J., and Yan, L., 1987, Case history of hydraulic fracture performance in the naturally fractured paludal zone: The transitory effects of damage; SPE-16397-MS, Society of Petroleum Engineers, Low-Permeability Reservoirs Symposium, Denver, Colorado, 18–19 May, 11 p. https://doi.org/10.2118/16397-MS

Breckels, I. M., and van Eekelen, H.A., 1982, Relationship between horizontal stress and depth in sedimentary basins; SPE-10336-PA, Society of Petroleum Engineers, Journal of Petroleum Technology, v. 34, p. 2191–2198. https://doi.org/10.2118/10336-PA

Brown, E.T., and Hoek, E., 1978, Trends in relationships between measured in-situ stresses and depth; International Journal of Mechanics and Mining Sciences and Geomechanics Abstracts, v. 15, p. 211–215.

Chong, K.P., and Boresi, A.P., 1990, Strain rate dependent mechanical properties of New Albany reference shale; International Journal of Rock Mechanics and Mining Sciences and Geomechanics Abstracts, v. 27, p. 199–205. https://doi.org/10.1016/0148-9062(90)94328-Q

Chopra, S., and Marfurt, K.J., 2005, Seismic – A historical perspective; Geophysics, v. 70, p. 3SO–28SO. http://dx.doi.org/10.1190/1.2098670

Ciezobka, J., Courtier, J., and Wicker, J., 2018, Hydraulic Fracturing Test Site (HFTS) – Project overview and summary of results; URTeC 2937168, Unconventional Resources Technology Conference, Houston, Texas, 23–25 July 2018, 9 p.

Clark, J.A., 1982, Glacial loading: A cause of natural fracturing and a control of the present stress state in regions of high Devonian shale gas production; SPE-10798-MS, Society of Petroleum Engineers, Unconventional Gas Recovery Symposium, 16–18 May, Pittsburgh, Pennsylvania, 11 p. https://doi.org/10.2118/10798-MS

Cobbold, P.R., and Rodrigues, N., 2007, Seepage forces, important factors in the formation of horizontal hydraulic fractures and bedding-parallel fibrous veins ('beef' and 'cone-in-cone'); Geofluids, v. 7, p. 313–322. https://doi.org/10.1111/j.1468-8123.2007.00183.x

Cobbold, P.R., Zanella, A., Rodrigues, N., and Loseth, H., 2014, Bedding-parallel fibrous veins (beef and cone-in-cone): Worldwide occurrence and possible significance in terms of fluid overpressure, hydrocarbon generation and mineralization; Marine and Petroleum Geology, v. 43, p. 1–20.

Compton, R.R., 1985, Geology in the Field, Wiley, 416 p.

Cook, N.G.W., 1992, Natural joints in rock: Mechanical, hydraulic and seismic behavior and properties under normal stress; International Journal of Rock Mechanics and Mining Science and Geomechanics Abstracts, v. 29, p. 198–223.

Cooper, M., 1992, The analysis of fracture systems in subsurface thrust structures from the foothills of the Canadian Rockies; in, McClay, K.E., ed., Thrust Tectonics; London, Chapman and Hall, p. 391–405.

Cooper, S.P., 2013, Natural fractures and strain accommodation in the Tensleep Formation at Beer Mug Anticline; in, C. Knight and J. Cuzella, eds., Application of Structural Methods to Rocky Mountain Hydrocarbon Exploration and Development; American Association of Petroleum Geologists Studies in Geology 65, p. 139–156.

Cooper, S.P., Goodwin, L.B., and Lorenz, J.C., 2006, Fracture and fault patterns associated with basement-cored anticlines: The example of Teapot Dome, Wyoming; American Association of Petroleum Geologists Bulletin, v. 90, p. 1903–1920. http://doi.org/10.1306/06020605197

Costin, L.S., 1987, Time-dependent deformation and failure; in, Atkinson, B.K., ed., Fracture Mechanics of Rock, Academic Press, Geology Series, p. 167–215.

Craddock, J.P., and van der Pluijm, B.A., 1989, Late Paleozoic deformation of the cratonic carbonate cover of eastern North America; Geology, v. 17, p. 416–419.

https://doi.org/10.1130/0091-7613 (1989) 017<0416:LPDOTC>E2.3.CO;2

Crosby, G.W., 1969, Radial movements in the western Wyoming salient of the Cordilleran overthrust belt; Geological Society of America Bulletin, v. 80, p. 1061–1078.

Cruz, L., Izadi, G., Barton, C., and Hoeink, T., 2018, Hydraulic fracture propagation in multi-layer fractured media; American Association of Petroleum Geologists, Search and Discovery article #42201, 30 April, 2018, 16 p.

Cundall, P.A., Pierce, M.E., and Mas Ivars, D., 2008, Quantifying the size effect of rock mass strength; in, Potvin, Y., Carter, J., Dyskin, A., and Jeffrey, R., eds., 1st Southern Hemisphere International Rock Mechanics Symposium SHIRMS 2008 Australian Centre for Geomechanics, Perth, v. 2, p. 3–15.

Dahi-Taleghani, A.D. and Olson, J.E., 2011, Numerical modeling of multi-stranded hydraulic fracture propagation: Accounting for the interaction between induced and natural fractures; SPE-124884-PA, Society of Petroleum Engineers Journal, v. 16, p. 575–581 https://doi.org/10.2118/124884-PA

Daneshy, A.A., 1974, Hydraulic fracture propagation in the presence of planes of weakness; SPE-4852-MS, Society of Petroleum Engineers European Spring Meeting, Amsterdam, 29–30 May, 1974, 8 p. https://doi.org/10 .2118/4852-MS

Davis, G.H., Bump, A.P., Garcia, P.E., and Ahlgren, S.G., 2000, Conjugate riedel deformation band shear zones; Journal of Structural Geology, v. 22, p. 169–190.

Davis, J.C., 2002, Statistics and Data Analysis in Geology, 3rd ed., Wiley and Sons, 638 p.

Daw, A., Howel, F.T., and Woodhead, F.A., 1974, The effect of applied stress upon the permeability of some Permian and Triassic sandstones of northern England; in, Proceedings of the Congress of the International Society for Rock Mechanics, v. 2, Part A, Advances in rock mechanics; reports of current research, p. 537–542.

De Keijzer, M., Hillgartner, H., Al Dhahab, S., and Rawnsley, K., 2007, A surface–subsurface study of reservoir-scale fracture heterogeneities in Cretaceous carbonates, North Oman; in, Lonergan, L., Jolly, R.J.H., Rawnsley, K., and Sanderson, D.J., eds., Fractured Reservoirs; The Geological Society, London, Special Publication 270, p. 227–244. https://doi.org/10.1144/ GSL.SP.2007.270.01.15

DeGraff, J.M., and Aydin, A., 1987, Surface morphology of columnar joints and its significance to mechanics and direction of joint growth; Geological Society of America Bulletin, v. 99, p. 605–617.

Delaney, P.T., Pollard, D.D., Ziony, J.I., and McKee, E.H., 1986, Field relations between dikes and joints: Emplacement processes and paleostress analysis; American Geophysical Union, Journal of Geophysical Research, v. 91, p. 4920–4938.

Dershowitz, W.S., and Herda, H.J., 1992, Interpretation of fracture spacing and intensity; in, 33rd United States Symposium on Rock Mechanics, Santa Fe, New Mexico, p. 757–766.

Dershowitz, W. S., and LaPointe, P. R., 1994, Discrete fracture approaches for oil and gas applications, in, Nelson, R.A., and Laubach, S.E., eds., Rock Mechanics; Rotterdam, Balkema, p. 19–30.

Desayes, C., Villemin, T., and Pêcher, A., 2000, Microfracture pattern compared to core-scale fractures in the borehole of Soultz-sous-forêts granite, Rhine graben, France; Journal of Structural Geology, v. 22, p. 723–733.

Dixit, N.C., Hanks, C.L., Wallace, W.K., Ahmadi, M., and Awoleke, O., 2017, In situ stress variations associated with regional changes in tectonic setting, northeastern Brooks Range and eastern North Slope of Alaska; American Association of Petroleum Geologists Bulletin, v. 101, p. 343–360. http://doi.org/10.1306/08051616013

Doblas, M., 1998, Slickenside kinematic indicators; Tectonophysics, v. 295, p. 187–197.

Doe T., and Dershowitz, B., 2008, Derivation of basic fracture properties, Golder Associates Ltd; University of Washington, Civil and Environmental Engineering, Engineering Geology 1 course presentation, 30 p. https://faculty.washington.edu/dersh/Files/Quiz2-2009FractureStudy.pdf

Du Bernard, X., Eichubl, P., and Aydin, A., 2002, Dilation bands: A new form of localized failure in granular media; American Geophysical Union, Geophysical Research Letters, v. 29, p. 29-1 to 29-4. https://doi.org/10.1029/ 2002GL015966

Dunn, D.E., LaFountain, L.J., and Jackson, R.E., 1973, Porosity dependence and mechanism of brittle fracture in sandstones; Journal of Geophysical Research, v. 78, p. 2403–2417.

Durham, W.B., and Bonner, B.P., 1994, Self-propping and fluid flow in slightly offset joints at high effective pressures; American Geophysical Union, Journal of Geophysical Research, v. 99, p. 9391–9399.

Dutton, S.P., and Willis, B.J., 1998, Comparison of outcrop and subsurface sandstone permeability distribution in Lower Cretaceous Fall River Formation, South Dakota and Wyoming; Journal of Sedimentary Research, v. 68, p. 890–900.

Dyke, C.G., 1992, Stress insensitive natural fracture permeability within hydrocarbon reservoirs; in, Hudson, J.A., ed., Proceedings of the International Society of Rock Mechanics and Rock Engineering, Eurock 1992 Symposium, Chester, England, 14–17 September, 1992, p. 281–286.

Elkins, L.F., 1953, Reservoir performance and well spacing, Spraberry trend area field of West Texas; American Institute of Mining, Metallurgical and Petroleum Engineers, Petroleum Transactions, v. 198, p. 177–196. https://doi.org/10.2118/953177-G

Elkins, L.F., 1963, Cyclic water flooding in the Spraberry utilizes "end effects" to increase oil production rate; SPE-545-PA, Society of Petroleum Engineers, Journal of Petroleum Technology, v. 15, p. 877–884. https://doi.org/10.2118/545-PA

Elkins, L.F., and Skov, A.M., 1960, Determination of fracture orientation from pressure interference; American Institute of Mining, Metallurgical and Petroleum Engineers, Petroleum Transactions, v. 219, p. 301–304.

Ellsworth, W.L., 2013, Injection-induced earthquakes; Science, v. 341, p. 142–149.

Engelder, T., 1985, Loading paths to joint propagation during a tectonic cycle: An example from the Appalachian Plateau, USA; Journal of Structural Geology, v. 7., p. 459–476.

Engelder, T., 1993, Stress Regimes in the Lithosphere; Princeton University Press, Princeton, NJ., 457 p.

Engelder, T., and P. Geiser, 1980, On the use of regional joint sets as trajectories of paleostress fields during the development of the Appalachian Plateau, New York; Journal of Geophysical Research, v. 85, p. 6319–6341.

English, J.M., 2012, Thermomechanical origin of regional fracture systems; American Association of Petroleum Geologists Bulletin v. 96, p. 1597–1625.

Evans, D.M., 1966, The Denver area earthquakes and the Rocky Mountain Arsenal disposal well; Rocky Mountain Association of Geologists, The Mountain Geologist, v. 3, p. 23–36.

Fairhurst, C., 2003, Stress estimation in rock: A brief history and review; International Journal of Rock Mechanics and Mining Sciences, v. 40, p. 957–973.

Fall, A., Eichhubl, P., Bodnar, R.J., et al., 2015, Natural hydraulic fracturing of tight-gas sandstone reservoirs, Piceance Basin, Colorado; Geological Society of America Bulletin, v. 127, p. 61–75. http://doi.org/10.1130/B31021.1

Far, M.E., Sayers, C.M., Thomsen, L., et al., 2013, Seismic characterization of naturally fractured reservoirs using amplitude versus offset and azimuth analysis; Geophysical Prospecting, v. 61, p. 427–447.

Fast, R.E., Murer, A.S., and Timmer, R.S., 1994, Description and analysis of cored hydraulic fractures – Lost Hills Field, Dern County, California; SPE-24853-PA, Society of Petroleum Engineers, Production and Facilities, May, 1994, p. 107–113. https://doi.org/10.2118/24853-PA

Faulkner, D.R., Mitchell, T.M. Rutter, E.H and Cembrano, J., 2008, On the structure and mechanical properties of large strike-slip fault zones; in, Wibberley C.A.J., Kurtz W., Imber J., Holdsworth R.E., and Collettini C., eds., The Internal Structure of Fault Zones: Implications for Mechanical and Fluid-Flow Properties; The Geological Society, London, Special Publication 299, p. 139–150. http://doi.org/10.1144/SP299.9

Fernández-Ibáñez, F., DeGraff, J. M., and Ibrayev, F., 2018, Integrating borehole image logs with core: A method to enhance subsurface fracture characterization; American Association of Petroleum Geologists, v. 102, p. 1067–1090.

Ferrill, D.A., Morris, A.P., Stamatakos, J.A., and Sims, D.W., 2000, Crossing conjugate normal faults; American Association of Petroleum Geologists Bulletin, v. 84, p. 1543–1549.

Fertl, W.H., 1976, Abnormal Formation Pressures: Developments in Petroleum Science, no. 2; Elsevier, Amsterdam, 383 p.

Finley S.J., and Lorenz, J.C., 1988, Characterization of Natural Fractures in Mesaverde Core from the Multiwell Experiment; SAND88-1800, Sandia National Laboratories Technical Report, 133 p. https://doi.org/10.2172/6631435

Fossen, H., 2010, Structural Geology, Cambridge University Press, 463 p.

Fossen, H., and Bale, A., 2007, Deformation bands and their influence on fluid flow; American Association of Petroleum Geologists Bulletin, v. 91, p. 1685–1700.

Fossen, H., Schultz, R.A., Shipton, Z.K., and Mair, K., 2007, Deformation bands in sandstone: A review; Journal of the Geological Society, London, v. 164, p 744–769.

Frechette, V.D., 1972, The fractography of glass; in, Pye, L.D., Stevens, H.J., and LaCourse, W.C., eds., Introduction to Glass Science; Plenum Press, p. 433–450.

Friedman, M., 1969, Structural analysis of fractures in cores from Saticoy Field, Ventura County, California; American Association of Petroleum Geologists Bulletin, v. 53, p. 367–389. http://doi.org/10.1306/5D25C613-16C1-11D7-8645000102C1865D

Friedman, M., 1972, Residual elastic strain in rocks; Tectonophysics, v. 15, p. 297–330.

Friedman, M., 1975, Fracture in Rock; in, Bell, P.M., ed., Reviews of Geophysics and Space Physics, v. 13, No. 3, July 1975, U.S. National Report 1971–1974, 16th General Assembly International Union of Geodesy and Geophysics, Grenoble, France, 24 August – 6 September, 1975, p. 352–358 and 383–389.

Gale, J.E., and Reardon, E.J., 1984, Effects of groundwater geochemistry on the permeability of grouted fractures; Canadian Geotechnical Journal, v. 21, p. 8–20. http://dx.doi.org/10.1139/t84-002

Gale, J.E., 1990, Hydraulic behavior of rock joints; in, Barton, W, and Stephansson, E., eds., Rock Joints; Proceedings of the International Symposium on Rock Joints, Loen, Norway, 4–6 June, 1990, p. 351–362.

Gale, J.F.W., 2004, Self-organization of natural mode-I fracture apertures into power-law distributions; ARMA-04-488, American Rock Mechanics Association, 6th North American Rock Mechanics Symposium Houston, Texas, 5–9 June, 2004, 13 p.

Gale, J.F.W., 2014, Natural fracture patterns and attributes across a range of scales; American Association of Petroleum Geologists Search and Discovery article #41486, 17 November, 2014.

Gale, J.F.W., Reed, R. M., and Holder, J., 2007, Natural fractures in the Barnett Shale and their importance for hydraulic fracture treatments; American Association of Petroleum Geologists Bulletin, v. 91, p. 603–622.

Gale, J.F.W., Laubach, S.E., Olson, J.E., et al., 2014, Natural fractures in shale: A review and new observations; American Association of Petroleum Geologists Bulletin, v. 98, p. 2165–2216.

Gale, J.F.W., Elliott, S.J., and Laubach, S.E., 2018, Hydraulic fractures in core from stimulated reservoirs: Core fracture description of HFTS slant core, Midland Basin, West Texas; URTeC 2902624, Unconventional Resources Technology Conference, Houston, Texas, USA, 23–25 July 2018, 18 p.

Gallagher, J.J., Friedman, M., Handin, J., and Sowers, G.M., 1974, Experimental studies relating to microfracture in sandstone; Tectonophysics, v. 21, p. 203–247.

Gay, N.C., 1970, The formation of step structures on slickensided shear surfaces; Journal of Geology, v. 78, p. 523–532.

Geiser, P.A., and Sansone, S., 1981, Joints, microfractures, and the formation of solution cleavage in limestone; The Geological Society of America, Geology, v. 9., p. 280–285.

Gilbertson, N. J., 2006, 3-D Geologic Modeling and Fracture Interpretation of the Tensleep Sandstone, Alcova Anticline, Wyoming: M.S. thesis, Colorado School of Mines, Golden, 265 p.

Goodman, R.E., 1989, Introduction to Rock Mechanics, 2nd ed.; Wiley and Sons, New York, 288 p.

Gramberg, J., 1965, Axial cleavage fracturing, a significant process in mining and geology; Engineering Geology, v. 1, p. 31–72.

Gramberg, J., 1989, A non-conventional view on rock mechanics and fracture mechanics, A.A. Balkema, Rotterdam, 250 p.

Grauls, D., 1998, Overpressure assessment using a minimum principal stress approach; Overpressures in Petroleum Exploration, Proceedings Workshop, Pau, April, 1998, Bulletin Centre Recherche Elf Exploration and Production, Memoir 22, p 137–147.

Gretener, P.E., 1979, Pore pressure: Fundamentals, General Ramifications and Implications for Structural Geology (revised edition); American Association of Petroleum Geologists, Continuing Education Course Note Series #4, 131 p.

GRI, 1999, Mounds drill cuttings injection project, final report; Wolhart, S. (manager), Gas Research Institute report on CD, GRI-99/0173, June, 1999.

Griggs, D., and Handin, J., 1960, Observations on fracture and a hypothesis of earthquakes; Chapter 13, in, Griggs, D., and Handin, J., eds., Rock Deformation, a Symposium; Geological Society of America Memoir 79, p. 347–373.

Gross, M.R., 1993, The origin and spacing of cross joints: Examples from the Monterey Formation, Santa Barbara Coastline, California; Journal of Structural Geology, v. 15, p. 737–751. https://doi.org/10.1016/0191-8141(93)90059-J

Gross, M.R., and Engelder, T., 1995, Strain accommodated by brittle failure in adjacent units of the Monterey Formation, U.S.A.: Scale effects and evidence for uniform displacement boundary conditions; Journal of Structural Geology, v. 17, p. 1303–1307, 1309–1318. https://doi.org/10.1016/0191-8141(95)00011-2

Gross, M.R., and Eyal, Y, 2007, Throughgoing fractures in layered carbonate rocks; Geological Society of America Bulletin, v. 119, p. 1387–1404.

Grover, P. W., 2011, Petrographic recognition of artificially induced microfractures versus natural microfractures in oil/gas shales (abstract); American Association of Petroleum Geologists, Search and Discovery Article 90124, AAPG Annual Convention and Exhibition, 10–13 April, 2011. http://www.searchanddiscovery.com/abstracts/html/2011/annual/abstracts/Grover.html

Guerriero, V., Mazzoli, S., Iannace, A., et al., 2013, A permeability model for naturally fractured carbonate reservoirs; Marine and Petroleum Geology, v. 40, p. 115–134.

Gutierrez, M., Oino, L.E., and Nygard, R., 2000, Stress-dependent permeability of a demineralised fracture in shale; Marine and Petroleum Geology, v. 7., p. 895–907.

Haimson, B., 1978, Effect of cyclic loading on rock; in, Silver, M.L., and Tiedmann, D. A., eds., Dynamic Geotechnical Testing, STP654, p. 228–245.

Hancock, P.L., 1985, Brittle microtectonics; principles and practice; Journal of Structural Geology, v. 7, p. 437–457.

Hancock, P.L., 1986, Joint spectra; in Nichol, I., and Nesbitt, R.W., eds., Geology in the Real World – the Kingley Dunham volume; London, Institution of Mining and Metallurgy, p. 155–164.

Hancock, P.L., and Bevan, T.G., 1987, Brittle modes of foreland deformation; in, Coward, M.P., Dewey, J.F., and Hancock, P.L, eds., Continental Extension Tectonics; The Geological Society, London, Special Publication 28, p. 127–137.

Handin, J., Hager, R.V., Friedman, M., and Feathers, J.N., 1963, Experimental deformation of sedimentary rocks under confining pressure: Pore pressure tests; American Association of Petroleum Geologists Bulletin, v. 47, p. 717–755.

Hanks, C.L., Lorenz, J.C., Teufel, L. and Krumhardt, A.P., 1997, Lithologic and Structural Controls on natural fracture distribution and behavior within the Lisburne Group, Northeastern Brooks Range and North Slope subsurface, Alaska; American Association of Petroleum Geologists Bulletin, v. 81, p. 1700–1720.

Hart, B.S., 2006, Seismic expression of fracture-swarm sweet spots, Upper Cretaceous tight-gas reservoirs, San Juan Basin; American Association of Petroleum Geologists Bulletin, v. 90, p. 1519–1534.

Hart, B.S., 2011, Introduction to Seismic Interpretation; American Association of Petroleum Geologists Discovery Series No. 16, 215 p.

Haxby, W.F., and Turcotte, D.L., 1976, Stresses induced by the addition or removal of overburden and associated thermal effects; Geology, v. 4, p. 181–184.

Healy, D., Blenkinsop, T.G., Timms, N.E., et al., 2015, Polymodal faulting; time for a new angle on shear failure; Journal of Structural Geology, v. 80, p. 57–71.

Healy, J.H., Rubey, W.W., Griggs, D.T. and Raleigh, C.B., 1968, The Denver earthquakes; Science, v. 161, p. 1301–1310.

Heap, M.J., Baud, P., Reuschlé, T., and Meredith, P.G., 2014, Stylolites in limestones: Barriers to fluid flow?; The Geological Society of America, Geology, v. 42, p. 1–54.

Heck, E.T., 1955, Fractures and Joints; Producers Monthly, v. 19, February 1955, p. 20, 22–23.

Heffer, K., Greenhough, J., Main, I.G., et al., 2010, Low-cost monitoring of inter-well reservoir communication paths through correlations in well rate fluctuations: Case studies from mature field in the North Sea; SPE-130734-MS, Society of Petroleum Engineers Europe/European Association of Geoscientists and Engineers, Annual Conference and Exhibition, Barcelona, Spain, 14–17 June, 2010, 12 p. https://doi.org/10.2118/130734-MS

Heinemann, Z.E., and Mittermeir, G., 2015, Natural Fractured Reservoir Engineering; Professor Heinemanns Doktorandengruppe, Verein zur Förderung von wissenschaftlichen Arbeiten in Reservoircharakterisierung und-simulation, Textbook Series v. 5, 138 p.

Heritage, G.L, Large, A.R.G., eds., 2009, Laser Scanning for the Environmental Sciences; Wiley-Blackwell, 278 p.

Hertz, H., 1881, Über die Berührung fester elastischer Körper; Journal für die reine und angewandte Mathematik, v. 92, p. 156–171.

Hill, C., 2000, Overview of the geologic history of cave development in the Guadalupe Mountains, New Mexico; Journal of Cave and Karst Studies, v. 62, p. 60–71.

Hill, R.E., 1989, Analysis of deformation bands in the Aztec Sandstone, Valley of Fire State Park, Nevada; MSc thesis, University of Nevada, Las Vegas, 68 p.

Hill, R.E., Peterson, R.E., Warpinski, N.R., et al., 1994, Techniques for determining subsurface stress direction and assessing hydraulic fracture azimuth: SPE-29192-MS; Society of Petroleum Engineers Eastern Regional Conference and Exhibition, Charleston, WV, 8–10 November, 1994, p. 308–319. https://doi.org/10.2118/29192-MS

Hillis, R.R., 1998, The influence of fracture stiffness and the in situ stress field on closure of natural fractures; Petroleum Geoscience, v. 4, p. 57–65.

Hillis, R.R., 2000, Pore pressure/stress coupling and its implications for seismicity; Exploration Geophysics, v. 31, p. 448–454.

Hillis, R.R., 2001, Coupled changes in pore pressure and stress in oil fields and sedimentary basins; Petroleum Geoscience, v., 7, p. 419–425.

Hillis, R.R., 2003, Pore pressure/stress coupling and its implications for rock failure; in, Van Rensberg, N.P., Hillis, R.R., Maltman, J.J., and Morley, C.K., eds., Subsurface Sediment Mobilization; The Geological Society, London, Special Publication 216, p. 359–368.

Hodgson, R.A., 1961, Regional study of jointing in Comb Ridge-Navajo Mountain area, Arizona and Utah; American Association of Petroleum Geologists Bulletin, v. 45, p. 1–38.

Holcomb, D., Rudnicki, J.W., Issen, K.A., and Sternlof, K., 2007, Compaction localization in the earth and the laboratory: State of the research and research directions; Acta Geotechnica, v. 2, p. 1–15. http://doi.org/10.1007/s11440-007-0027-y

Hooker, J.N., Gale, J.F.W., Gomez, L.A., et al., 2009, Aperture-size scaling variations in a low-strain opening-mode fracture set, Cozzette Sandstone, Colorado; Journal of Structural Geology, v. 31, p. 707–718.

Hooker, J.N., Laubach, S.E., and Marrett, R., 2014, A universal power-law scaling exponent for fracture apertures in sandstones; Geological Society of America Bulletin, v. 126, p. 1340–1362. https://doi.org/10.1130/B30945.1

Hooker, J.N., Abu-Mahfouz, I.S., Meng, Q., and Cartwright, J., 2019, Fractures in mudrocks: Advances in constraining timing and understanding mechanisms; Journal of Structural Geology, in press.

Hopkins, C.W., Holditch, S.A., Rosen, R.L., and Hill, D.G., 1998, Characterization of an induced hydraulic fracture completion in a naturally fractured Antrim Shale Reservoir; SPE-51068-MS, Society of Petroleum

Engineers Eastern Regional Meeting, Pittsburgh, PA, 7–11 October, 1998, p. 177–185.

Horrocks, R.D., and Szukalski, B.W., 2002, Using geographic information systems to develop a cave potential map for Wind Cave, South Dakota; Journal of Cave and Karst Studies, v. 64, p 63–70.

Huang, Q., and Angelier, J., 1989, Fracture spacing and its relation to bed thickness; Geological Magazine, v. 126, p. 355–362.

Hubbert, M.K., and Willis D.G., 1957, Mechanics of hydraulic fracturing; American Institute of Mining, Metallurgical, and Petroleum Engineers, Petroleum Transactions, v. 210, p. 153–166, and Authors' Reply to the Reynolds and Coffer Discussion, p. 167–168.

Hubbert, M.K., and Rubey, W.W., 1959, Role of fluid pressure in mechanics of overthrust faulting: I. Mechanics of fluid-filled porous solids and its application to overthrust faulting; Geological Society of America Bulletin, v. 70 p. 115–166.

Hudson, J.A., and Priest, S.D., 1983, Discontinuity frequency in rock masses; International Journal of Rock Mechanics and Mining Sciences & Geomechanics Abstracts, v. 20, p. 73–89.

Hunt, J.M., 1990, Generation and migration of petroleum from abnormally pressured fluid compartments; American Association of Petroleum Geologists Bulletin, v. 74, p. 1–12.

Ingraham, M.D., Issen, K.A., and Holcomb, D.J., 2013, Use of acoustic emissions to investigate localization in high-porosity sandstone subjected to true triaxial stresses; Acta Geotechnica, v. 8., p. 645–663.

Jaeger, J.C., Cook, N.G.W., and Zimmerman, R.W., 2007, Fundamentals of Rock Mechanics, 4th ed.; Blackwell Publishing, 475 p.

Jamison, W.R., 2016, Fracture system evolution within the Cardium sandstone, central Alberta Foothills folds; American Association of Petroleum Geologists Bulletin, v. 100, p. 1099–1134.

Jamison, W.R., and Stearns, D.W., 1982, Tectonic deformation of Wingate Sandstone, Colorado National Monument; American Association of Petroleum Geologists Bulletin, v. 66, p. 2584–2608.

Janach, W., 1977, Failure of granite under compression; International Journal of Rock Mechanics Mining Science and Geomechanics Abstracts, v. 14, p. 209–215.

Jeffrey, R.G., Zhang, X., and Bunger, A.P., 2010, Hydraulic fracturing of naturally fractured reservoirs; SGP-TR-188, Proceedings of the 35th Workshop on Geothermal Reservoir Engineering, Stanford University, Stanford, California, February 1–3, 2010, 9 p.

Jenkins, C., Ouenes, A., Zellou, A., and J. Wingard, J, 2009, Quantifying and predicting naturally fractured reservoir behavior with continuous fracture models; American Association of Petroleum Geologists Bulletin, v. 93, p. 1597–1608.

Jensen, J.L., Lake, L.W., Corbett, P.W.M., and Goggin, D.J., 2007, Statistics for Petroleum Engineers and Geoscientists, 2nd ed.; in, Cubitt, J., ed., Handbook of Petroleum Exploration and Production 2; Elsevier, 338 p.

Jizba, D.L., 1991a, Mechanical and acoustical properties of sandstones and shales; PhD dissertation, Stanford University, 275 p.

Jizba, D.L., 1991b, Uniaxial compressive strength in tight gas sandstones and shales; Gas Research Institute Topical Report, Tight Gas Sands Project Area Contract No. 5089-211-1842, 64 p.

Johnson, R.B. and DeGraff, J.V., 1988, Principles of Engineering Geology; Wiley and Sons Ltd, 512 p.

Kassis, S., and Sondergeld, C.H., 2010, Fracture permeability of gas shale: Effects of roughness, fracture offset, proppant, and effective stress; SPE-131376-PP, presented at the Society of Petroleum Engineers, CSP-SPE International Oil and Gas Conference and Exhibition in Beijing, China, 8–10 June, 2010, 20 p. https://doi.org/10.2118/131376-MS

Kazemi, H., 1969, Pressure transient analysis of naturally fractured reservoirs with uniform fracture distribution; SPE-2156-A, Society of Petroleum Engineers Journal, v. 9, p. 451–462. https://doi.org/10.2118/2156-A

Kelley, V.C., and Clinton, N.J., 1960, Fracture Systems and Tectonic Elements of the Colorado Plateau; University of New Mexico Press, 104 p.

Knipe, R. J., and Lloyd, G.E., 1994, Microstructural analysis of faulting in quartzite, Assynt, NW Scotland: Implications for fault zone evolution; Earth and Environmental Science, v. 143, p. 229–254.

Kranz, R.L., 1983, Microcracks in rock: A review; Tectonophysics, v. 100, p. 449–480.

Kulander, B.R., Barton, C.C., and Dean, S.L., 1979, The Application of Fractography to Core and Outcrop Fracture Investigations; METC/SP-79/3, U.S. Department of Energy, 174 p.

Kulander, B.R., and S.L. Dean, 1985, Hackle plume geometry and joint propagation dynamics, in Ove Stephansson, ed., Fundamentals of Rock Joints; Sweden, Centek Publishers, p. 85–94.

Kulander, B.R., Dean, S.L., and Ward, B.J., 1990, Fractured Core Analysis: Interpretation, Logging, and Use of Natural and Induced Fractures in Core; American Association of Petroleum Geologists Methods in Exploration Series, n. 8, 88 p.

Lacazette, A., 1991, New stereographic technique for the reduction of scanline survey data of geologic features; Computers and Geosciences, v. 17, p. 445–463.

Ladeira, F.L., and Price, N.J., 1981, Relationship between fracture spacing and bed thickness; Journal of Structural Geology, v. 3, p. 179–183.

Lajtai, E.Z. and Alison, J.R., 1979, A study of residual stress effects in sandstone; Canadian Journal of Earth Science, v. 16, p. 1547–1557.

Lama, R.D., and Vutukuri, V.S., 1978, Handbook on Mechanical Properties of Rocks, Volume 2, Testing Techniques and Results; Trans Tech Publications, Bay Village, Ohio, 481 p.

Lamont, N., and Jessen, F., 1963, The effects of existing fractures in rocks on the extension of hydraulic fractures; Society of Petroleum Engineers, Journal of Petroleum Technology, v. 15, p. 203–209.

Landry, C.J., Eichhubl, P., Prodanović, M., and Tokan-Lawal, A., 2014, Matrix-fracture connectivity in Eagle Ford Shale; URTeC paper no. 1922708, Unconventional Resources Technology Conference, Denver, Colorado, 25–27 August, 2014, 10 p.

Landry, C.J., Eichhubl, P., Prodanović, M., Wilkins, S., 2016. Nanoscale grain boundary channels in fracture cement enhance flow in mudrocks; Journal of Geophysical Research, Solid Earth, v. 121, p. 3366–3376. http://dx.doi.org/10.1002/2016JB012810

Landry, C.J., Eichhubl, P., Prodanović, M., and Tokan-Lawal, A., 2017, Completely-cemented natural fractures in mudrocks: Flow barrier or highway?; American Association of Petroleum Geologists Search and Discovery, abstract 2101692, AAPG Annual Meeting and Exhibition, Denver Colorado, 31 May–2 June, 2015.

LaPointe, P.R., and Hudson, J.A., 1985, Characterization and interpretation of rock mass joint patterns; Geological Society of America Special Paper 199, 37 p.

Lash, G.G., and Engelder, T., 2005, An analysis of horizontal microcracking during catagenesis: Example from the Catskill delta complex; American Association of Petroleum Geologists Bulletin, v. 89, p. 1433–1449.

Laubach, S.E., 1992, Fracture networks in selected Cretaceous sandstones of the Green River and San Juan basins, Wyoming, New Mexico, and Colorado; in, Schmoker, J.W., Coalson, E.B., and Brown, C.A., eds., Geological Studies Relevant to Horizontal Drilling: Examples from Western North America; Rocky Mountain Association of Geologists, p. 115–128.

Laubach, S.E., Fall, A., Copley, L.K., et al., 2016, Fracture porosity creation and persistence in a basement-involved Laramide fold, Upper Cretaceous Frontier Formation, Green River Basin, USA; Geological Magazine, v. 153, p. 887–910. http://doi.org/10.1017/S0016756816000157

Law, B.E., and Spencer, C.W., 1998, Abnormal pressures in hydrocarbon environments; in, Law, B.E., Ulmishek, G.F., and Slavin, V.I., eds., Abnormal pressures in hydrocarbon environments; American Association of Petroleum Geologists Memoir 70, p. 1–11.

Lee, K., Tomasso, M., Ambrose, W.A., and Bouroullec, R., 2007, Integration of GPR with stratigraphic and lidar data to investigate behind-the-outcrop 3D geometry of a tidal channel reservoir analog, upper Ferron Sandstone, Utah; Society of Exploration Geophysists, The Leading Edge, v. 26, p. 994–998.

Li, H., Li, H., Wang, K., and Liu, C., 2018, Effect of rock composition microstructure and pore characteristic on its rock mechanics properties; International Journal of Mining Science and Technology, v. 28, p. 303–308. https://doi.org/10.1016/j.ijmst.2017.12.008

Li, J.Z., Laubach, S.E., Gale, J.F.W., and Marrett, R.A. 2018, Quantifying opening-mode fracture spatial organization in horizontal wellbore image logs, core and outcrop: Application to Upper Cretaceous Frontier Formation tight gas sandstones, USA; Journal of Structural Geology, v. 108, p. 137–156.

Limpert, E., Stahel, W.A., and Abbt, M., 2001, Log-normal distributions across the Sciences: Keys and clues; Bioscience, v. 51, no. 5, p. 341–352. https://doi.org/10.1641/0006-3568(2001)051[0341:LNDATS]2.0.CO;2

Lin, W., Kwasniewski, M., Imamura, T., and Matsuki, K., 2006, Determination of three-dimensional in situ stresses from anelastic strain recovery measurements of cores at great depth; Tectonophysics, v. 426, p. 221–238. http://dx.doi.org/10.1016/j.tecto.2006.02.019

Liu, C., Zhang, R., Zhang, H., et al., 2013, Characteristics and origin of microfracture in Lower Cretaceous tight sandstone from Kuqa Foreland Basin, NW China; American Association of Petroleum Geologists Search and Discovery Article #10514, 26 August, 2013, 25 p.

Liu, C, Pollard, D.D., Gu, K., and Shi, B., 2015, Mechanism of formation of wiggly compaction bands in porous sandstone: 2. Numerical simulation using discrete element method; Journal of Geophysical Research, Solid Earth, v. 120, p. 8152–8168. https://doi.org/10.1002/2015JB012374

Lockman, D.F., George, R.P., and Hayes, M.J., 1997, A Systematic Technique for Describing and Quantifying Fractures in Core; American Association of Petroleum Geologists, Pacific Section, 33 p.

Long J. C. S., Remer, J. S., Wilson, C. R., and Witherspoon, P. A., 1982; Porous media equivalents for networks of discontinuous fractures; American Geophysical Union, Water Resources Research, v. 18, p. 645–658.

Loosveld, R.J.H., and Franssen, R.C.M.W., 1992, Extensional vs. shear fractures: Implications for reservoir characterization; SPE-25017-MS, Society of Petroleum Engineers, European Petroleum Conference held in Cannes, France, 16–18 November, 1992, p. 23–30. https://doi.org/10.2118/25017-MS

Lorenz, J.C., 1992, Well-bore geometries for optimum fracture characterization and drainage; West Texas Geological Society Bulletin, v. 32, p. 5–8.

Lorenz, J.C., 1999, Stress-sensitive reservoirs; SPE-50977-JPT, Society of Petroleum Engineers, Journal of Petroleum Technology, v. 51, p. 61–63. https://doi.org/10.2118/50977-JPT

Lorenz, J.C., 2001, The stimulation of hydrocarbon reservoirs with subsurface nuclear explosions; Oil-Industry History, v. 2, p. 56–63.

Lorenz, J.C., 2006, Residual stress; Houston Geological Society Bulletin, September, 2006, v. 49, p. 29–45.

Lorenz, J.C., Warpinski, N.R., Branagan, P.T., and Sattler, A.R., 1989, Fracture characteristics and reservoir behavior in stress-sensitive fracture systems in flat-lying lenticular formations; SPE-15244-PA, Society of Petroleum Engineers, Journal of Petroleum Technology, v. 41, p. 614–622. https://doi.org/10.2118/15244-PA

Lorenz, J.C., Finley, S.J., and Warpinski, N.R., 1990, Significance of coring-induced fractures in Mesaverde core, northwestern Colorado; American Association of Petroleum Geologists Bulletin, v. 74, p. 1017–1029.

Lorenz, J.C., Teufel, L.W., and Warpinski, N.R., 1991, Regional fractures I: A mechanism for the formation of regional fractures at depth in flat-lying reservoirs; American Association of Petroleum Geologists Bulletin, v. 75, p. 1714–1737.

Lorenz, J.C., and Finley, S.J., 1991, Regional fractures II: Fracturing of Mesaverde reservoirs in the Piceance Basin, Colorado; American Association of Petroleum Geologists Bulletin, v. 75, p. 1738–1757.

Lorenz, J.C., and Hill, R.E., 1992, Measurement and analysis of fractures in core; in Schmoker, J.W., Coalson, E.B., and Brown, C.A., eds., Geologic Studies Relevant to Horizontal Drilling: Examples from Western North America; Rocky Mountain Association of Geologists, p. 47–59.

Lorenz, J.C., Warpinski, N.R., Teufel, L.W., 1993, Rationale for finding and exploiting fractured reservoirs, based on the MWX/SHCT-Piceance Basin experience; SAND93-1342, Sandia National Laboratories Report, 144 p. https://doi.org/10.2172/10187135

Lorenz, J.C., and Hill, R.E., 1994, Subsurface fracture spacing: Comparison of inferences from slant/horizontal and vertical cores; SPE-21877-PA, Society of Petroleum Engineers, Formation Evaluation, v. 9, p. 66–72. https://doi.org/10.2118/21877-PA

Lorenz, J.C., and Laubach, S.E., 1994, Description and interpretation of natural fracture patterns in sandstones of the Frontier Formation along the Hogsback, southwestern Wyoming; Gas Research Institute, Tight Sands and Gas Processing Research Department, Topical Report GRI-94/0020, 89 p.

Lorenz, J.C., Sterling, J.L., Schechter, D.S., et al., 2002, Natural fractures in the Spraberry Formation, Midland basin, Texas: The effects of mechanical stratigraphy on fracture variability and reservoir behavior; American Association of Petroleum Geologists Bulletin, v. 86, p. 505–524.

Lorenz, J.C., and Cooper, S.P., 2003, Tectonic setting and characteristics of natural fractures in Mesaverde and Dakota reservoirs of the San Juan Basin; New Mexico Geology, New Mexico Bureau of Geology and Mineral Resources, v. 25, p. 3–14. https://geoinfo.nmt.edu/publications/periodicals/nmg/25/n1/nmg_v25_n1_p3.pdf

Lorenz, J.C., Cooper, S.P., Arnold, B.W., et al., 2004, Natural Gas Production Problems: Solutions, Methodologies, and Modeling; SAND2004-4859, Sandia National Laboratories Technical Report, 616 p.

Lorenz, J.C., Krystinik, L.F., and Mroz, T.H., 2005, Shear reactivation of fractures in deep Frontier sandstones: Evidence from horizontal wells in the Table Rock Field, Wyoming; in, Bishop, M.G., et al., eds., Gas in Low Permeability Reservoirs of the Rocky Mountain Region; Rocky Mountain Association of Geologists guidebook, p. 267–288.

Lorenz, J.C., and Cooper, S.P., 2013, Natural fractures in folded sandstones of the Tensleep Formation, Wyoming; in, Knight, C., and Cuzella, J., eds., Application of Structural Methods to Rocky Mountain Hydrocarbon Exploration and Development; American Association of Petroleum Geologists, Studies in Geology 65, p. 175–213.

Lorenz, J.C., and Cooper, S.P., 2018a, Atlas of Natural and Induced Fractures in Core; Wiley and Sons Ltd, 305 p.

Lorenz, J.C., and Cooper, S.P., 2018b, An outcrop example of variable fracture mode in a sandstone; LinkedIn article, 25 October, 2018. https://www.linkedin.com/pulse/outcrop-example-variable-fracture-mode-sandstone-john-lorenz/

Lorenz, J.C., and Cooper, S.P., 2018c, Dynamically compatible fractures in sandstone outcrops, LinkedIn article, 14 May, 2018. https://www.linkedin.com/pulse/dynamically-compatible-fractures-sandstone-outcrops-john-lorenz/

Lorenz, J.C., and Cooper, S.P., 2019, Natural Fractures; in, Sorkhabi, R., ed., Encyclopedia of Petroleum Geoscience, Springer Nature Switzerland AG, 19 p. https://doi.org/10.1007/978-3-319-02330-4_300-1

Loucks, R.G., 1999, Paleocave carbonate reservoirs: Origins, burial-depth modifications, spatial complexity, and reservoir implications; American Association of Petroleum Geologists Bulletin, v. 83, p. 1795–1834.

Loucks, R.G., and Reed, R.L., 2016, Natural microfractures in unconventional shale-oil and shale-gas systems: Real, hypothetical, or wrongly defined?; Gulf Coast Association of Geological Societies Journal, v. 5, p. 64–72.

Makel, G.H., 2007, The modelling of fractured reservoirs: Constraints and potential for fracture network geometry and hydraulics analysis; in, Jolley, S. J., Barr, D., Walsh, J. J., and Knipe, R. J., eds., Structurally Complex Reservoirs;

The Geological Society, London, Special Publication 292, 375–403. http://dx.doi.org/10.1144/SP292.21

Mandl, G., 2005, Rock Joints, the Mechanical Genesis; Springer-Verlag, 222 p.

Marchegiani, L., Van Dijk, J.P., Gillespie, P.A., et al., 2006, Scaling properties of the dimensional and spatial characteristics of faults and fracture systems in the Majella Mountain, central Italy; in, Cello, G. and Malamud, B.D., eds., Fractal Analysis for Natural Hazards; The Geological Society, London, Special Publication 261, p. 113–131. http://doi.org/10.1144/GSL.SP.2006.261.01.09

Márquez, X., and Mountjoy, E.W., 1996, Microfractures due to overpressures caused by thermal cracking in well-sealed upper Devonian reservoirs, deep Alberta Basin; American Association of Petroleum Geologists Bulletin, v. 80, p. 570–588.

Marrett, R., and Allmendinger, R.W., 1990, Kinematic analysis of fault-slip data; Journal of Structural Geology, v. 12, p. 973–986.

Marrett, R., and Allmendinger, R.W., 1991, Estimates of strain due to brittle faulting: Sampling of fault populations; Journal of Structural Geology, v. 13, p. 735–738.

Marshall, J.D., 1982, Isotopic composition of displacive fibrous calcite veins: Reversals in pore-water composition trends during burial diagenesis; Journal of Sedimentary Petrology, v. 52, p. 614–630.

Martin, M.A., and Davis, T.L., 1987, Shear-wave birefringence: a new tool for evaluating fractured reservoirs; Society of Exploration Geophysists, The Leading Edge, v. 6, p. 22–28. https://doi.org/10.1190/1.1439333

Mauldon, M., Dunn, W.M., and Rohrbaugh, M.B. Jr., 2001, Circular scanlines and circular windows: New tools for characterizing the geometry of fracture traces; Journal of Structural Geology, v 23, p. 247–258.

McQuillan, H., 1973, Small-scale fracture density in Asmari Formation of SW Iran and its relation to bed thickness and structural setting; American Association of Petroleum Geologists Bulletin, v. 57, p. 2367–2385.

Medlin, W.L., and Fitch, J.L., 1988, Abnormal treating pressures in massive hydraulic fracturing treatments; SPE-12108-PA, Society of Petroleum Engineers, Journal of Petroleum Technology, v. 40, p. 633–642. https://doi.org/10.2118/12108-PA

Meredith, P.G., 2013, Strength recovery and vein growth during self-sealing of experimentally-induced faults in Westerly granite (abstract); MR41B-08, American Geophysical Union, Fall Meeting, San Francisco, California, 9–13 December, 2013.

Meredith, P.G., 2015, Strength recovery and vein growth during self-sealing of faults in Westerly granite (abstract); Geology of Geomechanics Conference, The Geological Society, London, 28–29 October, 2015.

Milliken, K.L., and Land, L.S., 1994, Evidence of fluid flow in microfractures in geopressured shales: Discussion; American Association of Petroleum Geologists Bulletin, v. 78, p. 1637–1640.

Milliken, K., and Laubach, S.E., 2000, Brittle deformation in sandstone diagenesis as revealed by scanned cathodoluminescence imaging with application to characterization of fractured reservoirs; in, Pagel, M., Barbin, V., Blanc, P., and Ohnenstetter, D., eds., Cathodoluminescence in Geosciences; p. 225–243. http://doi.org/10.1007/978-3-662-04086-7_9

Mogi, K., 1967, Effect of the intermediate principle stress on rock failure; Journal of Geophysical Research, v. 72, p. 5117–5131.

Molina, O.M., and Zeidouni, M., 2017, Fault reactivation in compartmentalized reservoirs: Effect of fault transmissibility enhancement on pressure transient behavior; OTC-27957-MS, presented at the Offshore Technology Conference, Rio de Janeiro, Brazil, 24–26 October, 2017, 16 p. https://doi.org/10.4043/27957-MS

Moore, D.E., and Lockner, D.A., 1995, The role of microcracking in shear-fracture propagation in granite; Journal of Structural Geology, v. 17, p. 95–114.

Moore, J.C., and 22 others, 1988, Tectonics and hydrogeology of the northern Barbados Ridge: Results from Ocean Drilling Program Leg 110; Geological Society of America Bulletin, v. 100, p. 1578–1593.

Morrow, N.R., Brower, K.R., Ma, S., and Buckley, J.S., 1990, Fluid flow in healed tectonic fractures; Society of Petroleum Engineers Journal of Petroleum Technology, v. 42, p. 1310–1318.

Multiwell Experiment Project Groups at Sandia National Laboratories and CER Corporation

Multiwell Experiment Final Report: I, The Marine Interval of the Mesaverde Formation, 1987, SAND87-0327, Sandia National Laboratories Technical Report, 276 p. National Technical Reports Library Accession Number DE87009389

Multiwell Experiment Final Report: II, The Paludal Interval of the Mesaverde Formation, 1988, SAND88-1008, Sandia National Laboratories Technical Report, 554 p. National Technical Reports Library Accession Number DE88012442

Multiwell Experiment Final Report: III, The Coastal Interval of the Mesaverde Formation, 1988, SAND88-3284, Sandia National Laboratories Technical Report, 1131 p. National Technical Reports Library Accession Number DE89011324

Multiwell Experiment Final Report: IV, The Fluvial Interval of the Mesaverde Formation, Part A, 1989, SAND89-2612/A, Sandia National Laboratories

Technical Report, 465 p. National Technical Reports Library Accession Number DE90007948

Multiwell Experiment Final Report: IV, The Fluvial Interval of the Mesaverde Formation, Part B, 1989, SAND89-2612/B, Sandia National Laboratories Technical Report, 448 p. National Technical Reports Library Accession Number DE90007938

Murrell, S.A.F., 1965, The effect of triaxial stress systems on the strength of rocks at atmospheric temperatures; Geophysical Journal of the Royal Astronomical Society, v. 10, p. 231–281.

Myer, L., 1991, Hydromechanical and seismic properties of fractures; Proceedings of the 7th International Rock Mechanics Congress, Aachen, Germany, 1991, v. 1, p. 397–409.

National Research Council, 1996, Rock Fractures and Fluid Flow, Contemporary Understandings and Applications; National Academy Press, Washington DC, 551 p. https://doi.org/10.17226/2309

Narr, W., 1991, Fracture density in the deep subsurface: Techniques with application to point Arguello oil field; American Association of Petroleum Geologists Bulletin, v. 75, p. 1300–1323.

Narr, W., 1996, Estimating average fracture spacing in subsurface rock; American Association of Petroleum Geologists Bulletin, v. 80, p. 1565–1586.

Narr., W., and Suppe, J., 1991, Joint spacing in sedimentary rocks; Journal of Structural Geology, v. 13, p. 1037–1048.

Narr, W., Schechter, D.S., and Thompson, L.B., 2006, Naturally Fractured Reservoir Characterization; Society of Petroleum Engineers, 115 p.

Nelson, R.A., 1975, Fracture permeability in porous reservoirs, an experimental and field approach; Ph.D. dissertation, Texas A&M University, 171 p.

Nelson, R.A., 1981, Significance of fracture sets associated with stylolite zones; American Association of Petroleum Geologists Bulletin, v., 65, p. 2417–2425.

Nelson, R.A., 1985a, Geologic Analysis of Naturally Fractured Reservoirs; Contributions in Petroleum Geology and Engineering No. 1, Gulf Publishing Co., Houston, Texas, 320 p.

Nelson, R.A., 1985b, Characterization and interpretation of rock mass joint patterns: reviews of recent publications; American Association of Petroleum Geologists Bulletin, v. 69, p. 2039–2040.

Nelson, R.A., 2001, Geologic Analysis of Naturally Fractured Reservoirs, 2nd ed.; Gulf Professional Publishing, Boston, 332 p.

Nelson, R.A., 2020, Static Conceptual Fracture Modeling: Preparing for Simulation and Development; Wiley and Sons Ltd, 208 p.

Nelson, R.A., Lenox, L.C., and Ward, B.J., 1987, Oriented core: Its use, error, and uncertainty; American Association of Petroleum Geologists Bulletin, v. 71, p. 357–367.

Nelson, R.A., Moldovanyi, E.P., Matcek, C.C., et al., 2000, Production characteristics of the fractured reservoirs of the La Paz field, Maracaibo Basin Venezuela; American Association of Petroleum Geologists Bulletin, v. 84, p. 1791–1809.

Nermoen, A., Korsnes, R., Christensen, H.F., et al., 2013, Measuring the Biot stress coefficient and its implications on the effective stress estimate; ARMA 13-282, American Rock Mechanics Association, 47th US Rock Mechanics/Geomechanics Symposium, San Francisco, CA, June 23–26 2013, 9 p.

Nolte, K.G., 1987, Discussion of influence of geologic discontinuities on hydraulic fracture propagation; SPE-17011-PA, Society of Petroleum Engineers, Journal of Petroleum Technology, v. 39, p. 998. https://doi.org/10.2118/13224-PA

Nolte, K.G., 1993, Discussion of examination of a cored hydraulic fracture in a deep gas well; SPE-26302-PA, Society of Petroleum Engineers, Production and Facilities, v. 8, p. 159–164. https://doi.org/10.2118/22876-PA

Northrop, D.A., Warpinski, N.R., Schmidt, R.A., and Smith, C.W., 1978, Stimulation and mineback experiment project-the direct observation of hydraulic and explosive fracturing tests; SAND78-0682C-1, Sandia National Laboratories Technical Report, 15 p.

Odling, N.E., Gillespie, P., Bourgine, B., et al., 1999, Variations in fracture system geometry and their implications for fluid flow in fractured hydrocarbon reservoirs; Petroleum Geoscience, v. 5, p. 373–384. http://doi.org/10.1144/petgeo.5.4.373

Olsson, W.A., 1974, Microfracturing and faulting in a limestone; Tectonophysics, v. 24, p 277–285.

Olsson, W.A., 1992, The effect of slip on the flow of fluid through a fracture; American Geophysical Union, Geophysical Research Letters, v. 19, p. 541–543.

Olsson, W.A., and Brown, S.R., 1993, Hydromechanical response of a fracture undergoing compression and shear; International Journal of Rock Mechanics, Mining Science and Geomechanics Abstracts, v. 30, p. 845–851.

Olsson, W.A., Lorenz, J.C., and Cooper, S.P., 2004, A mechanical model for multiply-oriented conjugate deformation bands; Journal of Structural Geology, v. 26, p. 325–338. https://doi.org/10.1016/S0191-8141(03)00101-9

Onasch, C.M., 1990, Microfractures and their role in deformation of a quartz arenite from the central Appalachian foreland; Journal of Structural Geology, v. 12, p. 883–894. https://doi.org/10.1016/0191-8141(90)90061-3

Ortega, O., and R. Marrett, R., 2000, Prediction of microfracture properties using microfracture information, Mesaverde Group sandstones, San Juan Basin, New Mexico; Journal of Structural Geology, v. 22, p. 571–588. https://doi.org/10.1016/S0191-8141(99)00186-8

Ortega, O.J., Marrett, R.A., and Laubach, S.E., 2006, A scale-dependent approach to fracture intensity and average spacing measurement; American Association of Petroleum Geologists Bulletin, v. 90, p. 193–208.

Ouenes, A., 2000, Practical Application of Fuzzy Logic and Neural Networks to Fractured Reservoir Characterization; Computer and Geosciences, v. 26, p. 953–962. https://doi.org/10.1016/S0098-3004(00)00031-5

Ouenes, A., 2019, Developing unconventional reservoirs using limited natural fractures statistics – challenges and opportunities; Sixth European Association of Geologists and Engineers Shale Workshop, 28 April–1 May, 2019, Bordeaux, France, 6 p.

Ouenes, A., Anderson, T., Klepacki, D., et al., 2010, Integrated characterization and simulation of the fractured Tensleep Reservoir at Teapot Dome for CO_2 injection design; SPE-132404-MS, Society of Petroleum Engineers, Western Regional Meeting, Anaheim, California, USA, 26–30 May 2010, 15 p. https://doi.org/10.2118/132404-MS

Parker, J.M., 1942, Regional systematic jointing in slightly deformed sedimentary rocks; Geological Society of America Bulletin, v. 53, p. 381–408.

Paterson, M.S., 1958, Experimental deformation and faulting in Wombeyan marble; Geological Society of America Bulletin, v. 69, p. 465–476.

Paterson, M.S., 1978, Experimental Rock Deformation-The Brittle Field; Springer-Verlag, 254 p.

Paterson, M.S., and Wong, T.-f., 2005, Experimental Rock Deformation – The Brittle Field, 2nd ed.; Springer, 345 p.

Perez, J.M., Poston, S.W., and Edwards, C.M., 1993, The effect of microfractures on oil recovery from low-permeability reservoirs; SPE-25857-MS, Society of Petroleum Engineers Low-Permeability Reservoirs Symposium, Denver, Colorado, 26–28 April, 1993, 11 p. https://doi.org/10.2118/25857-MS

Perkins, T.K., and Kern, L.R., 1961, Widths of hydraulic fractures; SPE-89-PA, Society of Petroleum Engineers, Journal of Petroleum Technology, v. 13, p. 937–949. https://doi.org/10.2118/89-PA

Peterson, R.E., Warpinski, N.R., Lorenz, J.C., et al., 2001, Assessment of the Mounds Drill Cuttings Injection Disposal Domain; SPE-71378-MS, Society of Petroleum Engineers Annual Technical Conference and Exhibition, New Orleans, Louisiana, 30 September–3 October, 2001, 16 p. https://doi.org/10.2118/71378-MS

Petit, J.P., 1987, Criteria for the sense of movement on fault surfaces in brittle rocks; Journal of Structural Geology, v. 9, p. 597–608. https://doi.org/10.1016/0191-8141(87)90145-3

PetroWiki, 2019, Fluid flow in naturally fractured reservoirs; Society of Petroleum Engineers, PetroWiki, 13 p. https://petrowiki.org/Fluid_flow_in_naturally_fractured_reservoirs

Pollard, D.D., and Aydin, A., 1988, Progress in understanding jointing over the past century; Geological Society of America Bulletin, v. 100, p. 1181–1204.

Pollard, D.D. and Fletcher, R.C., 2005, Fundamentals of Structural Geology; Cambridge University Press, 500 p.

Potluri, N., Zhu, D., and Hill, A.D., 2005, Effect of natural fractures on hydraulic fracture propagation; SPE-94568-MS, Society of Petroleum Engineers, European Formation Damage Conference, Scheveningen, The Netherlands, 25–27 May, 2005, 6 p. https://doi.org/10.2118/94568-MS

Price, N.J., 1966, Fault and Joint Development in Brittle and Semi-Brittle Rock; Pergamon Press, Oxford, 176 p.

Price, N.J., 1974, The development of stress systems and fracture patterns in undeformed sediments; Proceedings of the 3rd Congress of the International Society of Rock Mechanics, v. 1-A, p. 487–496.

Priest, S.D., and Hudson, J.A., 1976, Discontinuity spacing in rock masses: International properties of fault zones; Journal of Structural Geology, v. 32, p. 1557–1575.

Pryak-Nolte, L.J., Nolte, D.D., Myer, L.R., and Cook, N.G.W., 1980, Fluid flow through single fractures; in, Barton, W., and Stephansson, E., eds., Rock Joints; Balkema, Rotterdam, p. 405–412.

Ramsey, J.M., and Chester, F.M., 2004, Hybrid fracture and the transition from extension fracture to shear fracture; Nature, v. 428, p. 63–66. https://doi.org/10.1038/nature02333

Randolf, P.L., 1974, Massive stimulation effort may double U.S. Gas Supply; World Oil, v. 34, p. 340–344.

Raterman, K.T., Farrell, H.E., Mora, O.S., et al., 2017, Sampling a stimulated rock volume: An Eagle Ford example; URTeC 2670034, Unconventional Resources Technology Conference, Austin, Texas, USA, 24–26 July, 2017, 18 p. https://doi.org/10.15530/urtec-2017-2670034

Rath, A., Exner, U., Tschegg, C., et al., 2011, Diagenetic control of deformation mechanisms in deformation bands in a carbonate grainstone; American Association of Petroleum Geologists Bulletin, v. 95, p. 1369–1381.

Rawnsley, K., De Keijzer, M., Wei, L., et al., 2007, Characterizing fracture and matrix heterogeneities in folded Devonian carbonate thrust sheets, Waterton tight gas fields, Western Canada; in, Lonergan, L., Jolly, R.J.H., Rawnsley, K., and Sanderson, D.J., eds., Fractured Reservoirs; The Geological Society, London, Special Publication 270, p. 265–279. https://doi.org/10.1144/GSL.SP.2007.270.01.17

Reches, Z., 1983, Faulting of rocks in three-dimensional strain fields II, theoretical analysis; Tectonophysics, v. 95, p. 133–156.

Reches, Z., and Dieterich, J.H., 1983, Faulting of rocks in three-dimensional strain fields I, failure of rocks in polyaxial, servo-control experiments; Tectonophysics, v. 95, p. 111–132.

Reches, Z., and Lockner, D.A., 1994, Nucleation and growth of faults in brittle rocks; Journal of Geophysical Research, v. 99, p. 18,159–18,173. https://doi.org/10.1029/94JB00115

Reiss, L.H., 1980, The reservoir engineering aspects of fractured formations; Institut Francais Du Petrole Publications, Editions Technip, Paris, 108 p.

Reynolds, J.J., and Coffer, H.F., 1957, Discussion of Hubbert, M.K., and D.G. Willis, 1957, Mechanics of hydraulic fracturing; American Institute of Mining, Metallurgical, and Petroleum Engineers, Petroleum Transactions, v. 210, p. 167.

Rhett, D., 2001, Pore pressure controls on the origin of regional fractures: Experimental verification of a model (abstract); American Association of Petroleum Geologists, Search and Discovery Article #90906, extended abstract on CD from American Association of Petroleum Geologists Annual Convention, Denver, Colorado, 3–6 June, 2001. http://www.searchanddiscovery.com/abstracts/html/2001/annual/abstracts/6151.htm?q=%2BtextStrip%3Arhett

Rice, J., and Cleary, M., 1976, Some basic stress diffusion solutions for fluid-saturated elastic porous media with compressible constituents; Reviews of Geophysics, v. 14, p. 227–241.

Rice, G.K., Berg, G.E., and Belt, J.Q. Jr., 2002, Soil-gas hydrocarbon pattern changes during a West Texas waterflood; in, Schumacher, D., and LeSchack, L.A., eds., Surface Exploration Case Histories Applications of Geochemistry, Magnetics, and Remote Sensing; American Association of Petroleum Geologists, Studies in Geology n. 48, and Society of Exploration Geophysicists, Geophysical References Series n. 11, p. 157–174.

Robertson, E.C., 1983, Relationship of fault displacement to gouge and breccia thickness; Mining Engineering, v. 10, p. 1426–1432.

Robertson, E.C., 1988, Thermal properties of rocks; U.S. Geological Survey Open-File Report 88–441, 106 p.

Robin, P.-Y.F., 1973, Note on effective pressure; Journal of Geophysical Research, v. 78, p. 2434–2437.

Robinson, L.H., 1959, The effect of pore and confining pressure on the failure process in sedimentary rock; Colorado School of Mines Quarterly, v. 53, p. 177–199.

Robinson, L.H., and Holland, W.E., 1969, Some interpretations of pore fluid effects in rock failure; in, 11th U.S. Symposium on Rock Mechanics, 16–19 June, Berkeley, California, Chapter 29, p. 585–597.

Rodrigues, N., Cobbold, P.R., Loseth, H., and Ruffet, G., 2009, Widespread bedding-parallel veins of fibrous calcite ("beef") in a mature source rock (Vaca Muerta Fm, Neuquén Basin, Argentina): Evidence for overpressure and horizontal compression; Journal of the Geological Society, London, v. 166, p. 695–709. https://doi.org/10.1144/0016-76492008-111

Roggenthen, W.M., and Doe, T.W. and the EGS Collab team, 2018, Natural fractures and their relationship to the Enhanced Geothermal Systems (EGS) Collab project in the Underground of the Sanford Underground Research Facility; ARMA-2018-1190, American Rock Mechanics Association, 52nd United States Rock Mechanics Symposium, Seattle, Washington, 17–20 June, 2018, 10 p.

Rohrbaugh, M.B. Jr., Dunne, W.M., and Mauldon, M., 2002, Estimating fracture trace intensity, density, and mean length using circular scan lines and windows; American Association of Petroleum Geologists Bulletin, v. 86, p. 2089–2104.

Roldan, J.D., Escobedo Cabrera, D.E., Patrocinio, A., et al., 2013, Integration of outcrop studies to naturally fractured subsurface models – Example of the Mogollon Formation, Block X, Talara Basin, Peru; American Association of Petroleum Geologists, Search and Discovery Article #10526, AAPG International Conference and Exhibition, Cartagena, Columbia, September 8–11, 2013, 32 p.

Rothery, E., 1988, En echelon vein array development in extension and shear; Journal of Structural Geology, v. 10, p. 63–71. https://doi.org/10.1016/0191-8141(88)90128-9

Ruehlick, B., 2015, From borehole images to fracture permeability and fracturing pressure; American Association of Petroleum Geologists, Search and Discovery Article, #41537.

Russell, J.E., and Hoskins, E.R., 1973, Residual stresses in rocks; ARMA-72-0001, in, Hardy, H.R., and Stefanko, R., eds., New Horizons in Rock Mechanics; American Rock Mechanics Association, 14th Symposium on Rock Mechanics, University Park, Pennsylvania, 11–14 June, 1972, p. 1–24.

Salz, L.B., 1977, Relationship between fracture propagation pressure and pore pressure; SPE-6870-MS, Society of Petroleum Engineers Annual Fall Technical Conference and Exhibition, 9–12 October, 1977, Denver, Colorado, 7 p. https://doi.org/10.2118/6870-MS

Sangha, C.M., and Dhir, R.K., 1972, Influence of time on the strength, deformation and fracture properties of a lower Devonian sandstone; International Journal of Rock Mining Science & Geomechanics Abstracts, v. 9, p. 343–352. https://doi.org/10.1016/0148-9062(72)90002-2

Sano, O., Ito, I., Terada, M., 1981, Influence of strain rate on dilatancy and strength of Oshima Granite under uniaxial compression; Journal of Geophysical Research, Solid Earth, v. 86, p. 9299–9311. https://doi.org/10.1029/JB086iB10p09299

Sattler, A.R., 1989, Core Analysis in a Low Permeability Sandstone Reservoir: Results from the Multiwell Experiment; SAND89-071, Sandia National Laboratories Report, 69 p. https://doi.org/10.2172/6203510

Schechter, D.S., 2002, Waterflooding and CO_2 Injection in the Naturally Fractured Spraberry Trend Area; PETSOC-02-10-DAS, Society of Petroleum Engineers, Journal of Canadian Petroleum Technology, v. 41, no. 10, p. 9–14. https://doi.org/10.2118/02-10-DAS

Schneider, S., Eichkitz, C.G., Schreilechner, M.G., and Davis, J.C., 2016, Interpretation of fractured zones using seismic attributes – Case study from Teapot Dome, Wyoming, USA; Society of Exploration Geophysicists, Interpretation, v. 4, p. T249–T259. https://doi.org/10.1190/INT-2015-0210.1

Scholz. C.H. 1968, Microfracturing and the inelastic deformation of rock in compression; American Geophysical Union, Journal of Geophysical Research, v. 73, 1417–1432. https://doi.org/10.1029/JB073i004p01417

Schumacher, D., 2012, Hydrocarbon microseepage – a significant but underutilized geologic principle with broad applications for oil/gas exploration and production; American Association of Petroleum Geologists, Search and Discovery Article #40943, 11 June, 2012. http://www.searchanddiscovery.com/pdfz/documents/2012/40943schumacher/ndx_schumacher.pdf.html

Secor, D.T., 1965, The role of fluid pressure in jointing; American Journal of Science, v. 263. p. 633–646. http://dx.doi.org/10.2475/ajs.263.8.633

Segall, P., and Pollard, D.D., 1983, Joint formation in granitic rock of the Sierra Nevada; Geological Society of America Bulletin, v. 94., p. 563–575.

Sibson, R.H., 1990, Conditions for fault-valve behavior; in, Knipe, R.J., and Rutter, E.H., Deformation Mechanisms, Rheology and Tectonics; The Geological Society, London, Special Publication 54, p. 15–28. https://doi.org/10.1144/GSL.SP.1990.054.01.02

Sibson, R.H., 2003, Brittle-failure controls on maximum sustainable overpressure in different tectonic regimes; American Association of Petroleum Geologists Bulletin, v. 87, p. 901–908.

Skopec, R.A., 1994, Proper coring and wellsite core handling procedures: The first step towards reliable core analysis; SPE-28153-PA, Society of Petroleum Engineers, Journal of Petroleum Technology, v. 64, p. 280. https://doi.org/10.2118/28153-PA

Solum, J.G., Brandenburg, J.P., Naruk, S.J., et al., 2010, Characterization of deformation bands associated with normal and reverse stress states in the Navajo Sandstone, Utah; American Association of Petroleum Geologists Bulletin, v. 94, p. 1453–1474.

Souque, C., Knipe, R.J., Davies, R.K., et al., 2019, Fracture corridors and fault reactivation: Example from the Chalk, Isle of Thanet, Kent, England; Journal of Structural Geology, v. 122, p. 11–26. doi.org/10.1016/j.jsg.2018.12.004

Stearns, D.W., and Friedman, M., 1972, Reservoirs in fractured rock; in. King, R.E., ed., Stratigraphic Oil and Gas Fields – Classification, Exploration Methods, and Case Histories; American Association of Petroleum Geologists Memoir 16, p. 82–106.

Stephenson, B.J., Koopman, A., Hillgartner, H., et al., 2007, Structural and stratigraphic controls on fold-related fracturing in the Zagros Mountain, Iran: Implications for reservoir development; in, Lonergan, L., Jolly, R.J.H., Rawnsley, K., and Sanderson, D.J., eds., Fractured Reservoirs; The Geological Society, London, Special Publication 270, p. 1–21. https://doi.org/10.1144/GSL.SP.2007.270.01.01

Sternlof, K.R., Chapin, J.R., Pollard, D.D., and Durlofsky, L.J., 2004, Permeability effects of deformation band arrays in sandstone; American Association of Petroleum Geologists Bulletin, v. 88, p. 1315–1329.

Stoneley, R., 1983, Fibrous calcite veins, overpressures, and primary oil migration; American Association of Petroleum Geologists Bulletin, v. 67, p. 1427–1428.

Swarbrick, R.E. and Osborne, M.J., 1998, Mechanisms that generate abnormal pressures: An overview; in, Law, B.E., Ulmishek, G.F., and Slavin, V.I., eds., Abnormal Pressures in Hydrocarbon Environments; American Association of Petroleum Geologists Memoir 70, p. 13–34.

Tavener, E., Flottmann, T., and Brooke-Barnett, S., 2017, In situ stress distribution and mechanical stratigraphy in the Bowen and Surat basins, Queensland, Australia; in, Turner, J.P., Healy, D., Hillis, R.R., and Welch, M.J., eds., Geomechanics and Geology, The Geological Society, London, Special Publication 458, p. 31–47.

Terzaghi, K., 1923, Die Berechnung der Durchlaeassigkeitsziffer des Tones aus dem Verlauf der hydrodynamischen spannungserscheiungen; Sitzungsber Akademie de Wissenschaften in Wien, Mathematish-Naturwissen-Schaftiliche Klasee, v. 132, p. 125–138.

Terzaghi, R.D., 1965, Sources of error in joint surveys; Géotechnique, v. 15, p. 287–304. https://doi.org/10.1680/geot.1965.15.3.287

Teufel, L.W., 1983, Determination of in-situ stress from anelastic strain recovery measurements of oriented core; SPE-11649-MS, Society of Petroleum Engineers/Department of Energy Low Permeability

Reservoirs Symposium, Denver, CO, 14–16 March, 1983, p. 421–430. https://doi.org/10.2118/11649-MS

Teufel, L.W., 1996, Influence of pore pressure and production-induced changes in pore pressure on in situ stress; SAND96-0534, Sandia National Laboratories Technical Report, 54 p. https://doi.org/10.2172/208315

Teufel, L.W., and Rhett, D.W., 1991, Geomechanical evidence for shear failure of chalk during production of the Ekofisk field; SPE-22755-MS, Society of Petroleum Engineers Annual Technical Conference and Exhibition, 6–9 October, Dallas, Texas, 8 p. https://doi.org/10.2118/22755-MS

Teufel, L.W., Rhett, D.W., and Farrell, H.E., 1991, Effect of reservoir depletion and pore pressure drawdown on in situ stress and deformation in the Ekofisk Field, North Sea; in, Rogiers, J.C. ed., Rock Mechanics as a Multidisciplinary Science; Balkema, Rotterdam, p. 63–72.

Teufel, L.W., and Farrell, H.E., 1992, Interrelationship between in-situ stress, natural fractures, and reservoir permeability anisotropy: A case study of the Ekofisk Field, North Sea; in, Fractured and Jointed Rock Masses; International Society of Rock Mechanics 2nd Conference on Fractured and Jointed Rock, Balkema, Lake Tahoe, California, 3–5 June, 1992, p. 573–578.

Teufel, L.W., Rhett, D.W., Farrell, H.E., and Lorenz, J.C., 1993, Control of fractured reservoir permeability by spatial and temporal variations in stress magnitude and orientation; SPE-26437-MS, Society of Petroleum Engineers 68th Annual Technical Conference and Exhibition, Houston, TX, 3–6 October, 1993, 11 p. https://doi.org/10.2118/26437-MS

Toussaint, R., Aharonov, E., Koehn, D., et al., 2018, Stylolites: A review; Journal of Structural Geology, v. 114, p. 163–195.

Tsvankin, I., and Grechka, V., 2011. Seismology of Azimuthally Anisotropic Media and Seismic Fracture Characterization; Society of Exploration Geophysics, Geophysical References Series 17, 510 p.

Turcotte, D.L., 1986, Fractals and fragmentation; American Geophysical Union, Journal of Geophysical Research, v. 91, p. 1921–1926. https://doi.org/10.1029/JB091iB02p01921

Turner, J.P., Healy, D., Hillis, R.R., and Welch, M.J., 2017, Geomechanics and geology: Introduction; in, Turner, J.P., Healy, D., Hillis, R.R., and Welch, M.J. eds., Geomechanics and Geology; The Geological Society, London, Special Publication 458, p. 1–5. https://doi.org/10.1144/SP458.15

Tutuncu, A.N., 2010, Anisotropy, compaction and dispersion characteristics of reservoir and seal shales; ARMA 10-344, American Rock Mechanics Association, 44th US Rock Mechanics Symposium and 5th U.S.-Canada Rock Mechanics Symposium, held in Salt Lake City, Utah, 27–30 June, 2010, 14 p.

Twiss, R.J., and Moores, E.M., 1992, Structural Geology; W.H. Freeman and Co., New York, 532 p.

Ukar, E., and Laubach, S.E., 2016, Syn- and postkinematic cement textures in fractured carbonate rocks: Insights from advanced cathodoluminescence imaging; Tectonophysics, v. 690, p. 190–205.

Ukar, E., Lopez, T.G., Gale, J.F.W., et al., 2017, New type of kinematic indicator in bed-parallel veins, Late Jurassic-Early Cretaceous Vaca Muerta Formation, Argentina: E-W shortening during Late Cretaceous vein opening; Journal of Structural Geology, v. 104, p. 31–47. https://doi.org/10.1016/j.jsg.2017.09.014

van der Pluijm, B.A., Craddock, J.P., Graham, B.R., and Harris, J.H., 1997, Paleostress in cratonic North America: Implications for deformation of continental interiors; Science, v. 277, p. 794–796. https://doi.org/10.1126/science.277.5327.794

van Golf-Racht, T.D., 1982, Fundamentals of Fractured Reservoir Engineering, Developments in Petroleum Science, Book 12; Elsevier, 732 p.

van Poollen, H.K., Ishteiwy, A.A., and Chancelor, R.E., 1977, Massive hydraulic fracturing well Federal no. 498-4-1, Rio Blanco County, Colorado; U.S., Department of Energy, Fossil Energy report NVO/0677-1, 121 p.

Vega Navarro, O.G., 2014, Closure of natural fractures caused by increased effective stress, a case study: Reservoir Robore III, Bulo Field, Bolivia; SPE-153609-MS, Society of Petroleum Engineers, Latin American and Caribbean Petroleum Engineering Conference, Mexico City, Mexico, 16–18 April 2012, 11 p. https://doi.org/10.2118/153609-MS

Vermilye, J.H., and Scholz, C.H., 1995, Relation between vein length and aperture; Journal of Structural Geology, v. 17, p. 423–434.

Vincent, M.C., 2009, Examining our assumptions – Have oversimplifications jeopardized our ability to design optimal fracture treatments?; SPE-119143-MS, Society of Petroleum Engineers, Hydraulic Fracturing Technology Conference, The Woodlands, Texas, 19–21 January, 2009, 51 p. https://doi.org/10.2118/119143-MS

Warpinski, N.R. 1989, Elastic and viscoelastic calculations of stresses in sedimentary basins; SPE-15243-PA, Society of Petroleum Engineers, Formation Evaluation, v. 4, 9 p. https://doi.org/10.2118/15243-PA

Warpinski, N.R., 1991, Hydraulic fracturing in tight, fissured media; SPE-220154-PA, Society of Petroleum Engineers, Journal of Petroleum Technology, v. 73, p. 146–152, 208-209. https://doi.org/10.2118/20154-PA

Warpinski, N.R., Tyler, L.D., Vollendorf, C., and Northrop, D.A., 1981, Direct observation of a sand-propped hydraulic fracture; SAND81-0225, Sandia National Laboratories Technical Report, 60 p.

Warpinski, N.R., and Teufel, L.W., 1987a, Influence of geologic discontinuities on hydraulic fracture propagation; SPE-13224-PA, Society of Petroleum Engineers, Journal of Petroleum Technology, v. 39, p. 209–220. https://doi.org/10.2118/13224-PA

Warpinski, N.R., and Teufel, L.W., 1987b, Authors' reply to discussion of influence of geologic discontinuities on hydraulic fracture propagation; SPE-17074-PA, Society of Petroleum Engineers, Journal of Petroleum Technology, v. 39, p. 999. https://doi.org/10.2118/13224-PA

Warpinski, N.R., Teufel, L.W., Lorenz, J.C., and Holcomb, D.J., 1993a, Core based stress measurements: A guide to their application; GRI-93/0270 and SAND 93-2239, Topical Report for July 1991–June 1993, Gas Research Institute, Contract no. 5089-211-2059, 150 p.

Warpinski, N.R., Lorenz, J.C., Branagan, P.T., et al., 1993b, Examination of a cored hydraulic fracture in a deep gas well; SPE-22876-PA, Society of Petroleum Engineering, Production and Facilities, v. 8, p. 150–158. https://doi.org/10.2118/22876-PA

Warpinski, N.R., Lorenz, J.C., Branagan, P.T., et al., 1993c, Authors' reply to discussion of examination of a cored hydraulic fracture in a deep gas well; SPE-26946-PA, Society of Petroleum Engineering, Production and Facilities, v. 8, p. 159–164. https://doi.org/10.2118/22876-PA

Warpinski, N.R., and Lorenz, J.C., 2008, Analysis of the Multiwell Experiment data and results –Implications for the basin-centered gas model; in, Cumella, S.P., Shanley, K.W., and Camp, W.K., eds., Understanding, Exploring, and Developing Tight-Gas Sands; American Association of Petroleum Geologists, 2005 Hedberg Conference, Vail, Colorado, 24–29 April, 2005, v. 3, p. 157–176.

Warren, J.E., and Root, P.J., 1963, The behavior of naturally fractured reservoirs; SPE-426-PA, Society of Petroleum Engineers Journal, v. 3, p. 245–255. https://doi.org/10.2118/426-PA

Watkins, H., Bond, C.E., Healy, D., and Butler, R.W.H., 2015, Appraisal of fracture sampling methods and a new workflow to characterize heterogeneous fracture networks at outcrop; Journal of Structural Geology, v. 72, p. 67–82.

Weger, R.J., Murray, S.T., McNeill, D.F., et al., 2019, Paleothermometry and distribution of calcite beef in the Vaca Muerta Formation, Neuquén Basin, Argentina; American Association of Petroleum Geologists Bulletin, v., 103, p. 931–950.

Wei, Y., and Anand, L., 2008, On micro-cracking, inelastic dilatancy, and the brittle-ductile transition in compact rocks: A micro-mechanical study; International Journal of Solids and Structures, v. 45, p. 2785–2798.

Weijermars, R., 1997, Principles of Rock Mechanics; Alboran Science Publishing, 360 p.

Weinberger, R., and Bahat, D., 2008, Relative fracture velocities based on fundamental characteristics of joint-surface morphology; Terra Nova, v. 20, p. 68–73.

Wennberg, P.O., Azizzadeh, M., Aqrawi, A.A.M., et al., 2007, The Khaviz Anticline: An outcrop analogue to giant fractured Asmari Formation reservoirs in SW Iran; in, Lonergan, L., Jolly, R.J.H., Rawnsley, K., and Sanderson, D.J., eds., Fractured Reservoirs; The Geological Society, London, Special Publication 270, p. 23–42. https://doi.org/10.1144/GSL.SP.2007.270.01.02

Wennberg, P.O., Casini, G., Jonoud, S., and Peacock, D.C.P., 2016, The characteristics of open fractures in carbonate reservoirs and their impact on fluid flow: A discussion; Petroleum Geoscience, v. 22., p. 91–104. http://doi.org/10.1144/petgeo2015-003

White, A.J., Traigott, M.O., and Swarbrick, R.E., 2002, The use of leakoff tests as means of predicting minimum in-situ stress; Petroleum Geoscience, v. 8, p. 189–193.

Winslow, M.A., 1983, Clastic dike swarms and the structural evolution of the foreland fold and thrust belt of the southern Andes; Geological Society of America Bulletin, v. 94, p. 1073–1080.

Wise 1964, D.U., 1964, Microjointing in basement, Middle Rocky Mountains of Montana and Wyoming; Geological Society of America Bulletin, v. 75, p. 287–306.

Witherspoon, P.A., Wang, J.S.Y., Iwai, K., and Gale, J.E., 1979, Validity of the cubic law for fluid flow in a deformable rock fracture; LBL-9557, SAC-23, UC-70, Lawrence Berkeley Labs Report, 33 p.

Withjack, M.O., Olson, J., and Peterson, E., 1990, Experimental models of extensional forced folds; American Association of Petroleum Geologists Bulletin, v. 74, p. 1038–1054.

Wong, L.N., Maruvanchery, V., and Liu, G., 2015, Water effects on rock strength and stiffness degradation; Acta Geotechnica, v. 11, p. 713–737.

Wu, H., and Pollard, D.D., 1995, An experimental study of the relationship between joint spacing and layer thickness; Journal of Structural Geology, v. 17, p. 887–905.

Zahm, C.K., and Hennings, P.H., 2009, Complex fracture development related to stratigraphic architecture: Challenges for structural deformation prediction, Tensleep Sandstone at the Alcova anticline, Wyoming; American Association of Petroleum Geologists Bulleting, v. 93, p. 1427–1446.

Zeeb, C., Gomez-Rivas, E., Bons, P.D., and Blum, P., 2013, Evaluations of sampling methods for fracture network characterization using outcrops; American Association of Petroleum Geologists Bulletin, v. 97, no. 9, p. 1545–1566.

Zeng, L., 2010, Microfracturing in the Upper Triassic Sichuan Basin tight-gas sandstones: Tectonic, overpressure, and diagenetic origins; American Association of Petroleum Geologists Bulletin, v. 94, p. 1811–1825.

Zeng, L., and Li, X.-Y., 2009, Fractures in sandstone reservoirs with ultra-low permeability: A case study of the Upper Triasic Yanchang Formation in the Ordos Basin, China; American Association of Petroleum Geologists Bulletin, v. 93, p. 461–477. http://doi.org/10.1306/09240808047

Zhao, Z., Wang, W., and Gao, X., 2014, Evolution of laws of strength parameters of soft rock at the post-peakcd considering stiffness degradation; Journal of Zhejiang University Science A, Applied Physics and Engineering, v. 15, p. 282–290. https://doi.org/10.1631/jzus.A1300314

Zimmerman, R.W., and Bodvarsson, G.S., 1996, Hydraulic conductivity of rock fractures; Transport in Porous Media, v. 23, p. 1–30.

Zoback, M.D., 2007, Reservoir Geomechanics; Cambridge Univ. Press, New York, 449 p.

Index

d

e

f

t

u

v

w

y